LABORATORY MANUAL

HUMAN ANATOMY

Second Edition

EricWise

Santa Barbara City College

Connect
Learn
Succeed™

LABORATORY MANUAL TO ACCOMPANY HUMAN ANATOMY, SECOND EDITION

Published by McGraw-Hill, a business unit of The McGraw-Hill Companies, Inc., 1221 Avenue of the Americas, New York, NY 10020. Copyright © 2008 by The McGraw-Hill Companies, Inc. All rights reserved. No part of this publication may be reproduced or distributed in any form or by any means, or stored in a database or retrieval system, without the prior written consent of The McGraw-Hill Companies, Inc., including, but not limited to, in any network or other electronic storage or transmission, or broadcast for distance learning.

Some ancillaries, including electronic and print components, may not be available to customers outside the United States.

This book is printed on acid-free paper.

1 2 3 4 5 6 7 8 9 0 WDQ/WDQ 1 0 9 8 7 6 5 4 3 2 1 0

ISBN 978–0–07–325052–6
MHID 0–07–325052–X

Vice President & Editor-in-Chief: *Marty Lange*
Vice President, EDP: *Kimberly Meriwether-David*
Director of Development: *Kristine Tibbetts*
Executive Editor: *Colin H. Wheatley*
Senior Developmental Editor: *Kristine A. Queck*
Marketing Director: *Michelle Watnick*
Marketing Manager: *Denise M. Massar*
Project Coordinator: *Mary Jane Lampe*
Senior Production Supervisor: *Laura Fuller*
Senior Designer: *David W. Hash*
(USE) Cover Image: *colored angiogram (x-ray) of a healthy kidney, ©James Cavallini/Photo Researchers, Inc.; colored scanning micrograph (SEM) of a kidney glomerulus, ©Susumu Nishinaga/Photo Researchers, Inc.; colored scanning electron micrograph (SEM) of kidney blood vessels, ©Susumu Nishinaga/Photo Researchers, Inc.*
Senior Photo Research Coordinator: *John C. Leland*
Compositor: *Carlisle Publishing Services*
Typeface: *10/12 Janson*
Printer: *World Color Press Inc.*

All credits appearing on page or at the end of the book are considered to be an extension of the copyright page.

Some of the laboratory experiments included in this text may be hazardous if materials are handled improperly or if procedures are conducted incorrectly. Safety precautions are necessary when you are working with chemicals, glass test tubes, hot water baths, sharp instruments, and the like, or for any procedures that generally require caution. Your school may have set regulations regarding safety procedures that your instructor will explain to you. Should you have any problems with materials or procedures, please ask your instructor for help.

www.mhhe.com

CONTENTS

LABORATORY MANUAL TO ACCOMPANY SALADIN, *HUMAN ANATOMY*, SECOND EDITION CORRELATION GUIDE

This lab manual can be used independently or with Saladin's *Human Anatomy* text. Below is a correlation guide listing the chapters in Saladin's text that correspond to the exercises in this lab manual.

Wise Exercises	Saladin Chapters
1. Organs, Systems, and Organization of the Body	1. The Study of Human Anatomy Atlas A
2. Microscopy	2. Cytology: The Study of Cells
3. Cell Structure	2. Cytology: The Study of Cells
4. Tissues	3. Histology: The Study of Tissues
5. The Integumentary System	5. The Integumentary System
6. Introduction to the Skeletal System	6. Bone Tissue
7. Appendicular Skeleton	8. The Appendicular Skeleton
8. Axial Skeleton: Vertebrae, Ribs, Sternum, Hyoid	7. The Axial Skeleton
9. Axial Skeleton: Skull	7. The Axial Skeleton
10. Articulations	9. Joints
11. Introduction to the Study of Muscles and Muscles of the Shoulder and Upper Extremity	10. The Muscular System 12. The Appendicular Musculature
12. Muscles of the Hip, Thigh, Leg, and Foot	12. The Appendicular Musculature
13. Muscles of the Head and Neck	11. The Axial Musculature
14. Muscles of the Trunk	11. The Axial Musculature
15. Introduction to the Nervous System	13. Nervous Tissue
16. Brain and Cranial Nerves	14. The Central Nervous System 15. The Peripherial Nervous System
17. Spinal Cord and Somatic Nerves	14. The Central Nervous System 15. The Peripherial Nervous System
18. Introduction to Sensory Receptors	16. Sense Organs
19. The Endocrine System	17. The Endocrine System
20. Blood Cells	18. The Circulatory System: Blood
21. The Heart	19. The Circulatory System: Heart
22. Introduction to the Blood Vessels and Arteries of the Upper Body	20. The Circulatory System: Blood Vessels
23. Arteries of the Lower Body	20. The Circulatory System: Blood Vessels
24. Veins and Special Circulations	20. The Circulatory System: Blood Vessels 21. The Lymphatic System
25. The Respiratory System	22. The Respiratory System
26. The Digestive System	23. The Digestive System
27. The Urinary System	24. The Urinary System
28. The Male Reproductive System	25. The Reproductive System
29. The Female Reproductive System and Development	25. The Reproductive System

INSTRUCTOR PREFACE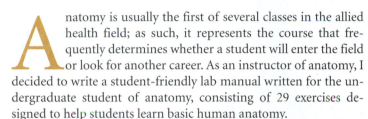

Anatomy is usually the first of several classes in the allied health field; as such, it represents the course that frequently determines whether a student will enter the field or look for another career. As an instructor of anatomy, I decided to write a student-friendly lab manual written for the undergraduate student of anatomy, consisting of 29 exercises designed to help students learn basic human anatomy.

The diversity of interests in today's anatomy students is due, in part, to the number of majors that either require or recommend the subject. This lab manual provides a framework for understanding anatomy for students interested in nursing, radiology, physical or occupational therapy, physical education, dental hygiene, or other allied health majors.

This lab manual can be used with Saladin's *Human Anatomy* text or it can be used independently. The illustrations are labeled; therefore, students do not need to bring their lecture texts to the lab. The lab manual accompanies the lecture text and lecture portion of the course and can be used in either a one-term or a full-year course. The illustrations are outstanding, and the balanced combination of line art and photographs provides effective coverage of the material. The amount of lecture material in the manual is limited, so there is little material included that is not part of the lab experience.

Practical lab experience is an invaluable opportunity to reinforce lecture concepts, enrich students' understanding of anatomy, and allow them to explore new dimensions in the subject area. The educational benefit of reinforcing lecture material with hands-on experiences and acquiring knowledge with a learn-by-doing philosophy makes the anatomy lab a very special educational environment. Many of us use lab experiences to provide those students who have different learning styles another avenue for learning.

The 29 exercises in this lab manual provide a comprehensive overview of the human body. Each exercise presents the core elements of the subject matter. You can tailor this manual to match your own vision of the course or use it in its entirety. The labs generally take between 2 and 3 hours to complete.

This lab manual was written for three types of anatomy courses. For courses that use the cat as the primary dissection animal, cat dissections are at the end of the lab manual or mammalian organ dissections follow the material on humans. For courses that use models or charts, numerous cadaver photographs are included, so that students can see the representative structures as they exist in the cadaver material. Finally, for courses that use cadavers, this lab manual can be used by studying the human material and omitting the cat dissection sections.

KEY FEATURES

1. **Dynamic art program.** All of the illustrations have been rendered by a state-of-the-art digital illustration company. They are extremely accurate, use bold and appealing colors, and offer a unique, three-dimensional look.

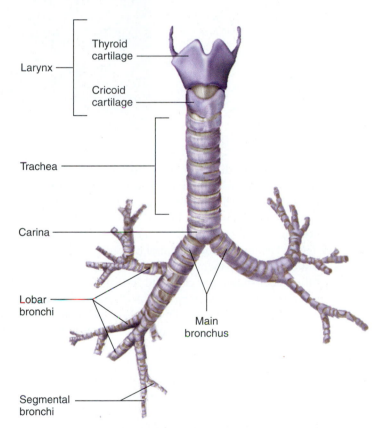

2. **Instructional photographs.** Numerous color photographs, including detailed histologic light micrographs and cadaver and cat dissections, show detailed structures.
3. **Labels.** Illustrations are labeled for students to learn the names and terminology by looking at real-life examples or models and by referring to the illustrations in the manual.
4. **Focus on the laboratory.** This manual focuses primarily on the material necessary for the laboratory and does not repeat the material presented in the lecture text, with the expectation that students can look up material in the lecture text when necessary.

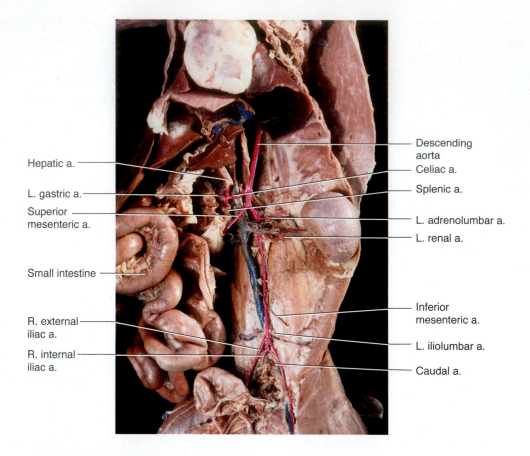

Hepatic a.

L. gastric a.

Superior
mesenteric a.

Small intestine

R. external
iliac a.

R. internal
iliac a.

Descending
aorta

Celiac a.

Splenic a.

L. adrenolumbar a.

L. renal a.

Inferior
mesenteric a.

L. iliolumbar a.

Caudal a.

5. **Use of the cat for a dissection specimen.** Cat dissections are located in a separate section at the end of the manual, so that laboratories focusing on the cat can view all of the cat anatomy photos in one location.

6. **Safety.** Safety guidelines appear in the inside front cover for reference. The international symbol for caution (⚠) is used throughout the manual to identify material that the reader should pay close and special attention to when preparing for or performing the laboratory exercise.

7. **Cleanup.** At the end of many laboratory exercises an icon for cleanup (✋) reminds the student to clean up the laboratory. Special instructions are given where appropriate.

8. **User-friendly format.** Each exercise begins on a right-hand page, and the pages are perforated to allow students to remove more easily the exercises to turn them in and to store them later.

9. **Study hints.** Study hints are located in selected chapters where additional information is provided to help students comprehend and retain the more difficult material presented.

10. **Key terms.** Current anatomic terminology is used throughout the lab manual. Key terms are boldfaced.

TEACHING AND LEARNING SUPPLEMENTS

In addition to this lab manual, an extensive array of supplemental materials is available for use in conjunction with Saladin's *Human*

Anatomy, second edition. Instructors can obtain teaching aids to accompany this lab manual by visiting www.mhhe.com/catalogs, by calling 800-338-3987, or by contacting a local McGraw-Hill sales representative. Students can order supplement study materials by calling 800-262-4729 or by contacting their campus bookstore.

Instructor's Manual for the Laboratory Manual

Visit www.mhhe.com/labcentral, to view and print the instructor's manual. This helpful preparation guide includes suggestions for coordinating laboratory exercises with the textbook, set-up instructions and materials lists, and answers to the laboratory review questions at the end of each exercise.

McGraw-Hill Connect *Human Anatomy* is a web-based assignment and assessment platform that gives students the means to better connect with their coursework, with their instructors, and with the important concepts that they will need to know for success now and in the future. With Connect *Human Anatomy,* instructors can deliver assignments, quizzes and tests easily online. Students can practice important skills at their own pace and on their own schedule. With Connect *Human Anatomy* Plus, students also get 24/7 online access to an eBook—an online edition of the text—to aid them in successfully completing their work, wherever and whenever they choose.

Now available on CD and online at aprevealed.com, this amazing multimedia tool is designed to help students learn and review human anatomy using cadaver specimens. Detailed cadaver photographs blended with a state-of-the-art layering technique provide a uniquely interactive dissection experience. This easy-to-use program features the following sections:

- **Dissection**—Peel away layers of the human body to reveal the structures beneath the surface. Structures can be pinned and labeled, just as in a real dissection lab. Each labeled structure is accompanied by detailed information and an audio pronunciation. Dissection images can be captured and saved.

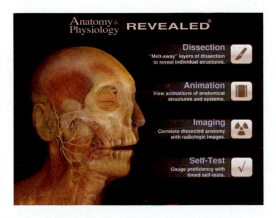

- **Animation**—Compelling animations demonstrate muscle action, clarify anatomic relationships, and explain difficult concepts.

- **Histology**—Labeled light micrographs presented with each body system allow students to study the cellular detail of tissues at their own pace.

- **Imaging**—Labeled X-ray, MRI, and CT images familiarize students with the appearance of key anatomic structures as seen through different medical imaging techniques.

- **Self-Test**—Challenging exercises let students test their ability to identify anatomic structures in a timed practical exam format or traditional multiple choice. A results page provides an analysis of test scores plus links back to all incorrectly identified structures for review.

- **Anatomy Terms**—This visual glossary of general terms includes directional and regional terms, as well as planes and terms of movement.

Virtual Anatomy Dissection Review

This unique multimedia program, created by John Waters and Melissa Janssen of Penn State University and Donna White of Collin County Community College, allows students to correlate cat and human structures. It contains 100 high-quality cat dissection photographs identifying over 600 anatomic structures, as well as corresponding illustrations of human anatomy. It also features audio pronunciations for each structure and brief video clips that demonstrate the various movements of the human body. It is available online or on CD-ROM.

Atlas of Skeletal Muscles

Prepared by Robert and Judith Stone of Suffolk Community College, this straightforward guide to the human skeletal muscles offers clear, precise illustrations of isolated muscles, along with a listing of the origin, insertion, action, and innervation of each muscle.

ACKNOWLEDGMENTS

Many people have been involved in the production of this lab manual, and I would like to thank the editorial and marketing teams at McGraw-Hill, including Marty Lange, Colin Wheatley, Kris Queck, Michelle Watnick, Nikki Puccio, and Jean Schmeider. Thanks also goes to the McGraw-Hill production team, including Mary Jane Lampe, John Leland, and David Hash for their input and encouragement.

Please feel free to write me or e-mail me with your comments, suggestions, and criticisms. I value your input and hope that your comments will lead to an even better third edition of this lab manual.

REVIEWERS

The writing of a laboratory manual is a collaborative effort and I was impressed by the dedication and insight given to me by the reviewers. I would like to gratefully acknowledge them, as they made significant contributions regarding improving the content and clarity of the manual. I appreciate their time and energy in making this a better lab manual; however, I alone take responsibility for any remaining errors.

Anne M. Burrows, Ph.D.
Duquesne University
Ruth E. Heisler
University of Colorado–Boulder
Karen M. Krabbenhoft
University of Wisconsin–Madison

Gavin C. O'Connor
Ozarks Technical Community College

Heather R. Roberts
Sierra College

Mohtashem Samsam, M.D., Ph.D.
University of Central Florida

Dr. Dean J. Scherer
Oklahoma State University–Oklahoma City

Mark A. Schlueter, Ph.D.
College of Saint Mary

Dr. Joseph Mark Shostell
Penn State University–Fayette

Dr. Laurie Walter
Chicago State University

Michael Yard
Indiana University–Purdue University at Indianapolis

Eric Wise
Santa Barbara City College
721 Cliff Drive
Santa Barbara, CA 93109
wise@sbcc.edu

This lab manual was written to help you gain experience in the lab as you learn human anatomy. The 29 exercises explore and explain the structure of the human body. You will be asked to study the structure of the body using the materials available in your lab, which may consist of models, charts, and mammal study specimens, such as cats, preserved or fresh internal organs of sheep or cows, and possibly cadaver specimens. You may also examine microscopic sections from various organs of the body. You should familiarize yourself with the microscopes in your lab very early, so that you can take the best advantage of the information they can provide.

As a student of anatomy you will be exposed to new and detailed information. The time it takes to learn the information will involve more than just time spent in the lab. You should maximize your time in the lab by reading the assigned laboratory exercises before you come to class. At the end of each exercise are review sheets that your instructor may wish to collect. The illustrations are labeled except on the review pages. All review materials can be used as study guides for lab exams or they may be handed in to the instructor.

Although the exercises are written for use with human cadavers, there is an entire section at the end of the manual for the respective cat anatomy. Get involved in your lab experience. Don't let your lab partner do all of the dissections; likewise, don't insist on doing everything yourself. Share the responsibility and you will learn more.

SAFETY

Safety guidelines appear in the inside front cover for reference. The international symbol for caution (⚠) is used throughout the lab manual to identify material that you should pay close and special attention to when preparing for or performing laboratory exercises.

CLEAN UP

Special instructions are provided for cleanup at the end of appropriate laboratory exercises and are identified by a unique icon (✋).

HOW TO STUDY FOR THIS COURSE

Some people learn best by concentrating on the visual, some by repeating what they have learned, and others by writing what they know over and over again. In this course, you will have to adapt your learning style to different study methods. You may use one study method to learn the muscles of the body and a completely different method for understanding the structure of the nervous system. Some students need only a few hours per week to succeed in this course, while others seem to study far longer with a much less satisfactory performance. You need to come to class. Come to lab on time. The beginning of the lab is when most instructors go over the material and point out what material to omit, what to change, and how to proceed. If you do not attend lab, you do not get the necessary information. Read the material ahead of time. The subject matter is very visual, and you will find an abundance of illustrations in this manual. Record on your calendar all of the lab quizzes and exams listed on the syllabus provided by your instructor. Budget your time so you study accordingly.

Work hard! There is absolutely no substitute for hard work to achieve success in a class. Some people do math easier than others, some people remember things easier, and some people express themselves better. Most students succeed because they work hard at learning the material. It is a rare student who gets a bad grade because of a lack of intelligence. Working at your studies will get you much further than worrying, ignoring your studies, or thinking that you can succeed by attending class some of the time and reviewing your notes the night before a lab exam. Be actively involved with the material and you will learn it better. Outline the material after you have studied it for awhile. Read your notes, go over the material in your mind, and then make the information your own. There are several ways that you can get actively involved.

Draw and doodle a lot. Anatomy is a visual science, and drawing helps. You do not have to be a great illustrator. Visualize the material in the same way you would draw a map to your house for a friend. You do not draw every bush or tree but, rather, create a schematic illustration that your friend could use to get the pertinent information. As you know, there are differences in maps. Some people need more practice than others, but anyone can do it. The head can be drawn as a circle, which can be divided into pieces representing the bones of the skull. Draw and label the illustration after you have studied the material and without the use of your text! Check yourself against the text to see if you really know the material. Correct the illustration with a colored pen, so that you highlight the areas you need work on. Go back and do it again until you get it perfect. This does take some time, but not as much as you might think.

Write an outline of the material. Take the mass of information to be learned and go from the general to the specific. Let's use the skeletal system as an example. You may wish to use these categories:

1. Bone composition and general structure
2. Bone formation
3. Parts of the skeleton
   a. Appendicular skeleton
      (1) Pectoral girdle
      (2) Upper extremity
      (3) Pelvic girdle
      (4) Lower extremity
   b. Axial skeleton
      (1) Skull
      (2) Hyoid
      (3) Ribs
      (4) Vertebral column
      (5) Sternum

An outline helps you organize the material in your mind and lets you sort the information into areas of focus. If you do not have an organizational system, then this course is a jumble of terms with no interrelationships. The outline can get more detailed as you progress, so you eventually place specific information into a framework of more general information.

Test yourself before an exam or a quiz. If you have practiced answering questions about the material you have studied, then you should do better on the real exam. As you go over the material, jot down possible questions to be answered later, after you study. If you compile a list of questions as you review your notes, then you can answer them later to see if you have learned the material well. You can also enlist the help of friends, study partners, or family (if they are willing to do this for you). You can also study alone. Some people make flash cards for anatomy. It is a good idea to do this for the muscle section of the class, but you may be able to get most of the information down by using the preceding technique. Flash cards take time to fill out, so use them carefully.

Use memory devices for complex material. A mnemonic device is a memory phrase that has a relationship to the study material. For example, there are two bones in the wrist right next to one another, the trapezium and the trapezoid. The mnemonic device used by one student was that "trapezium" rhymes with "thumb" and it is the bone under the thumb.

Use your study group as a support group. A good study group is very effective in helping you do your best in class. Hang out with people who will push you to do your best. If you get discouraged, your study partners can be invaluable support people. A good study group can help you improve your test scores, develop study hints, encourage you to do your best, and let you know that are not the only person who is living, eating, and breathing anatomy.

Just as a good study group can really help, a bad group can drag you down further than you might go on your own. If you are in a group that constantly complains about the instructor, that the class is too hard, that there is too much work, that the tests are not fair, that you don't really need to know this much anatomy for your own field of study and that this isn't medical school, and so on, then you need to get out of that group and into one that is excited by the information. Don't listen to people who complain constantly and make up excuses instead of studying. There is a tendency to start believing the complaints, and that begins a cycle of failure. Get out of a bad situation early and get with a group that will move forward.

Do well in the class and you will feel good about the experience. If you set up a study time with a group of people and they spend most of the time talking about parties, sports, or personal problems, then you aren't studying. There is nothing wrong with talking about parties, sports, or helping someone with personal problems, but you also need to address the task at hand, which is learning anatomy. Don't feel bad if you must get out of your study group. It is your education, and, if your partners don't want to study, then they don't really care about your academic well-being. A good study partner is one who pays attention in class, who has prepared for the study session, and who can explain information that you may have gotten wrong in your notes. You may want to get the phone number of two or three such classmates.

TEST-TAKING

Finally, you need to take quizzes and exams in a successful manner. By doing practice tests, you can develop confidence. Do well early in the semester. Study extra hard early—there is no such thing as overstudying! If you fail the first test or quiz, then you must work yourself out of an emotional ditch. Study early and consistently, and then spend the evening before the exam going over the material in a general way and learning those last few details. Some people do succeed under pressure and cram before exams; however, the information is temporarily stored and does not serve them well in their major field. If you study on a routine basis, then you can get up on the morning of a test, have a good breakfast, listen to some encouraging music, maybe review a bit, and be ready for the exam.

Your instructor is there to help you learn anatomy, and this lab manual was written with you in mind. Relate as much of the material as you can to your own body and keep an optimistic attitude.

Please feel free to write me or e-mail me with your comments, suggestions, and criticisms. I value your input and hope that your comments will lead to an even better third edition of this lab manual.

Eric Wise
Santa Barbara City College
721 Cliff Drive
Santa Barbara, CA 93109
wise@sbcc.edu

Organs, Systems, and Organization of the Body

INTRODUCTION

Human anatomy is the scientific study of the structure of the body. The term "anatomy" comes from Greek words meaning "to cut up" (***ana* = up, *toma* = to cut**). Anatomy forms the base of study for other courses in the health sciences. It is also an important course for athletic training, dance, and visual arts. The study of the human body requires an understanding of the details of the structure of the body, the orientation of the body or organ of study, and knowledge of body regions. In this exercise, you examine the levels of organization of the body (from the subatomic level to the whole organism), organ systems, reference terms, directional terms, planes of sectioning, and major regions of the body.

LEARNING OBJECTIVES

At the end of this exercise you should be able to
1. list the levels of structural hierarchy from smallest to largest;
2. list the 11 organ systems of the body and assign organs to each system;
3. put major organs, such as the heart, lungs, and stomach, in the proper organ system;
4. describe anatomic position;
5. give directional terms that are equivalent to up, down, front, back, toward the midline, and toward the surface of the body;
6. determine from an illustration whether a section is in the frontal, transverse, or sagittal plane;
7. assign select organs to respective body cavities;
8. identify the quadrants and nine regions of the abdomen.

MATERIALS

Models of human torso
Charts of human torso

PROCEDURE

Levels of Organization

The human body can be studied at different levels from the atomic (or even smaller levels) to the organismal. The earliest study involved **gross anatomy,** examining structures visible to the naked eye. As more sophisticated equipment was developed, other levels of organization became apparent. Today, the manipulation of atomic nuclei under magnetic fields has led to magnetic resonance imaging (MRI) studies that do not depend on dissection of the body (fig. 1.1). Examine the following list for the levels of study along with cited examples:

Level	Examples
Atomic	Oxygen, carbon, nitrogen
Molecular/chemical	Protein, lipid, carbohydrate
Organelle	Mitochondrion, ribosome
Cellular	Fibrocyte, red blood cell
Tissue	Epithelial, muscular
Organ	Stomach, kidney
Organ system	Digestive system, urinary system
Organism	*Homo sapiens*

Organ Systems

Anatomy can be studied in many ways. **Regional anatomy** is the study of areas of the body such as the head or leg. Another way to study anatomy is **systemic anatomy,** which is the study of **organ systems,** such as the skeletal system or the nervous system. Most undergraduate college courses in anatomy (and the format of this lab manual) use the systemic approach.

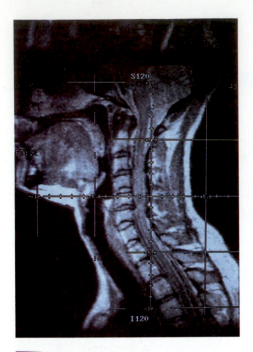

FIGURE 1.1

MRI   Image of the neck.

Although organ systems are studied separately, it is important to realize the connections between the systems. If the heart, part of the cardiovascular system, fails to pump blood, then the lungs (part of the respiratory system) do not receive blood for oxygenation and the intestines (in the digestive system) do not transfer nutrients to the blood as fuel. Organs throughout the body are no longer capable of functioning, and the result is death. The failure of one system has impacts on many other organ systems. Examine the torso models and charts in the lab and locate various organs. Using figure 1.2, find the following organ systems:

Reproductive	Respiratory	Digestive
Urinary	Skeletal	Endocrine
Nervous	Lymphatic	Cardiovascular
Muscular	Integumentary	

A quick way to remember all 11 systems is to remember this phrase: "Run Mrs. Lidec." Each letter of the phrase represents the first letter of one of the names of the organ systems.

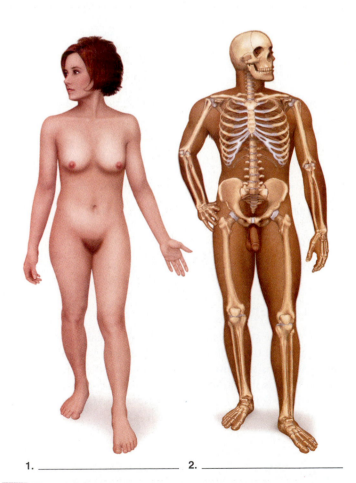
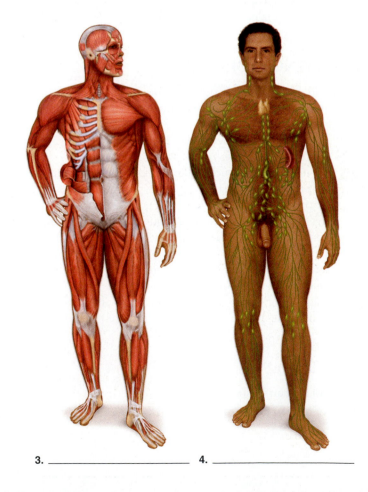

1. _____   2. _____   3. _____   4. _____

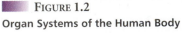

FIGURE 1.2

Organ Systems of the Human Body

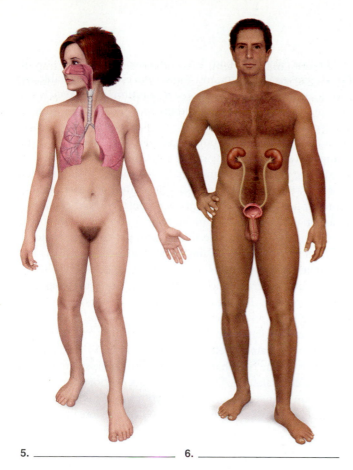

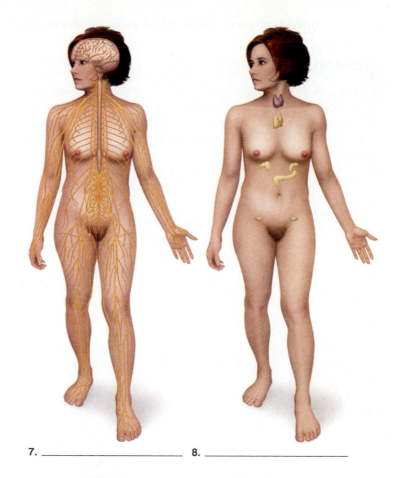

5. _____   6. _____   7. _____   8. _____

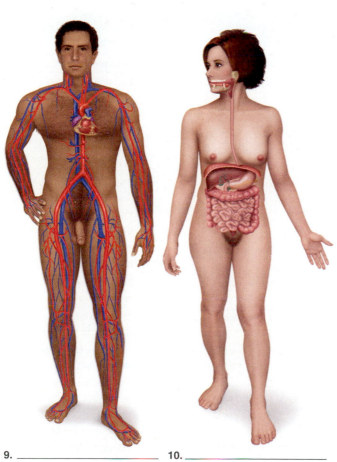

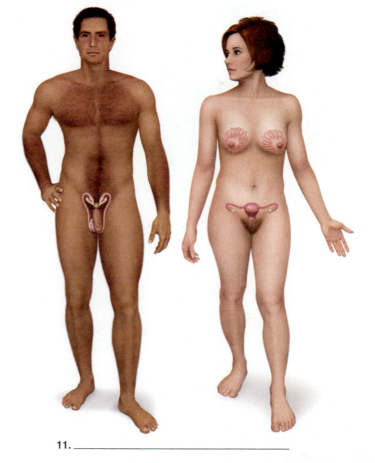

9. _____   10. _____   11. _____

FIGURE 1.2 *(continued)*

Anatomic Position

Anatomic position is the reference position for the human body. In clinical settings, it is important to have a proper orientation when dealing with patients. If two physicians are operating on a patient and one tells the other to make an incision to the left, the physician making the cut does not have to ask, "My left or your left?" because the cut is to the *patient's* left. When a person is in anatomic position, the head is erect, eyes open, arms straight and by the sides, palms facing forward, knees straight, feet together, and flat on the ground (fig. 1.3).

Directional Terms

With the body in the anatomic position, there are specific terms used to describe the location of one part with respect to another. **Superior** means that one part of the body is above another while

inferior is one part of the body below another. Table 1.1 lists the directional terms used for humans. You should examine the table and be able to use the terms easily. There are questions in the review section of this exercise that you should use to determine your understanding of these terms.

In quadrupeds (four-footed animals), the directional terms are somewhat different. Note in figure 1.4 that quadrupeds do not have a superior/inferior designation. In these animals, **dorsal** refers to the back, and **ventral** is the belly side. **Anterior** (or **craniad**) is the front, or head-end, of the animal and **posterior** (or **caudad**) is the rear, or tail-end, of the animal. There are other terms with unique meanings. For the digestive system, **proximal** refers to regions closer to the mouth, while **distal** is in reference to regions closer to the anus. **Parietal** is in reference to the body wall when compared with **visceral,** which refers to areas closer to the internal organs (figure 1.5). The heart, for example, has a visceral layer closer to the heart proper, while it also has a parietal layer farther from the heart. **Ipsilateral** refers to being on the same side of the body, and **contralateral** refers to being on the other half (left side/right side). The right hand and right arm are ipsilateral, while the ears are contralateral. Directional terms do not change if the body position changes. If you stand on your head, it is still superior to your feet because you reference the body as if it were in anatomic position.

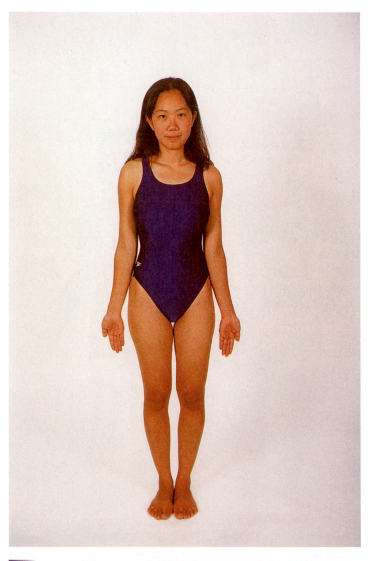

FIGURE 1.3
Anatomic Position

TABLE **1.1**		
Directional Terms Used in Anatomy		
Term	**Meaning**	**Example**
Superior	Above	The nose is superior to the chin.
Inferior	Below	The stomach is inferior to the head.
Medial	Toward the midline	The sternum is medial to the shoulders.
Lateral	Toward the side	The ears are lateral to the nose.
Superficial	Toward the surface	The skin is superficial to the heart.
Deep	Toward the core	The lungs are deep to the ribs.
Ventral (or anterior)	To the front	The toes are ventral/anterior to the heel.
Dorsal (or posterior)	To the back	The spine is dorsal/posterior to the sternum.
Proximal	For extremities, meaning near the trunk	The elbow is proximal to the wrist.
Distal	For extremities, meaning away from the trunk	The toes are distal to the knee.
Craniad	Toward the head	The nose is craniad to the chest in cats.
Caudad	Toward the "tail"	The tail is caudad to the chest in dogs.

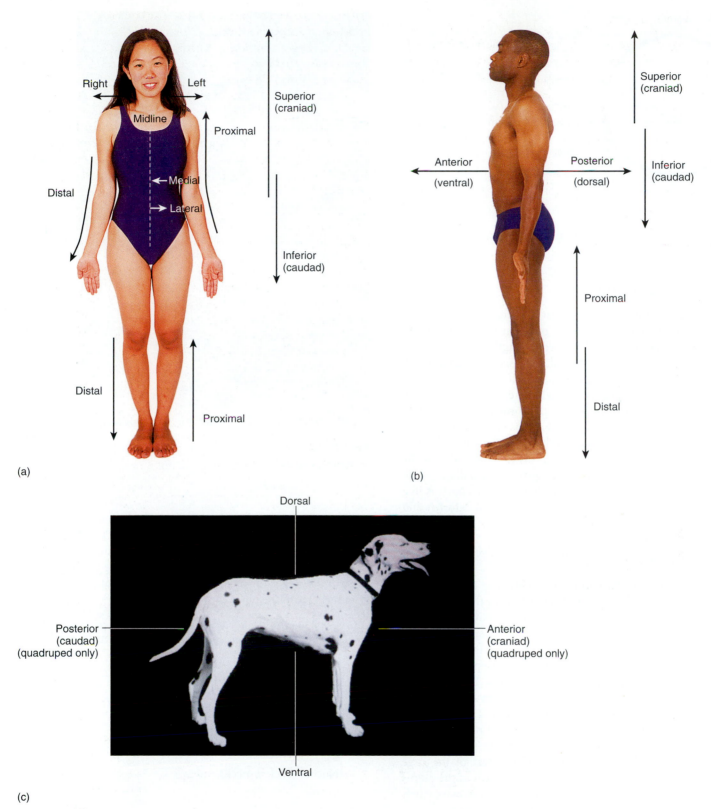

(a)

(b)

(c)

FIGURE 1.4

Directional Terms   (a) For humans, anterior view; (b) for humans, lateral view; (c) for quadrupeds, lateral view.

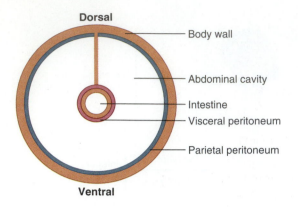

FIGURE 1.5

Idealized Cross Section of the Body with Relationships of Visceral and Parietal Terms

Planes of Sectioning

When viewing a picture of an organ that has been cut, it is very important to understand how the cut was made. Just as an apple looks different when cut crosswise as opposed to lengthwise, so do some organs. Examine figure 1.6 for the following **sectioning planes.** A cut that divides the body or organ into superior and inferior parts is in the **transverse plane** or **horizontal plane.** This is also known as a **cross section.** A cut that divides the body into anterior and posterior portions (separating front from back) is in the **frontal plane** or **coronal plane.** A cut that divides the body into left and right portions is in the **sagittal plane.** A cut that divides the body equally into left and right halves is called **midsagittal** (or **median**), while one that divides the body into unequal left and right parts is in a **parasagittal plane.** An organ, such as a kidney, cut into left and right sections is cut in the **median plane.**

Body Cavities

A cavity is an enclosed space inside the body. The **cranial cavity** houses the brain and the **vertebral canal** houses the spinal cord (fig. 1.7). Above the diaphragm is the **thoracic cavity,** which contains the heart and lungs. The lungs are located in specific cavities known as the **pleural cavities** and the heart is in the **pericardial cavity.** The **mediastinum** is the region between the pleural cavities consisting of the heart, esophagus, and vessels. Below the diaphragm is the **abdominopelvic cavity,** which can be subdivided into the **abdominal cavity** and the **pelvic cavity.** The abdominal cavity contains the stomach, small intestine, most of the large intestine and various digestive organs, such as the liver and the pancreas. The pelvic cavity contains the terminal part of the large intestine and some of the reproductive organs (such as the uterus and ovaries) of the female reproductive system. These two cavities are divided approximately at the superior portion of the hip bones.

Regions of the Body

Examine figure 1.8 for the various regions of the body. You will refer to these areas throughout this lab manual, so a complete study of these regions is essential. Anatomic names are listed first with the

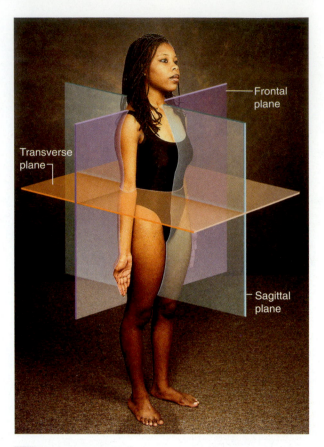

FIGURE 1.6

Sectioning Planes

common name (if appropriate) listed in parentheses. In anatomy there are some regions of the body that are described differently than what you might expect. In anatomic usage, the **arm** is the region between the shoulder and the elbow. Distal to the elbow is the forearm. The **leg** is the region between the knee and the ankle and from the knee to the hip is the thigh.

Cephalic (head)
    Frontal (forehead)
    Orbital (eye)
    Nasal (nose)
    Buccal (cheek)
    Oral (mouth)
    Mental (chin)
Cervical (neck)
Nuchal (back of neck)
Trunk
    Thoracic (chest)
      Pectoral
      Sternal
      Acromial (shoulder)
    Abdominal (belly)
    Inguinal

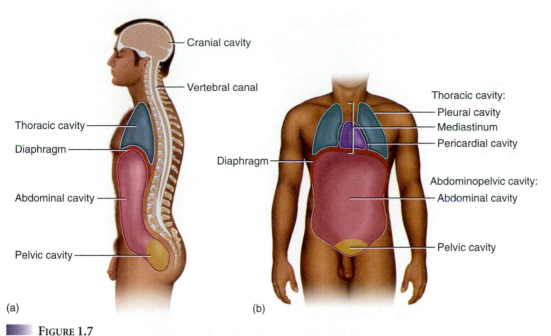

FIGURE 1.7

Body Cavities   (a) Lateral view; (b) anterior view.

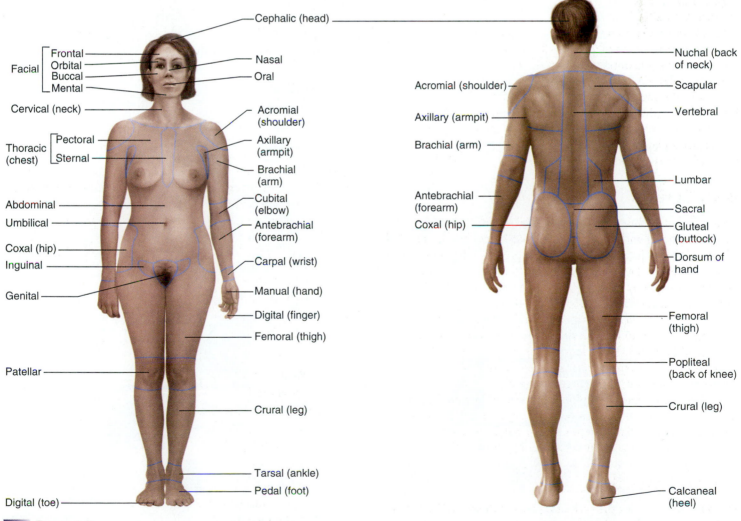

FIGURE 1.8

Regions of the Body

Genital (pubic)

Coxal (hip)

Upper extremity

Axillary (armpit)

Brachial (arm)

Cubital (elbow)

Antebrachial (forearm)

Carpal (wrist)

Manual (hand)

Digital (finger)

Lower extremity

Femoral (thigh)

Patellar (knee)

Popliteal (back of knee)

Crural (leg)

Tarsal (ankle)

Pedal (foot)

Digital (toe)

Find the following locations on your body and provide the appropriate anatomical description for these regions.

Shin _____

Elbow _____

Neck _____

Toes _____

Shoulder _____

Thigh _____

Knee _____

Abdomen

The abdomen can be divided into either four quadrants or nine regions. In clinical practice, the abdomen is divided into quadrants. In anatomic studies, the nine region approach is frequently used. A **subcostal line** occurs inferior to the ribs. An **intertubercular line** separates the umbilical region from the hypogastric region. Locate the midclavicular lines that form the divisions between the medial and lateral regions. One of the symptoms of appendicitis is pain in the lower right quadrant or right iliac region. Examine figure 1.9 and locate the lines and regions listed.

Abdominal Quadrants

Right upper quadrant

Left upper quadrant

Right lower quadrant

Left lower quadrant

Abdominal Regions

Right hypochondriac

Left hypochondriac

Epigastric

Right lumbar (lateral abdominal)

Left lumbar (lateral abdominal)

Umbilical

Hypogastric

Right iliac (inguinal)

Left iliac (inguinal)

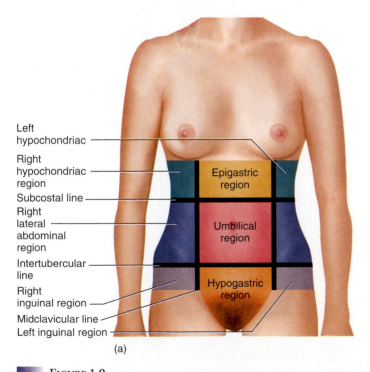

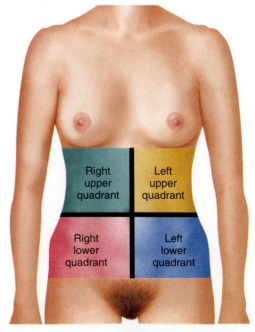

(a)   (b)

FIGURE 1.9

Abdomen   (a) Nine regions; (b) four quadrants.

Name _____ Date _____

1. The scientific study of the structure of the human body is known as _____.

2. Organs are associated into functionally related groups called _____.

3. In terms of reference, the body is placed in what position?

4. What body cavity lies directly inferior to the diaphragm?

5. The body cavity that is enclosed by the rib cage is known as the _____.

6. The body cavity surrounded by the hip bones is called the _____.

7. The term "arm" in anatomy refers to the region between the

    a. shoulder and elbow.   b. elbow and wrist.   c. shoulder and wrist.   d. shoulder and hand.

8. The term "leg" in anatomy refers to the region between the

    a. hip and knee.   b. knee and ankle.   c. hip and ankle.   d. ankle and foot.

9. A mitochondrion belongs to which level of organization?

    a. cellular   b. tissue   c. organelle   d. organ system

Use correct anatomic terminology to describe the following relationships.

10. In terms of up and down, the head is _____ to the toes.

11. In terms of nearness to the trunk, the fingers are _____ to the arm.

12. In terms of nearness to the surface, the brain is _____ to the scalp.

13. In terms of front to back, the nipples are _____ to the shoulder blades.

14. The lungs belong to the _____ system.

15. The stomach belongs to the _____ system.

16. If you sit on a horse's back, you are on the _____ aspect of the horse.

    a. anterior   b. ventral   c. posterior   d. dorsal

17. What is the difference between the abdomen and the abdominal cavity?

18. Complete the illustration by correctly placing the following terms:

    femoral   cervical

    crural   cephalic

    pedal   frontal

    genital   acromial

    coxal   axillary

    abdominal   brachial

    pectoral   antebrachial

    sternal   carpal

a. _____

b. _____

c. _____

d. _____

e. _____

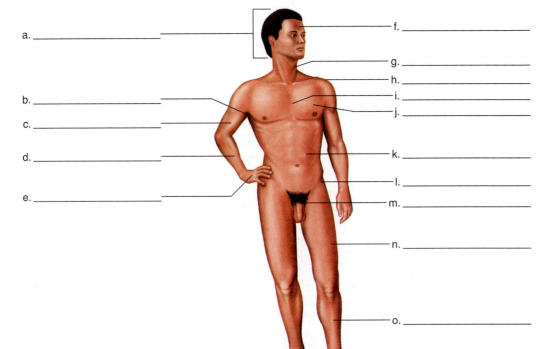

f. _____

g. _____

h. _____

i. _____

j. _____

k. _____

l. _____

m. _____

n. _____

o. _____

p. _____

19. In the following illustration, place the appropriate terms next to the sections that represent them.

    midsagittal   frontal   transverse

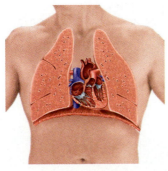

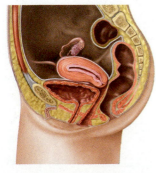

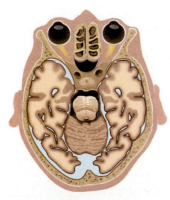

(a) _____

(b) _____

(c) _____

Microscopy

INTRODUCTION

This exercise covers the use of the light microscope, how to examine prepared slides under the microscope, and how to make simple slides for study. The study of anatomy was originally based on macroscopic, or gross, observation. With the invention of the light microscope much greater detail was observed, and the study of cells and tissues began. **Light microscopy** involves the use of visible light and glass lenses to magnify and observe a specimen. Electron microscopy, which uses electrons passing through or bouncing off of material, has revealed much greater detail than observable with light microscopy.

LEARNING OBJECTIVES

At the end of this exercise you should be able to
1. name the parts of the microscope presented in this exercise;
2. demonstrate the proper use of the light microscope;
3. place a microscope slide on the microscope and observe the material, in focus, under all magnifications of the microscope;
4. calculate the total magnification of a microscope based on the specific lenses used;
5. prepare a wet mount for observation;
6. list the rules for proper microscope use.

MATERIALS

Light microscope
Prepared slide with the letter *e* (or newsprint and razor blades)
Transparent ruler (or sections of overhead transparencies of rulers)
Glass microscope slides
Coverslips

Lens paper
Kimwipes or other cleaning paper
Lens cleaner
Small dropper bottle of water
1% methylene blue solution
Toothpicks
Histological slides of kidney, stomach, or liver
Silk threads prepared slide
Slide with grid etched on it (grid slide)

PROCEDURE

Care of the Microscope

Microscopes are expensive pieces of equipment, and you should always take great care handling them. There are a few rules concerning microscopes that you should observe:

1. When carrying the microscope, hold it securely with two hands—one hand under the base and one on the arm.
2. Keep the microscope upright at all times. Never tilt the microscope from an upright position, as lenses or filters may fall and break.
3. Keep microscope lenses clean with lens cleaner and softened lens paper. Do NOT use paper towels or your shirt.
4. Use only the fine-focus knob when using the high-power objective lens.
5. Remove slides from the microscope before putting it away.
6. Secure the cord with a rubber band, or wrap the cord carefully around the base of microscope.
7. Store the microscope with the low-power objective lens in place.
8. Put the microscope away in its proper location.

Microscope Set-Up Procedure

1. Remove the microscope from the storage area.
2. Take the microscope to your desk. Unwrap the electrical cord and familiarize yourself with the parts of the microscope. Compare the microscope that you have in lab with the one illustrated in figure 2.1. There may be differences between the microscope on your desk and the one in the figure in this lab manual, but you should be able to locate the parts listed in the checklist. Use the checklist to make sure that you find all of the listed parts. Place a check mark next to the appropriate space when you locate the part of the microscope.

Microscope Parts

____ Base	____ Arm	____ Objective lens
____ Condenser	____ Iris diaphragm lever	____ Body tube
____ Nosepiece	____ Coarse-focus knob	____ Fine-focus knob
____ Mechanical stage	____ Ocular (eyepiece) lens	____ Light source

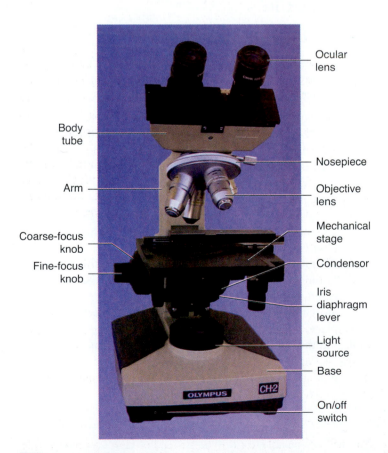

FIGURE 2.1
Compound Light Microscope

3. Plug in the microscope, making sure the ocular lens or lenses are facing you. The cord should not hang over the counter or in the aisle where someone might trip on the cord or pull the microscope off the counter. If the microscope has an illuminator (light source) dial, make sure that the setting is on the lowest level.
4. Rotate the nosepiece until the low-power objective lens (the one with the lowest number or the shortest barrel) clicks into place. When first looking at microscope slides always use the low-power objective lens.
5. Examine a prepared slide with the letter *e* or take a piece of newsprint and, using a razor blade or scissors, cut out a single letter from the paper. Remove a glass microscope slide from the box and lay it flat on your desk. Place the letter on the slide and add a drop of water to the piece of paper.
6. Place a thin coverslip on the slide by touching one edge of the coverslip to the water and lowering it slowly over the piece of newsprint (figure 2.2). If you drop the coverslip on top of the slide you will probably trap air bubbles, which may obscure some of your specimen.
7. Locate and turn on the light switch.
8. Place the slide on the microscope stage, so that you can read the letter. Focusing the microscope requires a little bit of patience. When first looking at microscope slides, always use the low-power objective lens. Make sure the specimen is on the stage and centered in the open circle on the stage. There should be light coming through the specimen. The coverslip should be very close to the objective lens. Look at the microscope stage from the side and adjust the coarse-focus knob so that the coverslip is almost touching the objective lens. Look through the ocular lens and rotate the coarse-focus knob slowly, so that the objective lens and the slide begin to move away from each other. This should bring the object into focus in the field of view. The **field of view** is the circle that you see as you look into the microscope. After you get the specimen in focus under low power, you should examine the material under higher powers. Do this by centering the image that you observe in the field of view and then switching the objective lens located on the rotating

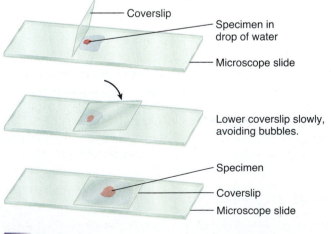

FIGURE 2.2
Preparation of a Wet Mount

nosepiece to the next higher power. Do not touch the focus knobs at this time. The next higher objective lens should clear the slide. Once you rotate the lens, adjust the focus by using the fine-focus knob. You can look at the subject under high power by the same procedure by turning the objective lens to the high-dry lens. If you cannot focus on the high power or have lost the image that you were looking for, *you should return to low power and try the process again.* If you still cannot find the object under high power, you should ask your instructor to help you.

9. Adjust the iris diaphragm for brightness.
10. Draw what the specimen looks like in the space provided.

Does the letter appear right-side-up or is the image inverted?

Is the letter oriented correctly or is the image flipped horizontally?

How much of the letter occupies the field of view?

11. Move the objective lens to the next higher power lens. Examine the specimen again. You may need to adjust the fine-focus knob so that the image becomes clearer. You may have to move the specimen a little to center the image. As magnification increases does the field of view increase or decrease in size?

Examination of a Prepared Slide

Examine three crossed threads under low power. As you focus the threads which color thread is on top? _____

Which color thread is in the middle? _____

Which color thread is on the bottom? _____

Under low power, how many threads or how much of one thread appears to be in focus? As you switch to the next higher power lens, how many threads or how much of one thread appears in focus? The depth of field is defined as the amount of material in focus under a particular magnification. What happens to the depth of field when you increase the magnification in the microscope?

Magnification and Field of View

You can determine the size of the object under observation if you know the diameter of the field of view. The field of view can be measured directly when using the low-power lens by using a clear ruler, an overhead transparency of a ruler, or a specialized slide that has a grid etched on it. If higher-power lenses are used, rulers won't work and you have to use a grid slide or calculate the field of view. There is an inverse relationship between the diameter of the field of view and the magnification used. You can first calculate the **total magnification** using the following procedure. Look at the barrels of your microscope and determine the magnification of each lens.

Eyepiece (ocular) magnification: _____

Low-power objective lens magnification: _____

Total magnification (= ocular magnification × objective magnification): _____

Place a transparent ruler, transparency, or grid slide on the stage of the microscope. The space between each dark line that runs vertically on a ruler is 1 millimeter (mm). Count the number of millimeters at the broadest part of the field of view and enter this number as the "diameter of the field of view" in the following space.

Diameter of the field of view (mm): _____

You can calculate the length of an object by determining how much of the diameter of the field of view it occupies. Let's say that the diameter of the field of view is 10 mm. If an object takes up one-half of the field of view, then you can estimate its size at 5 mm. If the object takes up only one-third of the field of view, how large is it? Record your answer in the following space.

Object size (mm): _____

As the magnification increases, the field of view decreases proportionally. Thus, if the diameter of the field of view is 10 mm at one magnification and you double the magnification by changing lenses, the field of view is reduced to a diameter of 5 mm. If you switch to a new lens and increase the magnification by 10 times, then the field of view is reduced to one-tenth of the original field of view. Look at figure 2.3 for a representation of this.

Keep the clear ruler, transparency, or grid slide under the microscope and increase the magnification to the next higher power by moving the next larger objective lens in place. Record the total magnification of your microscope with this objective lens.

Total magnification: _____

Examine the ruler, transparency, or grid slide under the microscope and record the diameter of the field of view in millimeters.

Diameter of the field of view in millimeters: _____

Has the increase in magnification produced a decrease in the field of view? If there is a decrease in the field of view, is it proportional to the magnification?

Now calculate the total magnification of the microscope using the high-power objective lens.

Magnification with high-power objective lens: _____

You will not be able to measure the field of view accurately under high power with a ruler or transparency; however, you should be able to use a grid slide or calculate the diameter of the field of view.

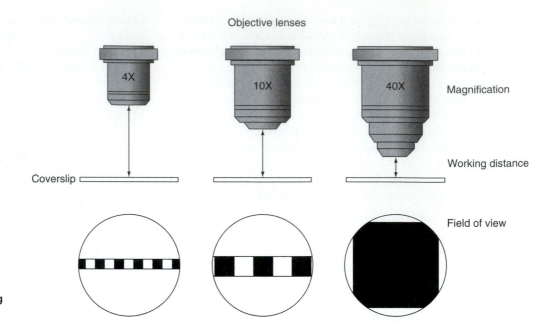

Objective lenses

4X

10X

40X   Magnification

Coverslip

Working distance

Field of view

FIGURE 2.3
Increasing Magnification and Decreasing Field of View

For example, if the diameter of the field of view is 2.5 mm at 40 power (40×), then it is 0.25 mm at 400× (10 times more magnified yet one-tenth the field of view). Calculate or measure the diameter of the field of view under high power.

Diameter of the field of view under high power: _____

As you move from one lens to another, the specimen should remain in the center of the field of view. This is known as being **parcentric**.

Proper Lighting

Too much or too little light makes a specimen difficult to see. You can adjust the light by either adjusting the brightness of the illuminator or adjusting the iris diaphragm lever. On some faintly stained specimens, the material may be difficult to see on low power. One trick is to locate the edge of the coverslip and turn the coarse-focus knob up and down until the edge is in sharp focus. This lets you know that you are in the approximate focal plane for examining the material on the slide. Move the slide to where the specimen should be, adjust the light, and reexamine it.

Preparation of a Wet Mount

You can make relatively quick and easy observations under the microscope as long as the material is thin enough and small enough. One technique for cell examination is to examine cells from the inside of the oral cavity. Do the following procedure.

1. With a toothpick, gently rub the inside of your cheek.
2. Smear the cheek material from the toothpick on a clean microscope slide.
3. Place a drop of methylene blue on the smear.
4. Place a coverslip on one edge of the drop and slowly lower it. Avoid trapping air bubbles in the process

(fig. 2.2). The small, oval structures inside the cells are the nuclei.
5. Draw your observations in the space provided.

Your illustration of a cheek cell:

Microscope Troubleshooting

If you are having a difficult time seeing anything or seeing your specimen in focus, there might be several reasons. Use the following troubleshooting list to help you.

Problem	Solution
Nothing is visible in the lens.	Plug in the microscope.
	Turn on the power supply.
	Rotate the objective lens so that it clicks into place.
	The bulb is burned out; replace the bulb.
You see a dark crescent.	The objective lens is not in proper position; click the lens into place.
All you see is a light circle.	The microscope is out of focus; adjust the coarse-focus knob.
	The light is up too high; turn down the light.
	The iris diaphragm is open too much; close it down slightly.

For another observation, remove a hair from your head (preferably one with split ends) and examine it by making a wet mount. Tie your hair in a knot, place it in the center of the slide, and add a drop of water. Draw what you see in the space below.

Observation of a Prepared Slide

Examine a prepared slide of tissue provided by your instructor. Examine the entire sample using the low-power objective lens. Scan the entire area, looking for areas you want to observe more closely. Move to the next higher power and adjust the focus using the fine-focus knob. Finally, examine the material with the high-power objective lens and draw what you see in the space provided.

Name of the sample (kidney, liver, intestine, etc.):

Your illustration of material from a prepared slide:

You should use the iris diaphragm lever to adjust the amount of light that strikes the specimen. Too much or too little light produces significant changes in the observed image.

Oil Immersion Lens

The objective lenses you have used so far are called "dry" lenses. Your lab may be equipped with microscopes that have oil immersion lenses. The techniques for using these lenses are somewhat different from those for dry lenses. Once you have examined the specimen using the high-power dry lens, find the spot you want to examine and center it in the field of view. Add a drop of immersion oil on top of the coverslip and carefully swing the oil immersion lens into place. Use the fine-focus knob only, or you may drive the oil immersion lens through the slide and break it. Once you have examined the slide, swing the lens away and remove the slide, carefully wiping away the immersion oil with a clean piece of lens paper (do not use your shirt or a paper towel; these can scratch the lens). Use only lens paper to clean the oil from the oil immersion lens. Use another lens paper to remove any remaining oil, if needed.

Cleaning the Microscope

Smudges on the images you view through the microscope may be due to several things. There may be makeup or dirt on the ocular lens (or lenses). There may be dirt, oil, salt, stains, or other material on the objective lenses. To clean a lens, place a small amount of lens-cleaning fluid on a clean sheet of lens paper. Make one circular pass on the lens and throw the paper away. If you continue to clean the lens with the same lens paper, you can grind dirt or dust into the lens. Use a fresh piece of lens paper and repeat the procedure if further cleaning is needed.

You may want to clean the microscope slide before you examine it. Use a cleaning paper, such as a Kimwipe, to clean oil or dust from the slide.

Finally, dust may have collected inside the microscope over the years or the lenses may be scratched. There is nothing you can do about this, though you may want to bring this to your instructor's attention.

Name _____ Date _____

1. If the ocular lens is 10×, what are the total magnifications for the following objective lenses?

   a. 7× _____

   b. 15× _____

   c. 20× _____

2. What is the name of the thin glass slide that is placed on top of a specimen?

3. The microscope that you use in lab is a(n)

   a. light microscope.   b. dissecting microscope.   c. electron microscope.

4. What is the name of the circle you see when you look through the ocular lens of the microscope?

5. What is the function of the iris diaphragm of the microscope?

6. If the diameter of the field of view is 5.6 mm at 40×, what is the diameter at 80×?

7. Label the parts of the microscope illustrated.

    fine-focus knob

    objective lens

    mechanical stage

    ocular lens

    coarse-focus knob

    body tube

    light source

    base

    arm

    condensor

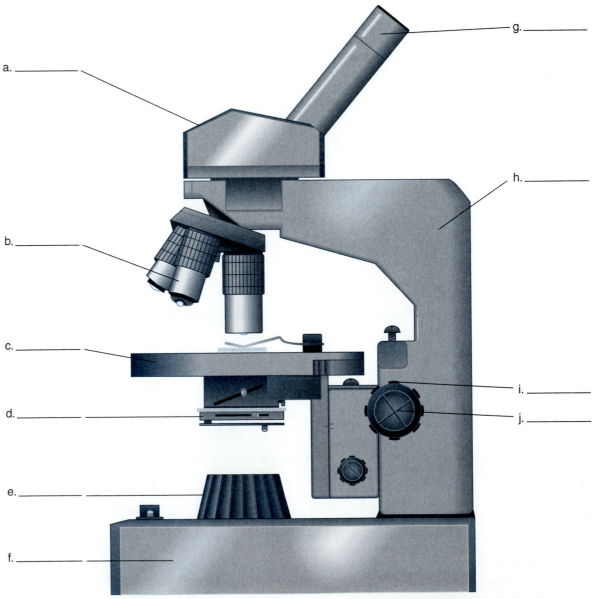

a. _____

b. _____

c. _____

d. _____

e. _____

f. _____

g. _____

h. _____

i. _____

j. _____

8. When you switch from a low-power objective lens (for example, 4×) to a higher-power objective lens (for example, 10×), what happens to the working distance between the lens and the coverslip?

9. When you change from a low-power objective lens to a high-power one, what happens to the field of view? Does it increase or decrease?

10. When should you use the low-power lens on the microscope?

11. How should you clean the lenses of a microscope?

12. What is the proper way to carry the microscope in lab?

13 Examine the following field of view and determine what the size of the object is.

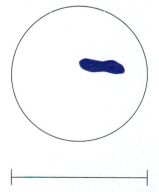

4.5 mm

Cell Structure

INTRODUCTION

The cell is the structural and functional unit of living organisms, including humans. **Cytology,** the scientific study of cells, is an essential part of the study of anatomy. Most diseases that produce obvious physical dysfunction can be traced to some type of cellular change. Cells grow, divide, acquire nutrients, release wastes, respond to local stimuli, and perform many functions, some of which are unique to the organ in which they are found.

The membrane of the cell is the dynamic interface between the internal, living environment of the cell and the external environment. In humans, most cells are found in a liquid medium, which provides the cells with nutrients, oxygen, hormones, water, ions, and other materials. From the cell's interior, it releases ammonia, carbon dioxide, and other metabolic products into this liquid medium. The cell membrane is vital in the exchange of materials between the cell's internal fluid and the cell's external environment. The exchange of materials between the cell and its environment maintains the homeostatic balance the cell must have in order to survive. Even small changes in the concentration of certain materials in the cell can lead to cellular death, so the constant adjustment of water, ions, and other metabolic products is extremely important. In large part, the cell membrane actively regulates what enters and what exits the cell.

In this exercise, you examine the structure of animal cells and learn how the cells of the body divide to make new cells.

LEARNING OBJECTIVES

At the end of this exercise you should be able to
1. describe the importance of cells to the body;
2. list the functions of the cell membrane;
3. list all of the organelles and their functions;
4. describe the three main events of the cell cycle;
5. name the four phases of mitosis and what happens during each phase.

MATERIALS

Models or charts of animal cells

Electron micrographs of cells or a textbook with electron micrographs

Prepared slides of whitefish blastula

Microscopes

Modeling clay (plasticine)—two colors

PROCEDURE

Overview of the Cell

Cells consist of the **plasma** (or **cell**) **membrane** and the **cytoplasm** (fig. 3.1). The plasma membrane is the outer boundary of the cell; although it cannot be seen when examining a slide using the light microscope, its location can usually be determined by the difference in color between the **cytosol,** or **intracellular fluid (ICF),** and the **extracellular fluid (ECF).**

The plasma membrane is composed of a **phospholipid bilayer.** This bilayer consists of outer and inner phospholipid molecules, each with a peripheral phosphate head and a hydrocarbon (lipid) tail directed toward the middle of the membrane (fig. 3.2). Other phospholipid bilayer membranes are found enclosing cytoplasmic structures, and these are known as **unit membranes.**

Inside the plasma membrane is the cytoplasm. The cytoplasm is the portion of the cell in which the cytosol, the cytoskeleton, organelles, and other cellular structures are found. The **organelles** (*organelle* = small organ), are small, cellular structures suspended in the cytosol. In the spaces provided, draw a sketch of the following structures and label them using the listed terms.

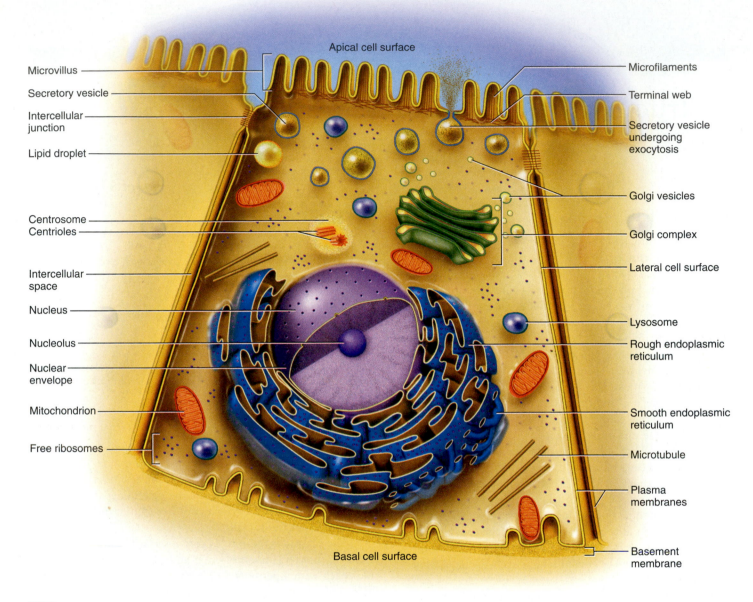

Microvillus

Secretory vesicle

Intercellular junction

Lipid droplet

Centrosome
Centrioles

Intercellular space

Nucleus

Nucleolus

Nuclear envelope

Mitochondrion

Free ribosomes

Apical cell surface

Microfilaments

Terminal web

Secretory vesicle undergoing exocytosis

Golgi vesicles

Golgi complex

Lateral cell surface

Lysosome

Rough endoplasmic reticulum

Smooth endoplasmic reticulum

Microtubule

Plasma membranes

Basement membrane

Basal cell surface

FIGURE 3.1
Overview of the Cell

Plasma membrane—phospholipid, phospholipid bilayer, transmembrane protein, peripheral protein, channel protein, cholesterol

Mitochondrion—cristae, intramembranous space, matrix, inner membrane, outer membrane

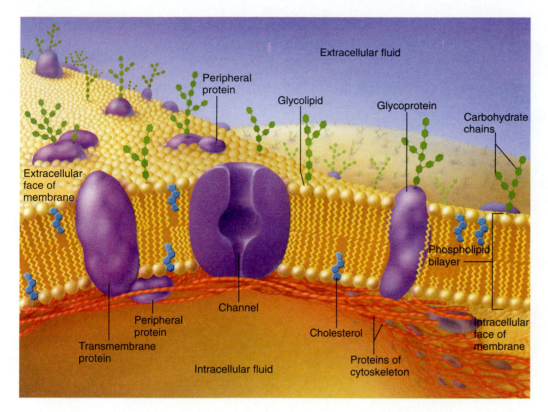

Figure 3.2
Plasma Membrane

Golgi apparatus—cisternae, vesicle, unit membrane

Nucleus—nucleoplasm, nucleolus, nuclear membrane, nuclear pore

Rough and **smooth endoplasmic reticulum**—cisternae, unit membrane, ribosome

Centrioles—microtubules

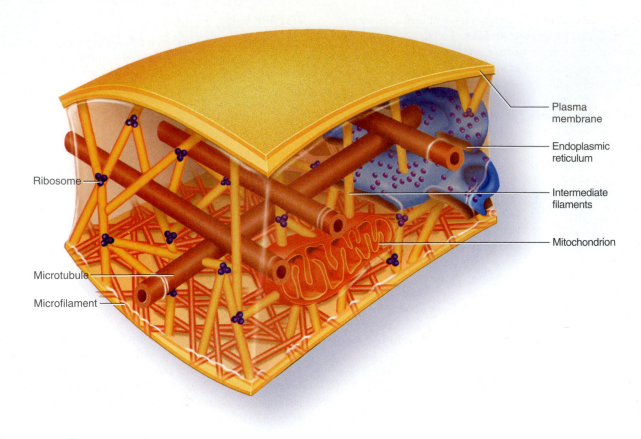

FIGURE 3.3

Cytoskeleton   The cytoskeleton consist of microtubules, microfilaments, and intermediate filaments.

Other Cellular Components

The **cytoskeleton** (fig. 3.3) is a network of small filaments and tubules. Examine the cell models or charts in lab and review material in your text, which has more detailed descriptions of cellular structures.

Microvilli are small extensions of the plasma membrane of some cells that increase the surface area of the cell. They are found in the cells of the digestive and urinary systems.

Cilia and **flagella** extend from the edges of the cell; they consist of bundles of microtubules and are covered by the plasma membrane. Most cells do not have cilia, and flagella are found only in human sperm cells. Cilia are found on cells that are involved in movement, such as the movement of mucus along the free edge of the respiratory passage or in the lumen of the uterine tubes. The general structure of cilia is the same as that of flagella, but cilia are shorter. Cilia are seen in figure 3.4.

Functions of Cellular Structures

Each organelle has at least one function in the cell. **Ribosomes,** the smallest of the organelles (about 25 nm in diameter), produce proteins. **Mitochondria** convert the energy in food to energy in molecules of adenosine triphosphate (ATP). **Rough endoplasmic reticulum** is near the nucleus and produces proteins for use out-

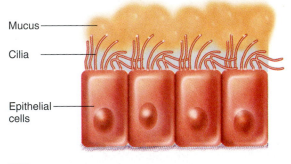

FIGURE 3.4

Cilia on Cells in the Respiratory Passage

side the cell. **Smooth endoplasmic reticulum** is a distal extensions of the rough endoplasmic reticulum; it produces lipid compounds (phospholipids, steroids, etc.) and detoxifies materials. The **Golgi complex** receives material from the endoplasmic reticulum and other parts of the cytoplasm, assembles large molecules, and transports the material out of the cell in **secretory vesicles.** The membrane of the secretory vesicle fuses with the plasma membrane and molecules are ejected without disrupting the plasma membrane.

Two specialized vesicles in the cytoplasm of the cell are lysosomes and peroxisomes. **Lysosomes** digest material with hydrolytic enzymes in a process known as phagocytosis; therefore, they are sometimes known as phagocytic vesicles. **Peroxisomes** use enzymes to convert hydrogen peroxide (a toxic material) to water and oxygen. Hydrogen peroxide is formed in cells by the metabolism of fatty acids and amino acids, as well as by the interaction of water with an unstable form of oxygen known as oxygen free radicals.

Centrioles are unique organelles. They form microtubules and a structure known as the spindle apparatus, which is involved in cellular division.

The **nucleus** is an organelle that has two major functions: one is to house the genetic information of the cell, and the other is to control the various tasks of the cell. **Nucleoli** (singular, **nucleolus**) make ribosomes, the protein-producing organelles in the cytoplasm of the cell.

The Cell Cycle

One of the great wonders of science is the mechanism by which a single cell, the result of the fusion of egg and sperm, develops into a complex, multicellular organism, such as a human. Various estimates put the number of cells in the human body in the trillions. All of these cells came from the first cell, or zygote. In this part of the lab exercise you examine the mechanism by which this occurs.

Most cells produce more cells by a process known as the cell cycle. The cell cycle can be divided into three events: **interphase, mitosis,** and **cytokinesis,** illustrated in figure 3.5. Most of the time spent in the life of the average cell is in interphase.

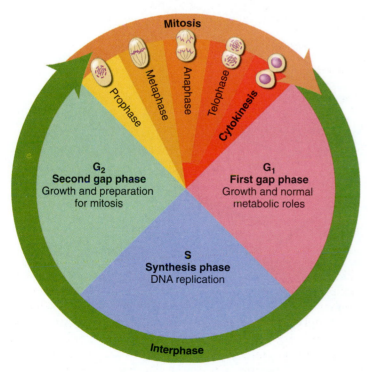

FIGURE 3.5

The Cell Cycle   The mitotic phase has been expanded to see the distinct sections. Most of the time, a cell is in interphase. Cytokinesis is a separate event, beginning in anaphase and ending in telophase.

INTERPHASE

Interphase is the time when a cell undergoes growth and duplication of DNA in preparation for the next cell division. If a cell is not going to divide any further (such as most neural cells and some muscle cells), then interphase is regarded as the time when a cell carries out normal cellular function.

Interphase has three separate phases, known as the G_1 phase, S phase, and G_2 phase. In the G_1 phase (G stands for "gap"), cells are in the process of growing in size and producing organelles. In the S phase (S stands for "synthesis"), the DNA of the cell is duplicated. The double helix of the DNA molecule unzips and two new, identical DNA molecules are produced. In the final phase of interphase, the G_2 phase, the cell continues to grow and prepares for the process of mitosis. Some cells do not undergo further division and are said to be in the G_0 (G zero) phase. Cells in interphase have a distinct nuclear membrane, and the genetic information is dispersed in the nucleus as chromatin.

MITOSIS

Mitosis is a continuous phenomenon that has been divided into four distinct phases. Mitosis is nuclear division and it involves the division of genetic information to produce two identical nuclei, which eventually occupy two cells. In order for mitosis to occur, the chromatin in the nucleus of the cell must condense into compact units called chromosomes (each species has its own number). Chromosomes consist of two chromatids held at the center by a centromere. Examine figure 3.6 for the structure of a chromosome. You should also note the structure of chromosomes as you study the cells undergoing mitosis.

The four phases of mitosis—prophase, metaphase, anaphase, and telophase—are described next. Refer to figure 3.7 as you read the descriptions.

Prophase   The first indication that a cell is undergoing mitosis is the condensation of chromatin into **chromosomes.** Each chromosome consists of two elongated arms known as **chromatids,** which are connected to each other by a **centromere.** See figure 3.7*a.*

In addition to the thickening of the chromosomes, the nucleolus disappears and the nuclear membrane begins to disassemble.

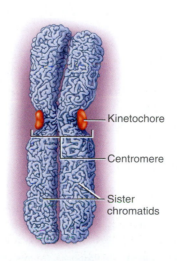

Kinetochore

Centromere

Sister chromatids

FIGURE 3.6

Structure of an Isolated Chromosome

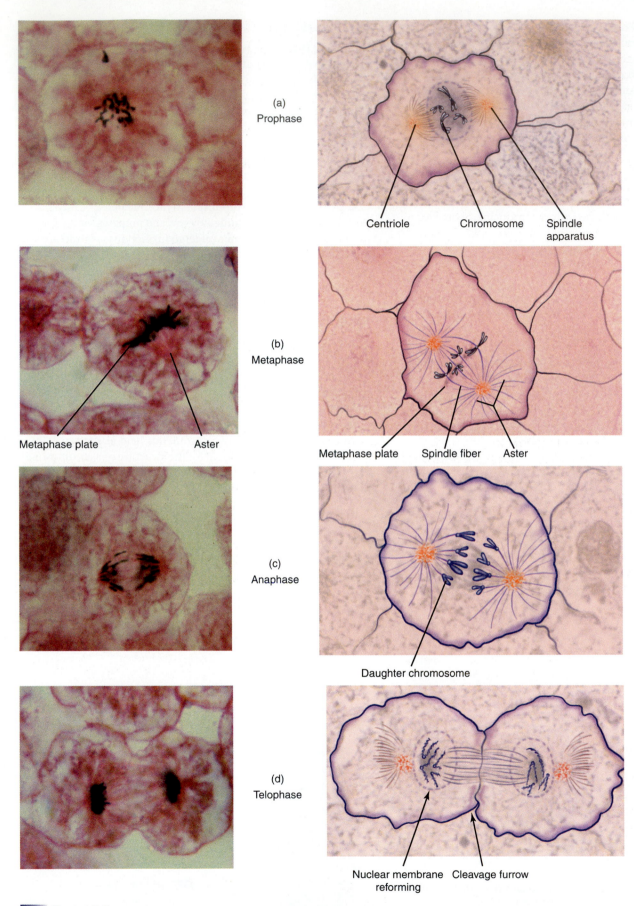

(a)
Prophase

Centriole   Chromosome   Spindle
apparatus

(b)
Metaphase

Metaphase plate   Aster

Metaphase plate   Spindle fiber   Aster

(c)
Anaphase

Daughter chromosome

(d)
Telophase

Nuclear membrane   Cleavage furrow
reforming

FIGURE 3.7

Phases of Mitosis Whitefish blastula (1,000×)   (a) Prophase; (b) metaphase; (c) anaphase; (d) telophase with cytokinesis.

In order for the chromosomes to separate and move apart, the nuclear membrane must not be present. Additionally, the mitotic apparatus appears. The mitotic apparatus consists of spindle fibers, which attach to the chromosomes at regions of the centromere known as the **kinetochores,** and two asters, which are points of radiating astral fibers at each end (pole) of the cell. In the center of the astral spindle fibers are two small structures known as centrioles forming the centrosome.

Metaphase   In this phase the chromosomes align between the poles of the cell in a region known as the metaphase plate (fig. 3.7*b*).

Anaphase   In anaphase, the chromatids separate at the centromere and each chromatid is now known as a new chromosome. The spindle fibers pull the new chromosomes toward opposite poles of the cell. The centromere region moves first and the arms of the chromosomes follow (fig. 3.7*c*).

Telophase   Once the new chromosomes reach the poles, telophase begins. The chromosomes begin to unwind into chromatin, the nucleolus reappears, and the nuclear membrane begins to reform. The mitotic apparatus disassembles, thus terminating mitosis (fig. 3.7*d*).

CYTOKINESIS

The splitting of the cell's cytoplasm into two parts is known as cytokinesis. Although cytokinesis is a distinct process, it frequently begins during late anaphase or early telophase. In late anaphase, as the chromosomes are moving to the poles, the plasma membrane begins to constrict, primarily by contracting actin filaments, at a region known as the **cleavage furrow.** This begins the process of dividing the cytoplasm (fig. 3.7*d*), ending as the cell splits into two separate daughter cells. The cytoplasm and the organelles are effectively divided into two parts.

Examine a slide of whitefish blastula and look for the various phases of the cell cycle in those cells. Most of the cells that you see are in a particular part of the cell cycle. What is this phase and why are most of the cells in this phase?

Compare a slide with figure 3.7. Draw representative cells in each phase of mitosis in the space below.

Simulation of Mitosis

Review the phases of mitosis in table 3.1. Using two colors of modeling clay, make chromosomes. You should have a long chromosome and a short chromosome of each color, for a total of four chromosomes. Each chromosome should have two chromatids, and the chromosomes should be joined in the middle. Draw a large circle on a sheet of paper to represent a cell. Manipulate the clay chromosomes to show how mitosis occurs. After you have done this, describe the process of mitosis in your own words below. List each stage and what happens in that stage.

TABLE 3.1	
Major Events of Mitosis	
Prophase	Chromatin condenses to form chromosomes.
	Nuclear membrane disappears.
	Spindle apparatus forms.
	Nucleolus disappears.
Metaphase	Chromosomes align on the metaphase plate.
Anaphase	Chromosomes split and daughter
	chromosomes migrate to poles;
	cytokinesis often begins.
Telophase	Chromosomes reach poles;
	nuclear membrane reforms.
	Chromosomes unwind to chromatin;
	cytokinesis divides the cytoplasm.
	Nucleolus reappears.

Name _____ Date _____

1. Cells in the body have a fluid surrounding them. What is the name of this fluid?

2. The cytoplasm has a liquid portion. What is it called?

3. What structure in a cell is mostly composed of a phospholipid bilayer?

4. Which organelle is responsible for ATP production?

5. Which organelle makes protein for use outside of the cell?

6. Which organelle in the cell produces lipids?

7. Which organelle contains DNA?

8. Which cellular structure is responsible for ribosome production?

9. Name the cellular structures in the illustration using the terms provided.

ribosomes   nucleolus

mitochondrion   Golgi complex

smooth endoplasmic reticulum   rough endoplasmic reticulum

centriole   plasma membrane

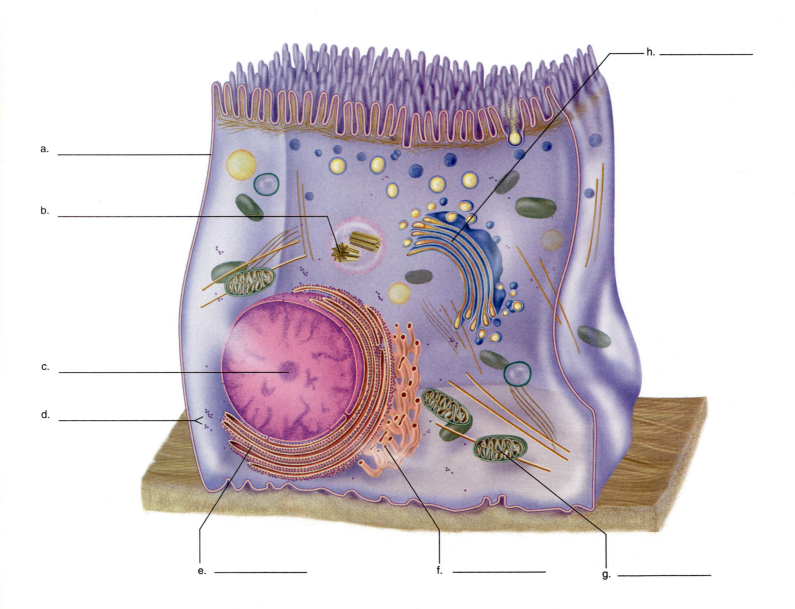

10. What are the three events of the cell cycle?

11. Of the three events of the cell cycle, in which one is DNA duplicated?

12. When in the cell cycle do chromosomes first split apart?

13. The division of the cytoplasm occurs in what part of the cell cycle?

14. Describe the four phases of nuclear (mitotic) division and what occurs during those phases.

15. Name the phases of the cell cycle as illustrated.

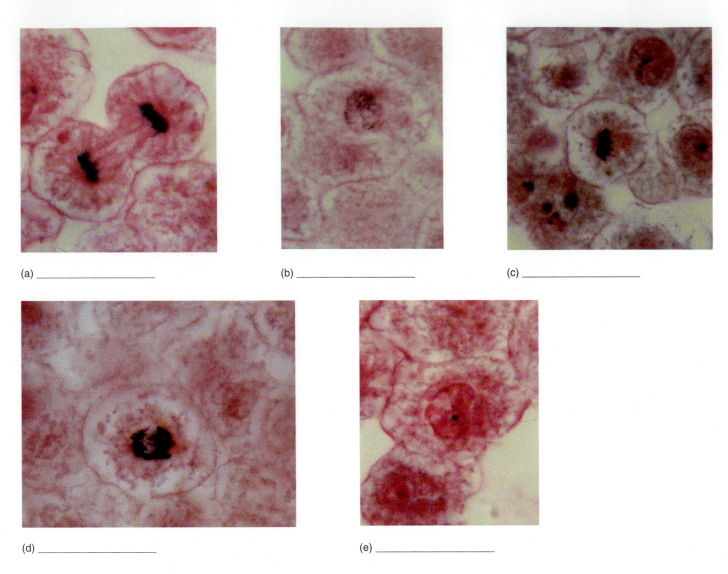

(a) _____

(b) _____

(c) _____

(d) _____

(e) _____

16 Name the phases in interphase and describe what happens during those phases.

LABORATORY

Tissues

INTRODUCTION

The study of tissues, called **histology,** is microscopic anatomy. Histology is very important because many organic dysfunctions of the human body are diagnosed at the tissue level. Surgical specimens are routinely sent to pathology labs, so that accurate assessment of the health of the tissue, and consequently the health of the individual, can be made.

Individual tissues have specific functions. For example, muscle contracts, and nervous tissue conducts impulses. These functions are due to the specialization of the cells that form the tissue and the extracellular material.

There are four main tissue types found in the human body—epithelial tissue, connective tissue, muscular tissue, and nervous tissue—and the organs of the body are formed by two or more of these tissues. These tissues vary by the type, function of the individual cells, and by the nature of the **matrix,** or extracellular material, present.

In this exercise you examine numerous slides of tissue and begin an introduction to histology. In later exercises you revisit histology as you examine various organ systems.

LEARNING OUTCOMES

At the end of this exercise you should be able to
1. recognize the various types of epithelium;
2. associate a particular tissue type with an organ, such as kidney or bone;
3. examine a slide under the microscope or a picture of a tissue and name the tissue represented;
4. distinguish between cartilage and other connective tissues;
5. list the three parts of a neuron;
6. describe the muscle cell types according to location and structure.

MATERIALS

Microscope
Colored pencils
Epithelial tissue slides
  Simple squamous epithelium
  Simple cuboidal epithelium
  Simple columnar epithelium
  Pseudostratified columnar epithelium
  Stratified squamous epithelium
  Transitional epithelium
Muscular tissue slides
  Skeletal muscle
  Cardiac muscle
  Smooth muscle
  All three muscle types
Nervous tissue slide
  Spinal cord smear
Connective tissue slides
  Loose (areolar) connective tissue
  Dense regular connective tissue
  Dense irregular connective tissue
  Elastic connective tissue
  Adipose tissue
  Reticular connective tissue
  Bone marrow
  Hyaline cartilage
  Fibrocartilage
  Elastic cartilage
  Ground bone
  Cancellous bone
  Blood

PROCEDURE

Before you begin this exercise you should be thoroughly familiar with the microscopes in your lab. If you need a review, go back to laboratory exercise 2. As you examine various tissues, look for distinguishing features that will identify the tissue. It is a good idea to examine more than one slide of a particular tissue, so that you see a range of samples of that tissue. Frequently, the material you see in the lab is from a slice of an organ and as such includes more than one tissue type. For example, a sample of cartilage taken from the trachea frequently contains epithelial tissue, adipose, and other connective tissues in addition to the cartilage you want to study. If you are looking for smooth muscle from the digestive tract, you may also find both epithelial tissue and connective tissue in the slide. Use the figures in this exercise to help you locate tissues on the prepared slides. Examine each slide by holding the slide up to the light and visually locating the sample. Then put the slide on the microscope and examine it on low power, scanning around the slide. Move to increasing magnifications after you have identified the tissue. If you cannot identify the tissue after some searching, ask your lab partner or your instructor for help.

Epithelial Tissue

Tissues consist of cells and extracellular material known as matrix. Epithelial tissue is a highly cellular tissue, meaning that it is composed mostly of cells with little matrix. It covers or lines parts of the body (such as the skin on the outside or the digestive tract on the inside) or is found in glandular tissue, such as the sweat glands or the pancreas. When you look at epithelial tissue under the microscope, examine the edge of the sample because epithelial tissue is frequently found as a lining. In most cases, epithelial tissue adheres to the underlying layers by way of a **basement membrane,** a noncellular adhesive layer. In the skin, the epidermis is made of epithelial tissue and the basement membrane connects the epidermis to the underlying dermis. Epithelial tissue is classified according to the shape of the cells and the number of layers present. The cell shapes are **squamous** (SKWAY-mus=flattened), **cuboidal,** and **columnar.** The number of layers are **simple** (cells in a single layer) or **stratified** (cells stracked in more than one layer). Examine tables 4.1 and 4.2 for an overview of epithelium. Epithelial tissue is listed by cell type in the following discussion. Tissues are usually stained, so that the details become visible. The most common stain is hematoxylin and eosin (H&E) stain. Hematoxylin stains the nucleus purple, and eosin colors the cytoplasm pink. The edge of the cell touching the basement membrane is the basal surface, and the upper edge is the apical surface.

SIMPLE EPITHELIA

The single layer of cells in simple epithelia is located on the basement membrane, which may adhere to connective tissue, muscle, or other epithelial tissue. Simple epithelium is classified according to the following shapes.

TABLE 4.1
Simple Epithelia

Simple Squamous Epithelium

Microscopic appearance: single layer of flat cells

Significant locations: lungs, inside of heart and blood vessels

Functions: diffusion and smooth lining, secretion of serous fluid

Simple Cuboidal Epithelium

Microscopic appearance: small cubes or wedge-shaped cells in single layer

Significant locations: kidney tubules and liver

Functions: absorption and secretion

Simple Columnar Epithelium

Microscopic appearance: tall cells in one layer, with nuclei typically in basal part of cell

Significant locations: from stomach to intestines

Functions: absorption and secretion

Pseudostratified Columnar Epithelium

Microscopic appearance: looks stratified but all cells arise from basement membrane, often ciliated

Significant locations: respiratory passages

Functions: secretion of mucus and trapping dust particles, moving them away from lung

TABLE 4.2
Stratified Epithelia

Stratified Squamous Epithelium

Microscopic appearance: many layers, cells cuboidal but flattened toward surface

Significant locations: epidermis, oral cavity, vagina

Functions: Resists abrasion, prevents microbial infection, retards water loss in skin

Transitional Epithelium

Microscopic appearance: many layers with tear-drop shaped cells that do not flatten toward surface

Significant locations: urinary bladder

Functions: allows stretching of urinary bladder

Simple Squamous Epithelium   This epithelial type consists of thin, flat cells that lie on the basement membrane like floor tiles. If these cells are seen from a side view, they look flat and their resemblance to floor tiles becomes apparent. Examine a prepared slide of simple squamous epithelium, usually seen as either a surface view or a side view. Simple squamous epithelium is found in the air sacs of lungs; it lines blood vessels and is called **endothelium.** It can be found as the surface layer of many membranes, where it is called **mesothelium.** It provides a smooth lining (as found on the inside of blood vessels), filters material in the

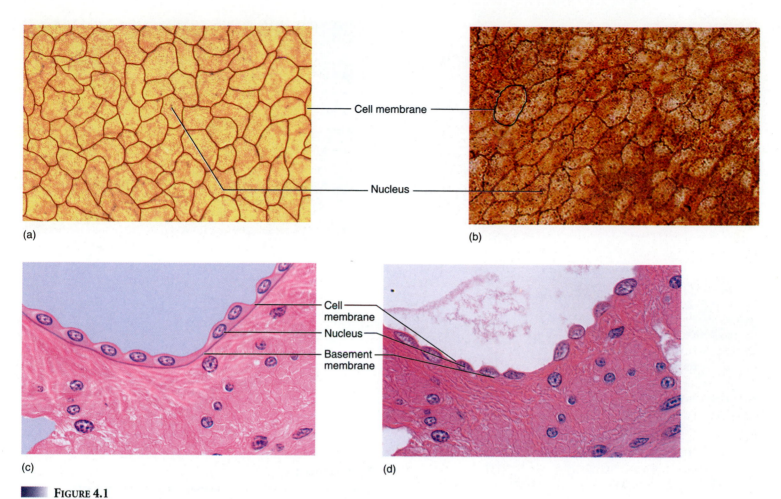

Cell membrane

Nucleus

(a)

(b)

Cell membrane
Nucleus
Basement membrane

(c)

(d)

FIGURE 4.1

Simple Squamous Epithelium Top view of mesothelium—(a) diagram; (b) photograph (400×). Side view of serosa of small intestine—(c) diagram; (d) photograph (400×).

kidney, or allows diffusion (as in the lungs). Compare your slide with the photograph in figure 4.1. Draw what you see under the microscope in the space provided. Note whether your slide shows the cell as a surface or side view.

Illustration of simple squamous epithelium:

of the glands and glandular organs of the body and form much of the kidneys. Simple cuboidal epithelium is frequently found lining tubules, such as sweat ducts. It is often involved in the secretion of fluids (sweat, oil) or in reabsorption (kidneys). Examine a prepared slide of simple cuboidal epithelium and compare it with figure 4.2. Draw what you see under the microscope in the space provided, labeling the nucleus of the cell and the basement membrane.

Illustration of simple cuboidal epithelium:

Simple Cuboidal Epithelium   This tissue consists of cubelike or wedge-shaped cells mostly equal on all sides. These cells form many

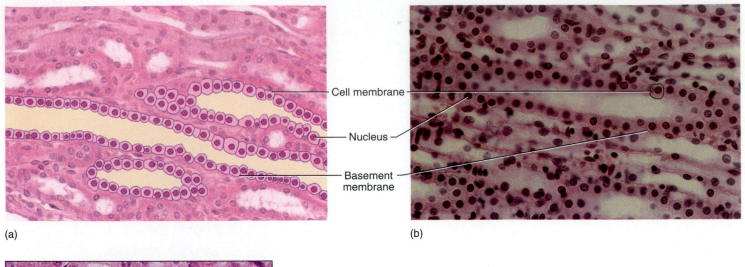

(a)

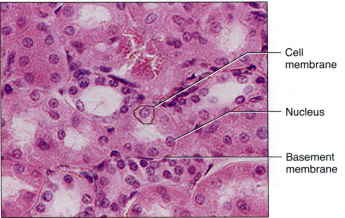

(c)

FIGURE 4.2

Simple Cuboidal Epithelium—Kidney Long section (a) diagram; (b) photograph (400×). Cross section (c) photograph (800×).

Simple Columnar Epithelium This epithelium resembles tall columns anchored at one end to the basement membrane. Simple columnar epithelium can be **ciliated** on the apical surface or it may be smooth. The nonciliated type lines the inner portion of the digestive tract and provides an absorptive area for digested food. Mucus-secreting **goblet cells** are frequently found in simple columnar epithelium. Simple columnar epithelium also lines the uterine tubes and is ciliated in this case. Examine a prepared slide of simple columnar epithelium and compare it with figure 4.3. Draw a representation of what you see in the space provided.

Illustration of simple columnar epithelium:

PSEUDOSTRATIFIED COLUMNAR EPITHELIUM

This tissue may appear as if it occurs in a few layers, but all of the cells originate from the basement membrane. The nuclei are at different levels. Pseudostratified columnar epithelium lines some portions of the respiratory passages, where it protects the lungs by trapping dust particles in a mucous sheet and moves the particles away from them. Examine a prepared slide of pseudostratified columnar epithelium and compare it with figure 4.4. Draw a representative sample of what you see in the space provided.

Illustration of pseudostratified columnar epithelium:

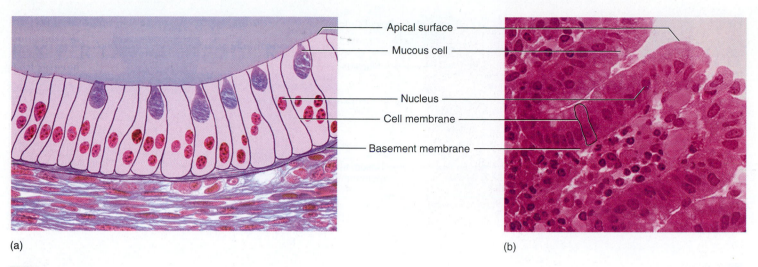

(a)   (b)

FIGURE 4.3
Simple Columnar Epithelium—Stomach (a) Diagram; (b) photograph (400×).

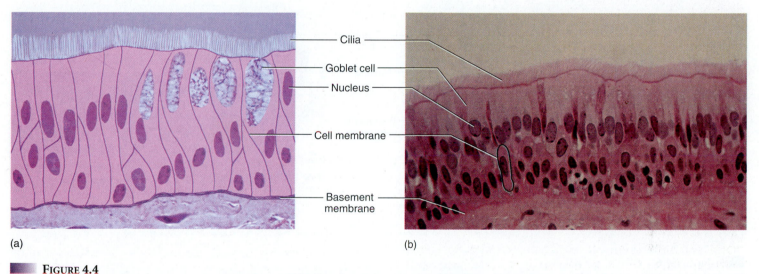

(a)   (b)

FIGURE 4.4
Pseudostratified Ciliated Columnar Epithelium—Trachea (a) Diagram; (b) photograph (400×).

STRATIFIED EPITHELIA

Stratified epithelium is so named because many epithelial cells occur in layers on a basement membrane. The term "stratified" comes from the word "strata" and refers to the layering of these cells. There are two common types belonging to this group.

Stratified Squamous Epithelium   This is a tissue that covers the outside of the body and forms the outermost layer of skin. It also lines the vaginal canal and mouth. The multiple layers of this tissue protect the underlying tissue from mechanical abrasion. This epithelium may have cuboidal-shaped cells at the basement layer, but it derives its name from the cell shape at the free surface. Stratified squamous epithelium comes in two distinct types, **keratinized** and **nonkeratinized** (keratin is a tough protein that hardens cells in the outer layer of the skin). Examine a prepared slide of stratified squamous epithelium and locate the basement membrane. Compare your slide with figure 4.5 and draw what you see under the micro-

scope in the space provided. Locate the basement membrane, and count the layers of cells between it and the surface of this cell type.

Illustration of stratified squamous epithelium:

Record the numbers of layers of cells here: _____

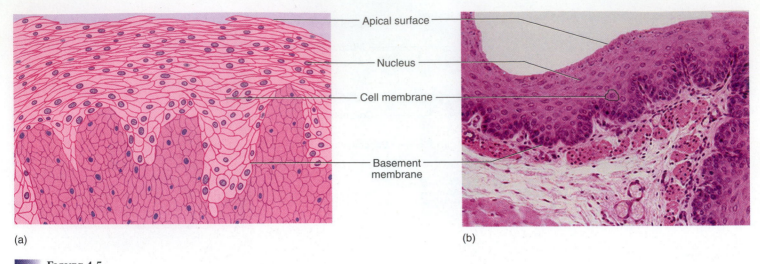

(a)   (b)

FIGURE 4.5

Stratified Squamous Epithelium—Esophagus (a) Diagram; (b) photograph (100×).

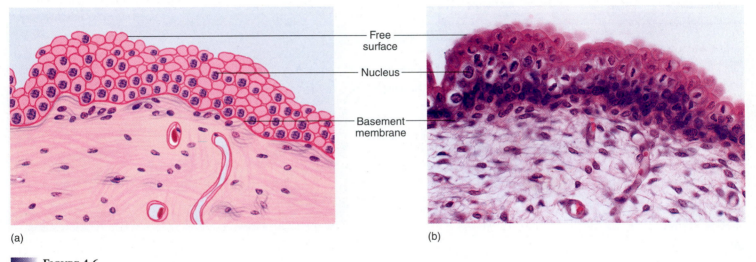

(a)   (b)

FIGURE 4.6

Transitional Epithelium—Allantoic Duct of Umbilical Cord (a) Diagram; (b) photograph (400×).

Transitional Epithelium   This is an unusual tissue in that it has some remarkable stretching capabilities. Transitional epithelium lines the ureter, urinary bladder, and proximal urethra and allows these organs to expand as urine collects within them. Examine a prepared slide of transitional epithelium. Look at figure 4.6 and draw what you see in the space provided.

Illustration of transitional epithelium:

Cell Connections

Cells are held together in many ways. Epithelial cells are held to other structures by the basement membrane, as discussed previously. Cells are frequently held together at specific regions known as **desmosomes. Gap Junctions** connect one cell to another, while allowing a direct cytoplasmic connection between cells. **Tight junctions** hold the cells and seal them off from the extracellular fluid (ECF) near the apical surface. These are illustrated, along with other types of connections, in figure 4.7.

Muscular Tissue

Like epithelial tissue, muscular tissue is a cellular tissue, having mostly cells and little matrix. These cells are contractile and shorten in length due to the sliding of protein filaments across one another. There are three types of muscle: skeletal, cardiac, and smooth muscle.

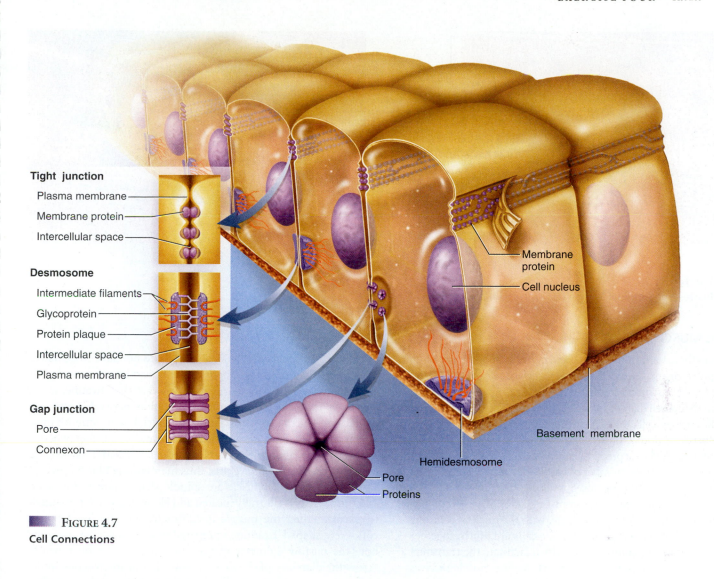

Tight junction
Plasma membrane
Membrane protein
Intercellular space

Desmosome
Intermediate filaments
Glycoprotein
Protein plaque
Intercellular space
Plasma membrane

Gap junction
Pore
Connexon

Membrane protein
Cell nucleus

Basement membrane

Hemidesmosome

Pore
Proteins

FIGURE 4.7
Cell Connections

SKELETAL MUSCLE

The muscles of the human body are organs made mostly of skeletal muscle. Skeletal muscle is sometimes known as striated muscle because it has obvious **striations** (stry-A-shuns) (they look like stripes) in the fiber. Skeletal muscle is voluntary because you have conscious control over this type of muscle. The individual muscle cells have many nuclei and are thus called **multinucleate** or **syncytial.** The nuclei are elongated and are found on the periphery of the cell. Examine a prepared slide of skeletal muscle under high power. Compare this slide with figure 4.8. Draw a representation of the muscle in the space provided. Examine the widths of muscle cells that fit across the diameter of your microscope when viewed at high power. The large diameter of the cells is one way that you can determine if the sample you have in your microscope is skeletal muscle.

Illustration of skeletal muscle:

How many muscle fiber widths do you count? _____

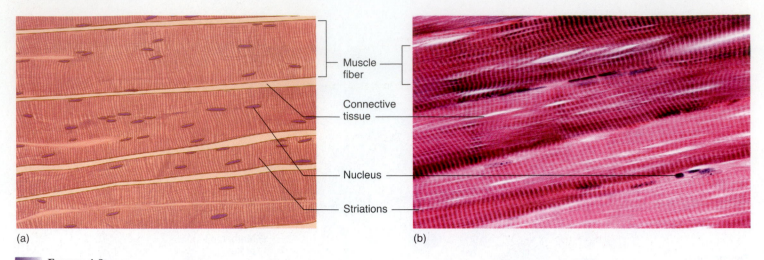

(a)   (b)

FIGURE 4.8
Skeletal Muscle—Longitudinal Section (a) Diagram; (b) photograph (400×).

CARDIAC MUSCLE

Cardiac muscle is found only in the heart, and it is the main tissue making up that organ. Cardiac muscle is somewhat similar to skeletal muscle in that it is striated, but the striations are much less obvious and the individual cell diameters are less than those of skeletal muscle cells. Cardiac muscle cells are called **myocytes** or **cardiocytes.** Another difference between cardiac muscle and skeletal muscle is that cardiac muscle is branched. Place a prepared slide of cardiac muscle under the microscope, and count the number of widths of cardiac muscle fibers across the diameter of the microscope under high power. Cardiac muscle is an involuntary tissue, as it contracts on its own.

These cells are mostly uninucleate, with the nucleus centrally located and oval in shape. Cardiac muscle cells are joined together by intercalated discs (gap junctions), which facilitate the transmission of the electrical impulses in the heart. As one muscle receives an impulse, it sends it on to the next cell. Look at a prepared slide of cardiac muscle and compare it with figure 4.9, noting the striations, intercalated discs, and nuclei of the cells. Draw the cells in the space provided.

Illustration of cardiac muscle:

SMOOTH MUSCLE

This muscle is nonstriated in that the cells do not have cross-hatchings perpendicular to the length of the fiber. Smooth muscle, like cardiac muscle, is involuntary and is found in the intestine, where it propels food along by a process known as peristalsis and segmentation. The uterine contractions during labor are smooth muscle contractions. Smooth muscle is also found in the skin, where it causes hair to stand on end. In blood vessels, smooth muscle regulates the diameter of the vessels, thus maintaining blood pressure. The cells of smooth muscle are spindle-shaped and uninucleate, and the nucleus is centrally located and elongated when the muscle is relaxed. When the muscle is contracted, the nucleus appears corkscrew-shaped. Examine a prepared slide of smooth muscle and note the narrow diameter of the fibers. Count the fiber widths across the diameter of the field of view of your microscope under high power. Compare your slide with figure 4.10 and draw an illustration of smooth muscle in the space provided.

Illustration of smooth muscle:

Record the number of cell widths that occupy the diameter of your field of view at high power: _____

Number of cell widths under high power: _____

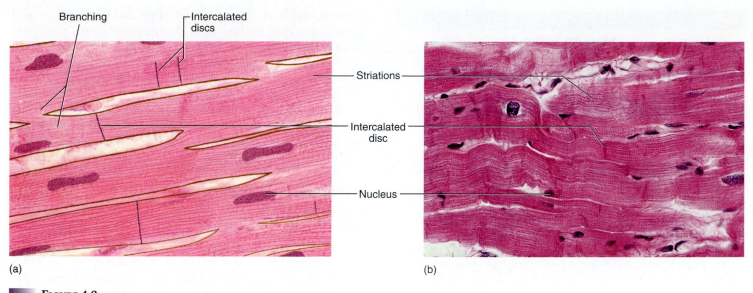

(a)   (b)

▰ **FIGURE 4.9**
Cardiac Muscle—Longitudinal Section (a) Diagram; (b) photograph (1,000×).

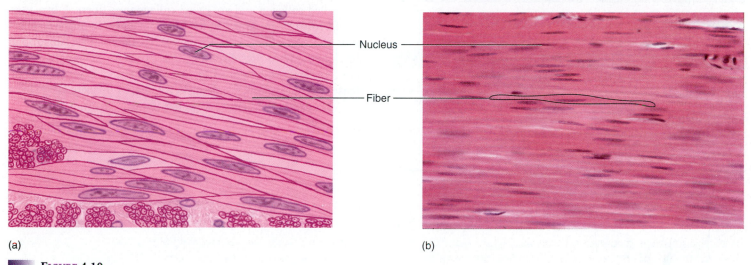

(a)   (b)

▰ **FIGURE 4.10**
Smooth Muscle—Small Intestine, Longitudinal Section (a) Diagram; (b) photograph (400×).

Compare the cell widths in the slides and distinguish among skeletal muscle, cardiac muscle, and smooth muscle.

Examine a prepared slide of all three muscle types, if available. Work with your lab partner and identify each muscle cell. Examine table 4.3 and compare the various muscle types and their characteristics.

Nervous Tissue

Nervous tissue, like epithelial and muscular tissues, is cellular. Nervous tissue is found in the brain, spinal cord, ganglia, and peripheral nerves of the body. The conductive cell of this tissue is the **neuron,** which receives and transmits electrochemical impulses. A neuron is a specialized cell with three major regions—the **dendrites,** the **nerve cell body** (**perikaryon** or **soma**), and the **axon.** Examine figure 4.11 for the regions of a typical neuron and review table 4.4.

Examine the neurons of a smear of the spinal cord of an ox. Look for purple star-shaped structures, which are the nerve cell bodies. Compare them with figure 4.12.

Special cells of the nervous system are called **glial cells** or **neuroglia** (= nerve glue). These cells guide developing neurons to synapses, remove some neurotransmitters from the synapse, and perform numerous other functions. You will study glial cells in greater detail in laboratory exercise 15, "Introduction to the Nervous System."

Connective Tissue

The tissues that you have seen so far, epithelial, muscular, and nervous, are composed mostly of cells with very little matrix between them. This is not the case with connective tissue. There is usually more matrix than cells in connective tissue. **Matrix** is extracellular

TABLE 4.3
Muscular Tissue

Skeletal Muscle

Microscopic appearance: large cell with many nuclei and obvious striations

Significant locations: skeletal muscles of body (biceps brachii, rectus abdominis, etc.)

Functions: voluntary contractions

Cardiac Muscle

Microscopic appearance: smaller, branched cell with one nucleus, intercalated discs and less obvious striations

Significant location: heart

Functions: rhythmic contraction of heart

Smooth Muscle

Microscopic appearance: small, slender cell with one central nucleus and no striations

Significant locations: digestive tract, uterus

Functions: sustained contractions, propulsion of food or delivery of infant

TABLE 4.4
Nervous Tissue

Microscopic appearance: large, star-shaped cells (neurons in brain and spinal cord) with smaller cells (glial cells) nearby

Significant locations: brain, spinal cord (nerves and ganglia)

Functions: transmission of information, assimilation

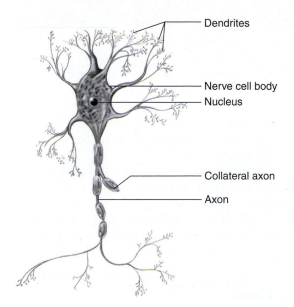

Dendrites

Nerve cell body

Nucleus

Collateral axon

Axon

FIGURE 4.11
Regions of a Typical Neuron

material, and it usually consists of fibers and fluid, gel, or solid ground substance. Because of the abundance of matrix, connective tissue is classified by *specific tissue*.

Connective tissue is diverse, and specific tissues do not seem to have much in common with one another; however, all arise from an embryonic tissue known as **mesenchyme**. Connective tissue can be divided into several subgroups for easier identification. These subgroups are *fibrous connective tissue*, *supportive connective tissue*, and *fluid connective tissue*. These are described in more detail next and are outlined in table 4.5.

FIBROUS CONNECTIVE TISSUE

Dense Connective Tissue   This tissue is also known as white fibrous connective tissue and is composed of collagenous fibers, which can either be parallel—in which case, it is known as dense regular connective tissue—or run in many directions—in which case, it is known as dense irregular connective tissue. **Dense regular connective tissue** is found in tendons and ligaments. **Dense irregular connective tissue** is found in the deep layers of the skin and the white of the eye. The fibers in these tissues are made of a protein called **collagen** and are called **collagenous fibers. Fibrocytes** and **fibroblasts** are found in the tissue, in addition to the collagenous fibers. The fibroblasts actively secrete fibers and subsequently develop into fibrocytes. Examine a slide of dense connective tissue and compare it with figure 4.13. Draw a section of dense regular or dense irregular connective tissue in the space provided.

Illustration of dense connective tissue:

Elastic Connective Tissue   The cells in this tissue are *fibroblasts* and *fibrocytes*. The fibers in this tissue are made of collagenous and **elastic fibers**. Elastic fibers are made of the protein **elastin**.

Elastic tissue is found in the muscular walls of the tunica media of arteries and can be identified as dark, squiggly lines there. Elastic tissue is also found in the vocal cords. Examine figure 4.14 as you look at a prepared slide of elastic tissue and review the tissues in table 4.6. Draw it in the space provided.

Illustration of elastic connective tissue:

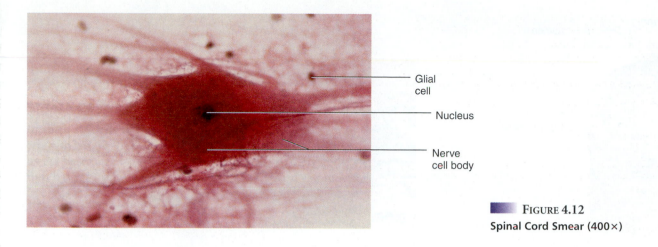

Glial cell

Nucleus

Nerve cell body

FIGURE 4.12
Spinal Cord Smear (400×)

TABLE 4.5
Connective Tissue

I. Fibrous connective tissue

  A. Dense connective tissue

    1. Dense regular connective tissue

    2. Dense irregular connective tissue

    3. Elastic connective tissue

  B. Loose connective tissue

    1. Reticular connective tissue

    2. Areola connective tissue

    3. Adipose tissue

II. Supportive connective tissue

  A. Bone marrow

  B. Cartilage

    1. Hyaline cartilage

    2. Fibrocartilage

    3. Elastic cartilage

  C. Bone

III. Fluid connective tissue

  A. Blood

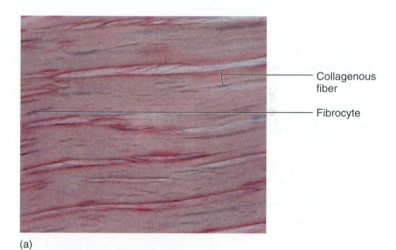

Collagenous fiber

Fibrocyte

(a)

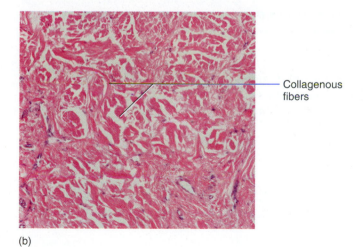

Collagenous fibers

(b)

FIGURE 4.13
Dense Connective Tissue (a) Dense regular connective tissue—tendon (400×); (b) dense irregular connective tissue—skin (100×).

Reticular Connective Tissue This tissue has fibroblasts and fibrocytes with **reticular fibers** (made of collagen with a glycoprotein coat). Reticular connective tissue is found in soft internal organs, such as the liver, spleen, and lymph nodes. This tissue provides an internal framework for the organ. If your slide is stained with a silver stain, the reticular fibers appear black. If your slide is stained with Masson stain, they appear blue. In either case, look for fibers that appear branched among small, round cells, which are the cells of the organ (such as spleen, lymph node, or tonsil) that the reticular connective tissue is holding together. Examine a prepared slide of reticular tissue and

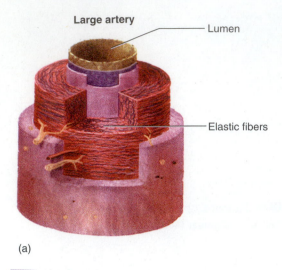

Large artery

— Lumen

— Elastic fibers

(a)

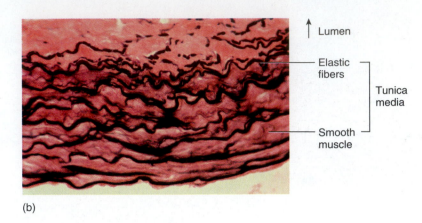

↑ Lumen

— Elastic fibers ⎤
                  ⎥ Tunica
                  ⎥ media
— Smooth ⎦
  muscle

(b)

FIGURE 4.14

Elastic Connective Tissue (a) Overview of artery; (b) photomicrograph (400×).

TABLE 4.6
Dense Connective Tissue

Dense Regular Connective Tissue

Microscopic appearance: closely packed, wavy collagen fibers (or elastic fibers in the case of elastic connective tissue)

Significant locations: tendons, ligaments (vocal cords in elastic connective tissue)

Functions: binds bones together or muscle to bone (provides elastic structure)

Dense Irregular Connective Tissue

Microscopic appearance: randomly appearing collection of densely clustered collagen fibers

Significant locations: dermis, sheaths around cartilage and bone

Functions: provides strength and resists stress and strain against tearing

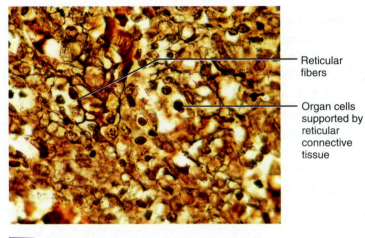

— Reticular fibers

— Organ cells supported by reticular connective tissue

FIGURE 4.15

Reticular Connective Tissue (400×)

compare it with figure 4.15. Review table 4.7. Draw a sample of what you see in the space provided.

Illustration of reticular connective tissue:

TABLE 4.7
Loose Connective Tissue

Reticular Connective Tissue

Microscopic appearance: reticular fibers forming meshwork around organ cells

Significant locations: spleen, thymus, lymph nodes

Functions: provides internal skeleton (framework) for soft organs

Areolar Connective Tissue

Microscopic appearance: scattered arrangement of collagenous fibers with elastic and reticular fibers along with many cells (many with protective immune functions)

Significant locations: attaches epithelia to lower layers, around many internal organs

Functions: binds epithelia to lower layers, insulates organs from infections

Adipose Tissue

Microscopic appearance: large, pale, open cells with nuclei near periphery of cell

Significant locations: under skin, breast tissue, outside of heart and kidney

Functions: energy storage, physical protection

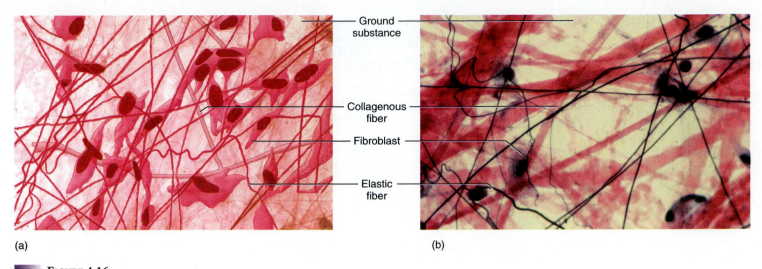

Ground substance

Collagenous fiber

Fibroblast

Elastic fiber

(a)   (b)

FIGURE 4.16

Areolar Connective Tissue (a) Diagram; (b) photograph (400×).

Areolar Connective Tissue Areolar connective tissue (also known as loose connective tissue) is a complex collection of fibers and cells that has a very distinctive look and is one of several specific tissues that make up fibrous connective tissue. Areolar connective tissue is found as a wrapping around organs, as sheets of tissue between muscles, and in many areas of the body where two different tissues meet (such as the boundary layer between fat and muscle). Areolar connective tissue consists of large, pink-stained **collagenous fibers;** smaller, dark **elastic** and **reticular fibers;** and a collection of cells that includes **fibroblasts, fibrocytes, mast cells,** and **macrophages**. Mast cells and macrophages have an immune function; that is, they protect the body from infections. Examine a prepared slide of areolar connective tissue and compare it with figure 4.16. Review table 4.7. Make a drawing of the slide in the space provided.

Illustration of areolar connective tissue:

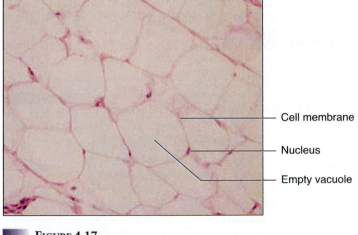

Cell membrane

Nucleus

Empty vacuole

FIGURE 4.17

Adipose Tissue (400×)

see in the microscope with figure 4.17 and make a drawing in the space provided. You may need to close down the iris diaphragm or reduce the amount of light shining on the specimen to see this tissue best.

Illustration of adipose tissue:

Adipose Tissue Adipose tissue (fat) is unusual as a connective tissue in that it is cellular. The cells that make up adipose tissue are called **adipocytes** (*fat cells*). Adipocytes store lipids in large vacuoles, while the nucleus and cytoplasm remain on the outer part of the cell near the plasma membrane. In prepared slides of adipose tissue, the fat has been dissolved in the preparation process. You should be able to locate the large, empty cells, with the nuclei on the edge of some of the cells. Review table 4.7. Compare what you

SUPPORTIVE CONNECTIVE TISSUE

Bone Marrow   Bone marrow is connective tissue that contains numerous cells. **Yellow bone marrow** contains mostly adipose tissue and is found in the shafts of long bones of adults. Another type of marrow is known as **red marrow,** which, in addition to adipocytes, contains stem cells (precursor cells of red blood cells and white blood cells). The function of stem cells is described in more detail in your text. Examine a prepared slide of red marrow and compare it with figure 4.18. Draw a sample of what you see in the space provided.

Illustration of bone marrow:

Cartilage (Cartilage and Perichondrium)   Cartilage is a type of connective tissue in which the matrix is composed of a pliable material that allows for some degree of movement. The ground substance of cartilage is made of chondroitin sulfate and forms a semisolid gel enclosing both fibers and cells. In prepared slides, some of the cells may have come out of the matrix and left a cavity. This cavity is known as a **lacuna** (plural, **lacunae**) and is diagnostic of cartilage tissue. The **perichondrium** is dense irregular connective tissue on the surface of cartilage. There are three types of cartilage in the human body, and they vary by the number of and type of protein fibers. These are hyaline cartilage, fibrocartilage, and elastic cartilage.

Hyaline Cartilage   The most common cartilage in the body is hyaline (HI-ah-lin) cartilage, which is found at the apex of the nose, at the ends of many long bones at joints, between the ribs and the sternum, and in other locations. Hyaline cartilage is clear and glassy in fresh tissue but usually appears light pink in prepared, stained slides. The cells in hyaline cartilage are **chondrocytes,** and the fibers are **collagenous fibers.** Find the chondrocytes in a prepared slide of hyaline cartilage. The fibers do not appear distinct because they blend into the color of the ground substance. The chondrocytes of hyaline cartilage frequently occur in pairs. One way to look at them is as pairs of eyes. Look around the edge of the sample of hyaline cartilage and locate the perichondrium. Review table 4.8. Compare the material under the microscope with figure 4.19. Draw what you see in the space provided.

Illustration of hyaline cartilage:

Fibrocartilage   This tissue is similar to hyaline cartilage in that it has chondrocytes and collagenous fibers, but it differs from hyaline cartilage by having more collagenous fibers. These fibers are visible in the prepared slides. Fibrocartilage is found in areas where more stress is placed on the cartilage, such as in the intervertebral discs, the symphysis pubis, and the menisci of each knee. Examine a prepared slide of fibrocartilage, and locate the collagenous fibers and chondrocytes. Compare the slide with figure 4.20 and make a drawing of fibrocartilage in the space provided.

Illustration of fibrocartilage:

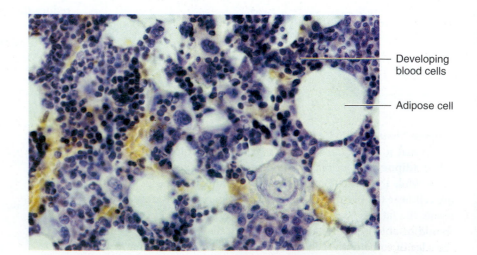

▆ **FIGURE 4.18**
Bone Marrow Photomicrograph of red marrow (400×).

Developing blood cells

Adipose cell

TABLE 4.8
Cartilage

Hyaline Cartilage

Microscopic appearance: usually light blue or pink stained matrix, frequently with pairs of cells appearing like "eyes"

Significant locations: ends of long bones, ribs, larynx, trachea

Functions: reduces friction at joints, keeps air passages open

Fibrocartilage

Microscopic appearance: numerous collagen fibers, cells frequently in rows of four or five

Significant locations: symphysis pubis, intervertebral discs

Functions: protects from wear and tear at weight-bearing or stressed joints

Elastic Cartilage

Microscopic appearance: netlike pattern of fibers around chondrocytes

Significant locations: external ear, epiglottis

Functions: provides flexible framework

Elastic Cartilage   Elastic cartilage is unique because it contains **elastic fibers,** which give the tissue its flexible nature. Elastic cartilage is found in the external ear and in the epiglottis of the larynx. If the prepared slide is stained with a silver stain, the elastic fibers appear black. If the slide is not stained in this way, the fibers may be hard to distinguish. Examine a prepared slide of elastic tissue and look for the chondrocytes and elastic fibers. Compare your slide with figure 4.21 and make a drawing of elastic cartilage in the space provided.

Illustration of elastic cartilage:

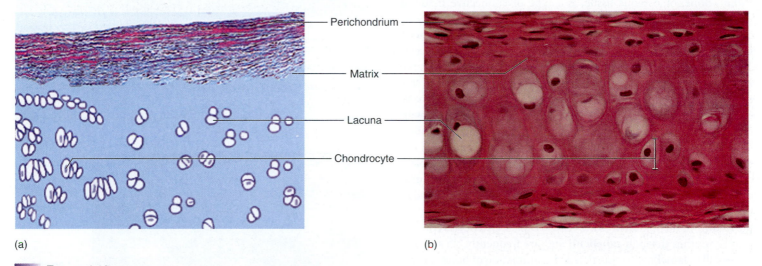

(a)   (b)

FIGURE 4.19

Hyaline Cartilage—Trachea (a) Overview; (b) photomicrograph (400×).

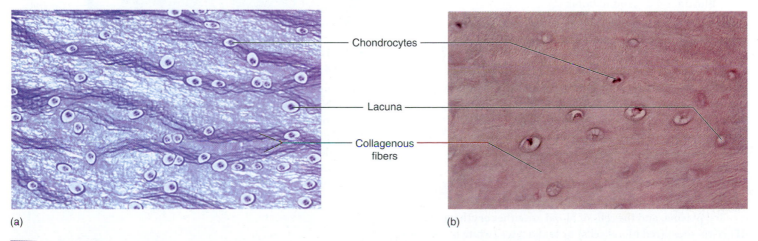

(a)   (b)

FIGURE 4.20

Fibrocartilage (a) Diagram; (b) photomicrograph (400×).

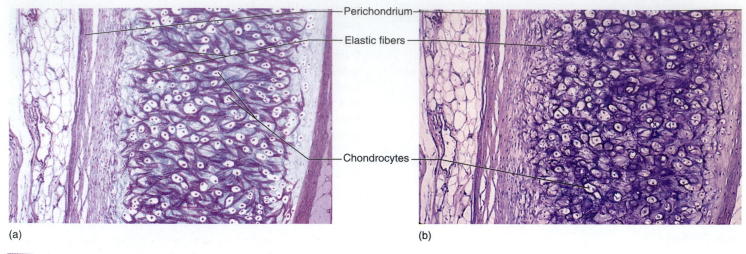

(a)

(b)

FIGURE 4.21
Elastic Cartilage (a) Diagram; (b) photograph (200×).

Bone   Bone, or osseous tissue, is a type of connective tissue in which the ground substance is made of a mineral component (calcium salts) called hydroxyapatite crystal, or bone salt. **Collagenous fibers** are found in the tissue along with osteocytes or bone cells. Two types of bone are common in the body. Compact bone is the more common one, as it makes up the dense material in a long section of bone. Cancellous (spongy or trabecular) bone is found in the end regions of long bones and has threads of bone interspersed with bone marrow. Review table 4.9. Look at a prepared slide of ground bone (which is compact bone) and compare it with figure 4.22. Locate the large, circular regions called **osteons** (which resemble a cross section of a tree trunk with the growth rings), the **central canal,** and the small spaces (lacunae) where **osteocytes** are found in the tissue. Osteons are the long, thin units of bone with the space in the middle of them known as the central canals. Osteocytes connect to one another through **canaliculi** and are frequently clustered in plates called **lamellae.** A more detailed examination of bone appears in laboratory exercise 6, "Introduction to the Skeletal System." After you study the slide, draw an osteon in the space provided.

Illustration of ground bone:

FLUID CONNECTIVE TISSUE

Blood   Blood is a connective tissue in which the ground substance is fluid and no fibers are visible. The ground substance of the blood is called **plasma,** and the cells of blood are either **erythrocytes** (eh-RITH-ro-sites) (red blood cells) or **leukocytes** (white blood cells). Examine a prepared slide of blood and look for the numerous, biconcave, pink discs in the sample. These are the erythrocytes. You

TABLE 4.9
Bone

Microscopic appearance: cells enclosed in calcium salts in concentric circles

Significant locations: skeleton

Functions: protection of soft organs, locomotion along with muscles, scaffold for body

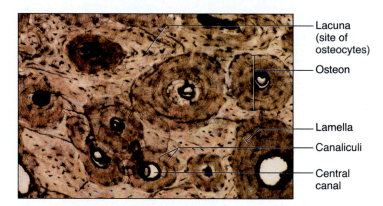

FIGURE 4.22
Ground Bone (150×)

may find larger cells with lobed, purple nuclei. These are the leukocytes. Numerous, small fragments of cells are in the slide. These are called platelets and they come from megakaryocytes. The details of blood are studied in laboratory exercise 20, "Blood Cells." Examine the slide under high power and compare it with figure 4.23. Review table 4.10. Make a drawing of blood in the space provided.

Illustration of blood:

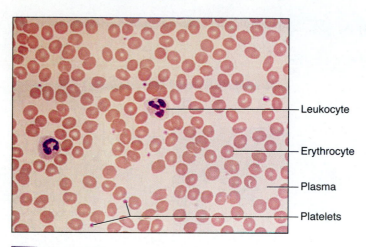

- Leukocyte
- Erythrocyte
- Plasma
- Platelets

FIGURE 4.23
Blood (400×)

Table 4.10
Blood

Microscopic appearance: many cells, most with no nucleus, suspended in plasma

Significant locations: heart, blood vessels

Functions: transport of gases, nutrients, hormones, water, and other material throughout the body

Study Hints

Examine as many different images of tissues as you can. Drawing the material observed in the lab is a good memory tool. Try to make associations between the abstract image of tissue and something common. For example, you could think of bone as looking like tree rings. Adipose tissue may look like angular balloons. If you have difficulty distinguishing between two specimens, try to find a characteristic that separates the two. For example, smooth muscle and dense connective tissue may look similar in certain stained preparations. You will note that smooth muscle has nuclei inside the fibers, while dense connective tissue has nuclei of the fibrocytes between adjacent fibers.

Name _____ Date _____

1. List the four main tissues of the body.

2. What is the name of the noncellular layer that attaches epithelial tissue to other layers?

3. Simple epithelium has _____ layer(s).

4. Name the three general cell shapes of epithelial tissue.

5. A single, flattened layer of cells represents what type of epithelium?

6. What functional problem might occur if the digestive tract were lined with stratified squamous epithelium instead of simple columnar epithelium?

7. A multiple layer of flattened epithelial cells represents what cell type?

8. Can you make the hairs on your arm "stand on end"? Based on your results, what kind of muscle is responsible for doing this?

9. Squamous cells have what general shape?

10. What cell type lines the inside of the urinary bladder?

11. What muscle cell types have the following features?

    nonstriated and involuntary _____

    striated and involuntary _____

    striated and voluntary _____

12. What muscle cell type has intercalated discs?

13. The intestine is mostly composed of what kind of muscle?

14. The heart is made of what type of muscle?

15. What type of muscle makes up most of the muscle of your arm?

16. What cell type is responsible for the transmission of electrochemical impulses?

17. What is the ground substance of blood called?

18. Name the three types of fibers found in areolar connective tissue.

19. What type of connective tissue is found in the middle walls of arteries?

20. What is the cell type found in adipose tissue?

21. Name the outer connective tissue layer that wraps around cartilage.

22. What kinds of fibers are found in fibrocartilage?

23. Describe the functional difference between fibrocartilage and hyaline cartilage.

24. What is another name for calcium salts in bone?

25. In a cross section of a bone, you can usually see two types of bone tissue. What are these called?

26. How do you distinguish between epithelial cells and connective tissue?

27. Label the following photomicrographs by tissue type and by using the terms provided.

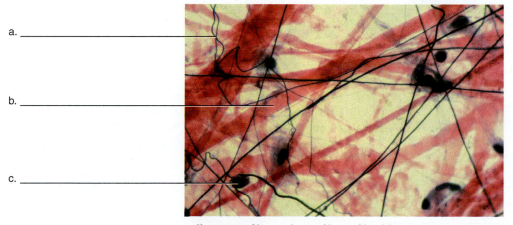

a. _____

b. _____

c. _____

collagenous fibers, elastic fibers, fibroblast

Tissue Name _____

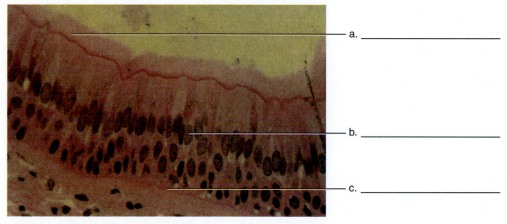

a. _____

b. _____

c. _____

cilia, basement membrane, nucleus

Tissue Name _____

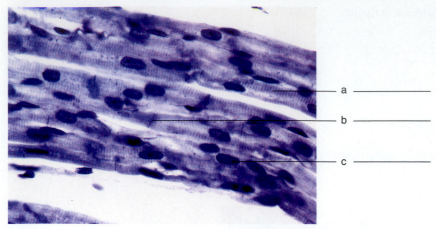

a _____

b _____

c _____

intercalated discs, striations, nucleus

Tissue Name _____

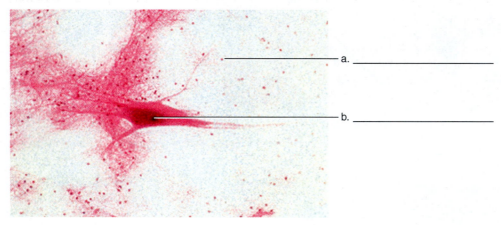

a. _____

b. _____

nerve cell body, glial cells

Tissue Name _____

The Integumentary System

Integumentary System

INTRODUCTION

The integumentary system consists of the skin and associated structures, such as hair, nails, and various glands. The integument consists primarily of a cutaneous membrane composed of an epidermis and a dermis. Below these two layers is an underlying layer called the hypodermis, which anchors the integument to deeper structures. The skin is the largest organ of the body. It protects against water loss, protects against the entrance of microorganisms, and regulates temperature. In this exercise you examine the structure of skin, hair, and nails and name the layers of the epidermis and dermis.

LEARNING OBJECTIVES

At the end of this exercise you should be able to
1. list the two main layers of the integument;
2. list all of the layers of the epidermis from deep to superficial in both thick and thin skin;
3. describe the structure and function of sweat glands and sebaceous glands;
4. draw a hair in a follicle in longitudinal section;
5. discuss the protective nature of melanin;
6. list the layers of the dermis from deep to superficial.

MATERIALS

Models and charts of the integumentary system
Microscopes
Prepared microscope slides of

Thick skin (with Pacinian corpuscles)
Hair follicles
Pigmented and nonpigmented skin
Thin skin

PROCEDURE

Overview of the Integument

Examine models or charts of the integumentary system in the lab and locate the two major regions of the **integument,** the **epidermis,** and the **dermis.** The dermis consists primarily of collagenous fibers and the epidermis is the most superficial layer; it can be determined by the epithelial cells of that layer. Also locate the **hypodermis,** (**subcutaneous layer,** or **superficial fascia**), not a layer of the integument but a structure anchoring the integument to underlying bone or muscle. The hypodermis can be distinguished by the presence of significant amounts of adipose tissue found there. Compare the models in the lab with figures 5.1 and 5.2.

Locate associated structures, such as the **hair follicles** in the dermis, **sebaceous (oil) glands,** and **sudoriferous (sweat) glands.** The sweat glands are connected to the surface by **sweat ducts.**

Dermis

The dermis is responsible for the structural integrity of the integumentary system. It is composed primarily of closely apposed fibers along with blood vessels, nerves, sensory receptors, hair follicles, and glands. The dermis consists of two major regions, the superficial **papillary layer** and a deeper **reticular layer.** The papillary layer is so named for the **papillae** (bumps) that interdigitate with the epidermis, providing good adhesion between the two layers. The reticular layer is the largest layer of the dermis. It is primarily composed of irregularly arranged collagenous fibers along with some elastic and reticular fibers. Examine a slide of thick skin and locate the two layers of the dermis, as seen in figure 5.2.

The majority of the fibers of the dermis are **collagenous** fibers along with lesser numbers of **elastic** and **reticular** fibers. The collagenous fibers provide strength and flexibility. The elastic fibers, as their name implies, provide elasticity to the skin.

Blood vessels are found in the dermis, bringing nutrients to both the dermis and epidermis. The blood vessels do not enter the

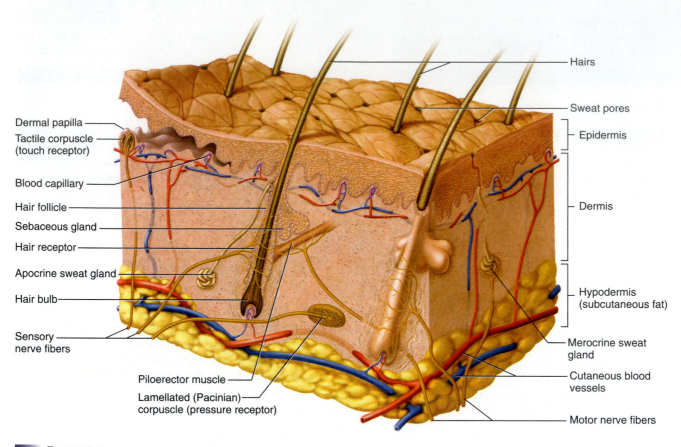

Dermal papilla
Tactile corpuscle
(touch receptor)
Blood capillary
Hair follicle
Sebaceous gland
Hair receptor
Apocrine sweat gland
Hair bulb
Sensory
nerve fibers
Piloerector muscle
Lamellated (Pacinian)
corpuscle (pressure receptor)

Hairs
Sweat pores
Epidermis
Dermis
Hypodermis
(subcutaneous fat)
Merocrine sweat
gland
Cutaneous blood
vessels
Motor nerve fibers

■ **FIGURE 5.1**
Diagram of the Integument

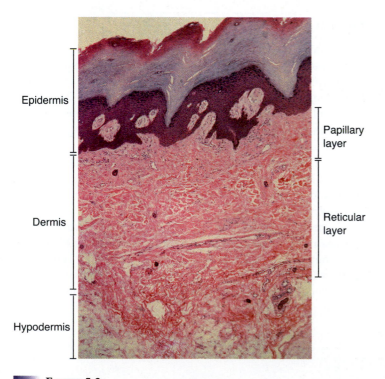

Epidermis

Dermis

Hypodermis

Papillary
layer

Reticular
layer

■ **FIGURE 5.2**
Photomicrograph of the Integument (40×)

epidermis, but the nutrients and oxygen diffuse from the dermis into the cells of the epidermis. Blood vessels are also important because they release heat to the external environment. As the internal temperature of the body rises, vasodilation of blood vessels of the dermis occurs and allows heat to radiate from the skin. In cold weather vessels constrict, decreasing the amount of blood flowing to the dermis, keeping heat in the body core.

There are many **nerves** and neural derivatives present in the dermis. Sensory information, such as light touch, changes in temperature, and the feeling of pain, is transmitted from receptors in the dermis to the central nervous system. These are covered in laboratory exercise 18, "Introduction to Sensory Receptors."

Epidermis

The epidermis is composed of **keratinized stratified squamous epithelium** (**keratinocytes**) hardened with the protein keratin. Keratin is a tough material that protects the lower layers of the epidermis from abrasion.

MICROSCOPIC EXAMINATION OF THE INTEGUMENT

Examine a prepared slide of thick and thin skin under low power. Locate the hypodermis, dermis, and epidermis. The layer with a significant amount of adipose tissue is the hypodermis. In slides stained with hematoxylin and eosin (H&E) stain, the dermis is often colored pink and the superficial region with multicolored layers is the epidermis. The colors may vary in the slides used in your lab. Compare these layers in the prepared slide with figures 5.3 and 5.4.

STRATA OF THE EPIDERMIS

In thick skin the deepest layer of the epidermis is known as the **stratum basale.** This single layer of cells is attached to the dermis by the **basement membrane,** a noncellular layer. Cells in the stratum basale divide repeatedly and push up toward the superficial layers as the cells age. This process is illustrated in figure 5.3.

Melanocytes are specialized cells in the stratum basale. These cells produce a brown pigment known as **melanin.** Melanin protects the stratum basale from the damaging effects of ultraviolet radiation. All people have approximately the same number of melanocytes in their skin, but the amount of melanin produced by these cells varies. People with darker pigmentation produce more melanin, while people with lighter pigmentation have melanocytes that produce less melanin. People with light skin are more susceptible to the effects of ultraviolet radiation and have higher rates of skin cancer.

The **stratum spinosum** is a layer of numerous cells superficial to the stratum basale. In prepared slides, the cells appear star-shaped

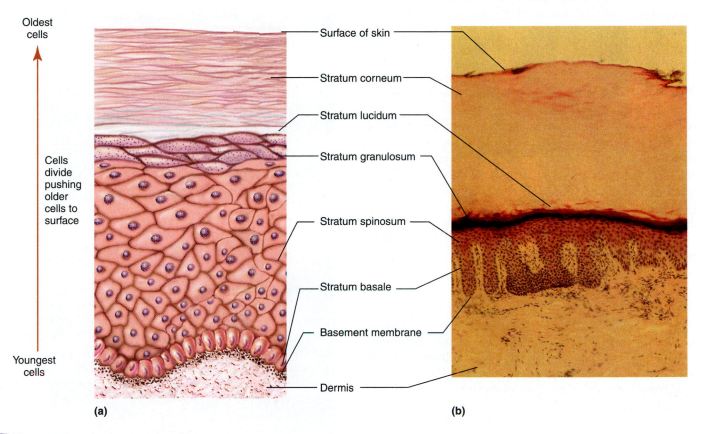

FIGURE 5.3
Proliferation of the Epidermis (a) Diagram; (b) photograph (100×).

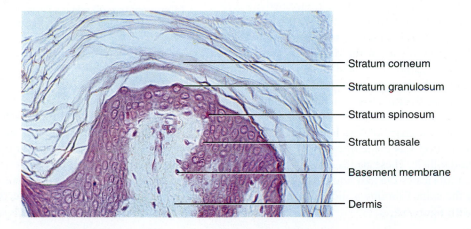

FIGURE 5.4
Strata of the Epidermis Thin skin.

because the cell membrane pulls away from the other cells except in areas where they are joined together at microscopic junctions called desmosomes. The stratum basale and stratum spinosum are sometimes referred to as the **stratum germinativum.** A layer of granular cells is present in the next layer. This is known as the **stratum granulosum.** The cells have purple-staining granules in their cytoplasm. The granules are precursors of keratin found in the outermost layer of the epidermis.

Superficial to the stratum granulosum is the **stratum lucidum,** so named (*lucid* = light) because this layer appears translucent in fresh material. This layer generally appears clear or pink in prepared slides stained with hematoxylin and eosin. It is found only in the skin of the palms of the hand and soles of the feet.

The most superficial layer of the integument is the **stratum corneum,** which is a tough layer consisting of dead cells that are flattened at the surface. Cells of the stratum corneum have been toughened by complete **keratinization,** and thus the epidermis is made of **keratinized stratified squamous epithelium.** Examine a prepared slide of thick skin and locate the various strata of the epidermis as illustrated in figure 5.3.

Cells in the epidermis go through three major events. In the deepest layer, the stratum basale, cells are involved in rapid cell division. Superficial to this layer, cells undergo a process of producing precursor molecules that are toxic to the cells. The final event is the completion of this process, which causes the death of the epithelial cells. The most superficial cells of the epidermis are tough. They protect the skin from microorganisms and desiccation and eventually slough off. The epidermis renews itself about every 6 weeks.

Integumentary Glands

A significant number of glands are found in the dermis and hypodermis. These include **sudoriferous (sweat) glands, lactiferous (milk) glands, sebaceous (oil) glands,** and **ceruminous (earwax) glands.** The sweat glands are composed of simple cuboidal epithelium and come in two different types. **Merocrine sweat glands,** the most common, are sensitive to temperature and produce normal body perspiration. Perspiration reduces the temperature of the body by evaporative cooling. **Apocrine sweat glands** secrete water and a higher concentration of organic acids than merocrine glands. The organic acids, along with bacterial action, produce the characteristic pungent body odor. These glands are concentrated in the region of the axilla and groin and are associated with hair follicles. Examine a prepared slide of skin and look for clusters of cuboidal epithelial cells with purple nuclei in the dermis or hypodermis. These are the sudoriferous glands illustrated in figure 5.5. Note that the glands are tubules and when these are cut in section on your slide you are mostly seeing the tubules cut in cross section.

Sebaceous glands are associated with hair follicles and are larger than sweat glands. You may not see a hair follicle with the sebaceous gland if the section was made through the gland and not through the gland and follicle. These glands secrete an oily material called **sebum,** which lubricates the hair and decreases the wetting action of water on the skin. Examine a slide of hair follicles and compare the slide with figure 5.5.

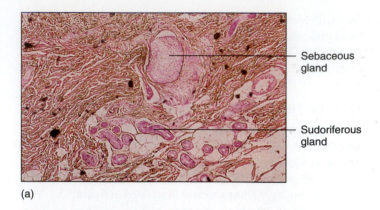
(a)
Sebaceous gland
Sudoriferous gland

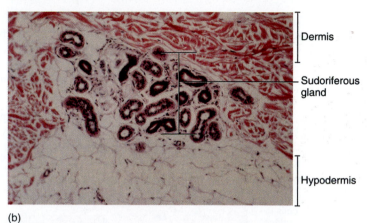
(b)
Dermis
Sudoriferous gland
Hypodermis

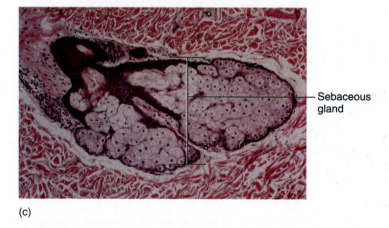
(c)
Sebaceous gland

FIGURE 5.5

Integumentary glands (a) Overview of glands (20×); (b) sudoriferous glands (100×); (c) sebaceous glands (100×).

Hair

Hair is considered an accessory structure of the integumentary system; it consists of keratinized cells that are produced in hair follicles. The **follicle** encloses the hair in the same way a bud vase encloses the stem of a rose. Hair follicles are actually projections of the epidermal layers into the dermis. Hair can thus be considered an epidermal derivative. The hair consists of the **shaft,** a portion that erupts from the skin surface, the **root,** which is enclosed by the follicle, and the **hair bulb,** which is the actively growing portion of the hair. There are two types of hair. **Determinate hair** grows to a specific length and then stops. Determinate hair is found in many places in the body, including the axilla, groin, arms, legs, eye lashes and eyebrows. **Indeterminate hair** continues to grow without regard to length. It is found on the scalp and in beards in males. Examine models and charts in the lab and compare them with figure 5.6.

Look at a prepared slide of hair and locate the root, follicle, and hair bulb. At the center of the bulb is a small structure known as the **dermal papilla**, a region with blood vessels and nerves that reach the hair bulb. The follicle is composed of an outer dermal layer of connective tissue and an inner epithelial layer. These form the **root sheath** of the hair. Locate these structures and compare them with figures 5.6 and 5.7.

PILOERECTOR MUSCLE

When the hair "stands on end," it does so by the contraction of **Piloerector (arrector pili) muscles.** A piloerector muscle is a cluster of parallel smooth muscle fibers that connects the hair follicle to the upper regions of the dermis. In response to cold conditions the piloerector muscles contract, producing goose bumps. The piloerector also contracts when a person is suddenly frightened. Find a piloerector

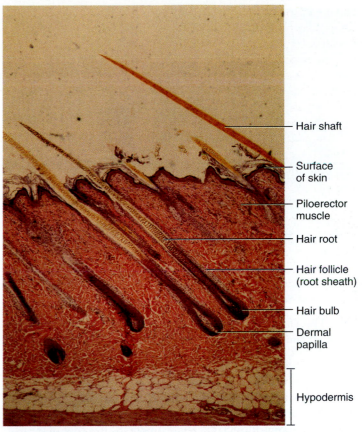

FIGURE 5.7
Photomicrograph of Hair in a Follicle (40×)

muscle in a prepared slide and compare it with figures 5.6 and 5.7. If you have trouble finding piloerector muscle, look for slender slips of smooth muscle that angle from the follicle to the superficial regions of the dermis. You may have to look at more than one slide to find them.

CROSS SECTION OF HAIR

The central portion of the hair is known as the **medulla,** which is enclosed by an outer **cortex.** The cortex may contain a number of pigments, which give the hair its particular color. The brown and black pigments are due to varying types and amounts of melanin. Blond hair has pheomelanin as a pigment. Iron-containing pheomelanin pigments produce red hair. Gray hair is the result of a lack of pigment in the cortex and air in the medulla. The layer superficial to the cortex is the **cuticle** of the hair and looks rough in microscopic sections. Figure 5.8 shows a cross section of hair.

Nails

Nails are sheets of keratinized cells of the stratum corneum on the distal ends of the fingers and toes. The nail itself is called the **nail body** and, as it grows, the **free edge** of the nail is the part you clip. The **cuticle** of the nail is known as the **eponychium,** and the part of the nail that generates the nail body is the **nail root,** which is underneath the eponychium. The **nail bed** is the layer deep to the **nail**

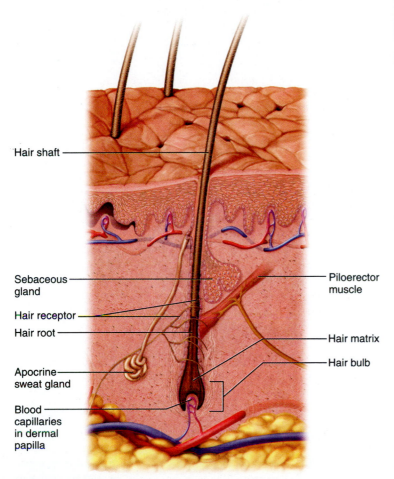

FIGURE 5.6
Diagram of Hair in a Follicle

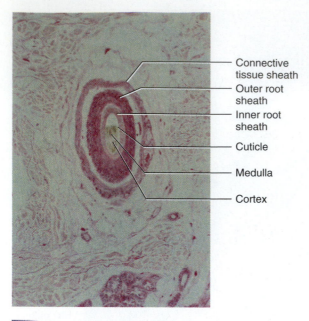

FIGURE 5.8
Hair Root and a Follicle, Cross Section (100×)

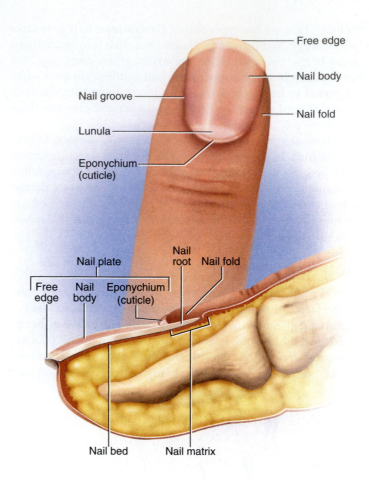

FIGURE 5.9
Fingernail, Top, and Midsagittal Views

body, while the **lunula** is the small, white crescent at the base of the nail. The **nail matrix** undergoes cell division to produce the nail root. The **hyponychium** is the region under the free edge of the nail. Examine figure 5.9 and compare your fingernails with the diagram.

On each side of the nail is the **nail groove,** a depression between the body of the nail and the skin of the fingers. The ridge above the groove is the **nail fold.**

Name_____ Date _____

1. Label the following illustration using the terms provided.

hair root	epidermis
piloerector	sweat gland
hypodermis	hair follicle
stratum corneum	dermis
hair shaft	sebaceous gland
stratum basale	stratum spinosum

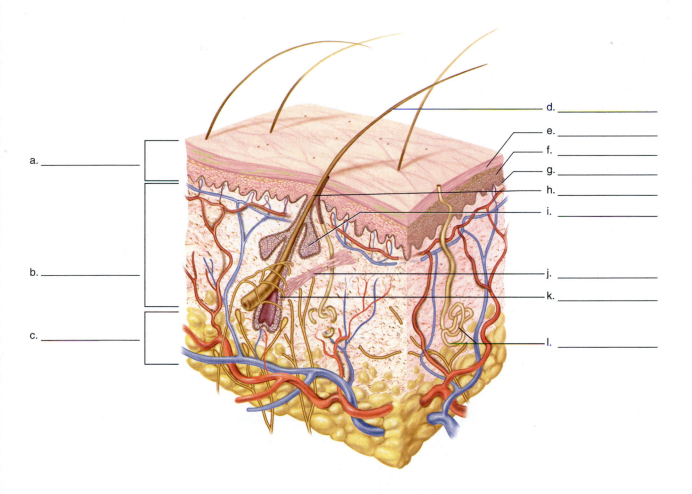

a. _____

b. _____

c. _____

d. _____

e. _____

f. _____

g. _____

h. _____

i. _____

j. _____

k. _____

l. _____

2. The main organ of the integumentary system is the skin. Name three structures that are associated structures in the integumentary system and discuss their function.

3. Which one of these three layers (epidermis, dermis, hypodermis) is not considered part of the integument?

4. What cell type produces a pigment that darkens the skin? What is the function of the pigment?

5. What specific protein makes the epidermis tough?

6. The most superficial layer of skin, (composed of keratinocytes), is dead and provides a protective barrier. If these cells were alive at the surface, how would that compromise the protective function of the integument?

7. Tatoos consist of ink that is injected into the skin. Do you think the ink is injected into the epidermis or the dermis? What support do you have for your answer?

8. Approximately how long does it take for the epidermis to renew itself?

9. The dermis has two main layers. Which one of these is the most superficial?

10. What is the most common connective tissue fiber found in the dermis?

11. The release of heat from the body occurs by blood vessels in what main layer of the integument?

12. Which kind of sweat gland (merocrine or apocrine) is involved in evaporative cooling?

13. What integumentary gland secretes sebum?

14. Hair of the axilla is considered determinate/indeterminate hair (circle the correct answer). Why?

15. Electrolysis is the process of hair removal using electric current. Explain how this might destroy the process of hair growth in relation to the hair bulb.

16. Since hair color is determined by pigment in the cortex and the hair shaft is dead, explain the fallacy of a person's hair turning white overnight.

17. What part of the hair is found on the outside of the skin?

18. What does the piloerector muscle do?

19. The outermost portion of a cross section of hair is known as the

    a. cuticle.   b. lunula.   c. root.   d. medulla.

20. What is another name for the cuticle of a fingernail?

    a. lunula   b. eponychium   c. hyponychium   d. base

Introduction to the Skeletal System

Skeletal System

INTRODUCTION

The skeletal system has numerous functions—including support of the body; protection of soft tissues, such as the brain and lungs; provision of a structure for movement, storage of minerals, and blood cell formation.

The skeletal and muscular systems are intimately related. The interaction between the skeletal and muscular systems is one in which the skeletal system provides the lever system on which the muscles can work, resulting in movement. This is described in more detail in laboratory exercise 10, "Articulations." In this exercise you examine the structure, composition, and development of the skeletal system.

LEARNING OBJECTIVES

At the end of this exercise you should be able to
1. describe the composition of bone tissue;
2. discuss how the skeletal system provides for efficient movement;
3. name the features on specific bones;
4. list the two major types of bone material found in long bones;
5. draw or describe the microscopic structure of compact bone.

MATERIALS

Chart of the skeletal system
Cut sections of bone showing compact and spongy bone
Articulated human skeleton
Articulated cat skeleton (if available)
Disarticulated human skeleton
Prepared slide of ground bone
Prepared slide of decalcified bone
Bone heated in oven at 350ºF for 3 hours or more
Bone placed in vinegar for several weeks or 1 N nitric acid for a few days
Microscopes

PROCEDURE

Divisions of the Skeletal System

The skeletal system can be sorted into two main divisions according to location. The **axial skeleton** is so named because it is in the vertical axis of the body. It consists of the skull, hyoid bone, vertebral column, ribs, and sternum. The **appendicular skeleton** is that part of the skeleton that is "attached to" the axial skeleton. It can be divided further into the **pectoral girdle**, the **upper extremity,** the **pelvic girdle,** and the **lower extremity**. Locate the major bones in lab and compare them with figure 6.1 and table 6.1.

Major Bones of the Body

Examine figure 6.1 for the major bones or bone regions of the body. You should be familiar with the major bones of the body before you study the specifics of the individual bones in later exercises. It helps to take a bone from a disarticulated skeleton and compare it with an articulated skeleton. Locate the major bones in the figure as you look at bones in the lab.

A young adult has about 206 bones, with more in younger individuals and fewer in older individuals. Many developing bones fuse as a person ages, forming larger bones and reducing the overall number. Not everyone has the same number of bones. Bones that develop in tendons have a shape like a sesame seed and are called **sesamoid bones.**

BONE SHAPE

Bones are classified according to four main shapes: long, short, flat, and irregular. **Long bones** are those that are longer than broad. Bones in the arm, forearm, fingers, thigh, and leg are long bones.

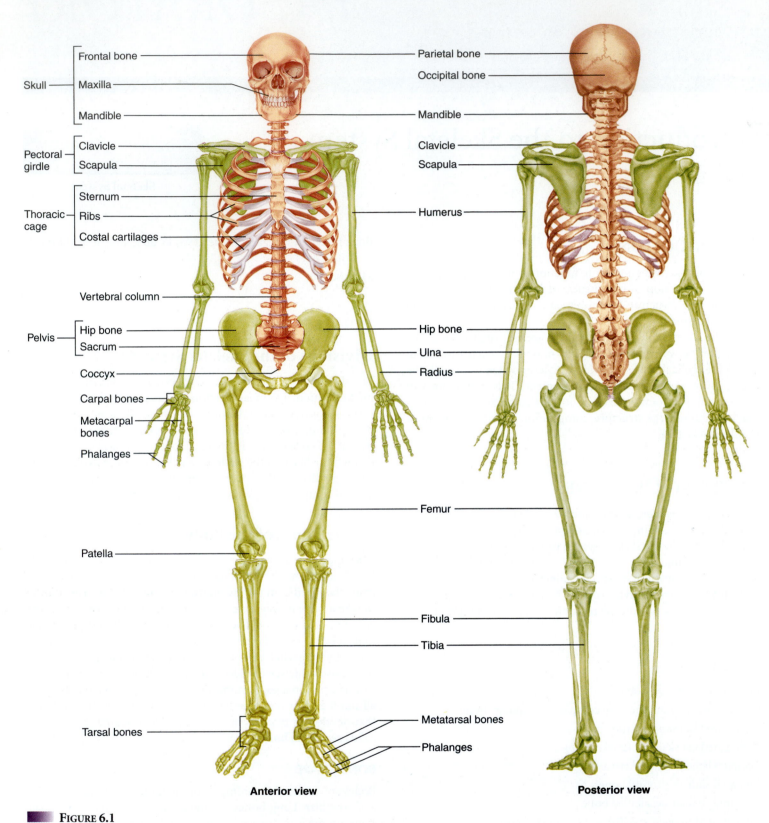

Frontal bone

Skull — Maxilla

Mandible

Pectoral girdle — Clavicle

Scapula

Sternum

Thoracic cage — Ribs

Costal cartilages

Vertebral column

Pelvis — Hip bone

Sacrum

Coccyx

Carpal bones

Metacarpal bones

Phalanges

Patella

Tarsal bones

Parietal bone

Occipital bone

Mandible

Clavicle

Scapula

Humerus

Hip bone

Ulna

Radius

Femur

Fibula

Tibia

Metatarsal bones

Phalanges

Anterior view

Posterior view

FIGURE 6.1

Major Bones of the Body Axial skeleton in beige, appendicular skeleton in green.

TABLE 6.1	
Axial and Appendicular Skeleton	
Axial Skeleton	
Skull	Ribs
Hyoid bone	Sternum
Vertebral column	
Appendicular Skeleton	
Pectoral girdle	**Pelvic girdle**
Clavicle	hip bone
Scapula	**Lower extremity**
Upper extremity	Femur
Humerus	Patella
Radius	Tibia
Ulna	Fibula
Carpals	Tarsals
Metacarpals	Metatarsals
Phalanges	Phalanges

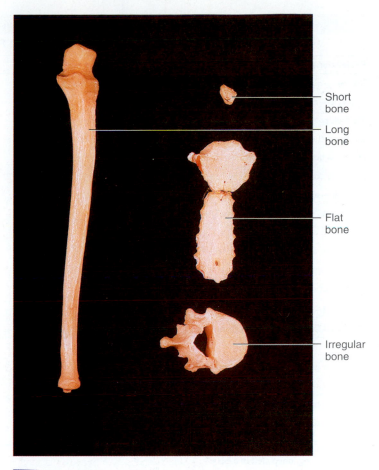

Short bone

Long bone

Flat bone

Irregular bone

FIGURE 6.2
Bone Shapes

Short bones are approximately the same length in all dimensions. Bones of the wrist and ankle are examples of short bones. **Flat bones,** such as the sternum, ribs, scapula, hip bones, and some of the bones of the cranium, are compressed. Small, flat bones sometimes occur in the cranium and these are known as **sutural bones** or **wormian bones.** Bones that do not fit into any of these shapes are known as **irregular bones.** Bones in the floor of the skull, facial bones, vertebrae, and clavicle are considered irregular bones. Examine bones in the lab and compare them with figure 6.2.

Look through the bones in lab and find an example that represents each of the bone shapes. Write the names of the bones in the following spaces:

Long bone _____

Short bone _____

Flat bone _____

Irregular bone _____

BONE FEATURES

There are specific terms that describe bumps, hollows, or spaces inside bones. **Projections** arise from the surface of bones, **depressions** occur on the surface of bones, and **cavities** are enclosed spaces in bones. Projections on bones reflect different functions. Some projections, such as condyles and heads, are surfaces where the bone articulates with another. Other projections, such as tubercles, spines, and trochanters, are attachment points for muscles or ligaments. Depressions, such as foramina, canals, and fissures, are often passageways from blood vessels and/or nerves. These are presented in table 6.2. You should study these terms and be able to find them on bones in the lab.

COMPOSITION OF BONE TISSUE

Bone consists of organic matter and inorganic matter. The **organic matter** is composed mostly of cells and collagenous fibers and makes up about one-third of the bone by weight. The **inorganic matter** makes up the remaining two-thirds of the bone and mostly consists of **hydroxyapatite,** a complex salt consisting of calcium phosphate. There is also a lesser amount of calcium carbonate in the inorganic portion of bone. If available in the lab, examine bones that have been soaked in acid. The acid dissolves the mineral salts, leaving the collagenous fibers as a flexible framework. Then examine bones that were baked in an oven for a few hours. Heating bones denatures the proteins; although the bones remain hard because of the mineral salts, they are brittle due to the destruction of the protein fibers. Notice how easily these bones break.

BONE STRUCTURE

The general anatomy of a long bone consists of the proximal and distal ends of the bone, the **epiphyses** (eh-PIF-uh-seez)(singular, **epiphysis**), and the shaft of the bone, the **diaphysis** (die-AH-fuh-sis) (fig. 6.3). The epiphyses of bones have **articular cartilage** at the ends of the bone. The articular cartilage is composed of hyaline cartilage and helps reduce friction as the joint moves.

TABLE 6.2	
Bone Features	
Projections from Bones	**Example**
Process—a general term for a projection from the surface of the bone	Styloid process of ulna
Tubercle—a relatively small bump on a bone	Greater tubercle of humerus
Tuberosity—a relatively large rough area on a bone	Deltoid tuberosity of humerus
Spine—a short, sharp projection	Vertebral spine
Head—a hemispheric projection that articulates with another bone	Head of femur
Neck—a constriction below the head	Neck of rib
Condyle—an irregular, smooth surface that articulates with another bone	Lateral condyle of femur
Epicondyle—an elevation proximal to a condyle	Medial epicondyle of humerus
Crest—an elevated ridge of bone	Crest of ilium
Line—a smaller elevation than a crest	Gluteal line of ilium
Facet—a smooth, flat face	Articular facets of vertebrae
Trochanter—a large bump (on femur)	Greater and lesser trochanter of femur
Ramus—a branch	Ramus of mandible
Depressions or Cavities in Bone	**Example**
Foramen—a shallow hole	Foramen magnum of occipital bone
Meatus or canal—a tunnel or deep hole	External auditory meatus of temporal bone
Sinus—a cavity in a bone	Maxillary sinus
Fossa—a shallow surface depression in a bone	Iliac fossa
Notch—a deep cutout in a bone	Greater sciatic notch of ilium
Groove or sulcus—an elongated depression	Intertubercular groove of humerus
Fissure—a long, deep cleft in a bone	Inferior orbital fissure

In young people, there is a hyaline cartilage pad between the epiphysis and the diaphysis. This is the **epiphyseal (growth) plate.** The cartilage increases in thickness by the action of chondrocytes. As the cartilage grows, the individual is pushed upwards by the increase in the cartilage plate. Remodeling of the cartilage takes place, and the cartilage is eventually replaced by bone. Once a person stops growing, the epiphyseal plate is replaced by a thin line of bone called the **epiphyseal** (EP-ih-FIZZ-ee-ul) **line.**

Examine a cut section of bone. Note the hard, **compact (cortical) bone** on the outside of the bone and the inner **cancellous,** or **spongy, bone.** The cancellous bone is made of **trabeculae,** which are thin plates of bone that run in the same direction as the stress applied to the bone. Stress on the bone may be in the form of gravity, or it may occur due to a force applied to the limb.

The innermost section of bone is hollow and is called the **medullary** (MED-you-lerr-ee) **(marrow) cavity. Marrow** is the material that occupies this cavity and can be of two types, a **hematopoietic** (blood cell–forming) **red marrow** and an adipose-containing **yellow marrow.** In flat bones the inner and outer compact bone encase a material called **diploe** (DIP-lo-ee), which is spongy bone. Compare the cut bones in the lab with figure 6.4 and note the features listed in the discussion.

Examine a bone that has been cut lengthwise and see how the compact bone is thicker in the middle of the bone than at the ends. Bones tend to break in the middle and this thickening reduces the breakage.

On the surface of the diaphysis, locate the **nutrient foramina** (for-AM-ih-nuh). These are small holes that allow for the passage of blood vessels into and out of the bone. They can be found in various locations in bones. The nutrient foramina lead to **perforating (Volkmann's) canals** that pass through compact bone. The central canals typically run vertically in bones, while perforating canals carry nutrients horizontally.

The outer surface of the bone is covered with a dense connective tissue sheath called the **periosteum** (PEAR-ee-OS-tee-um). It is the site where nerves and blood vessels occur on the outer surface of the bone. The periosteum is an anchoring point for tendons and ligaments. **Tendons** attach muscles to bone at the periosteum, and this attachment is strengthened by **perforating (Sharpey's) fibers** that provide greater surface area between the tendon and the periosteum. **Ligaments** are parallel straps of connective tissue that connect one bone to another, and they are secured to the bone by the periosteum.

The inner surface of long bones, near the medullary cavity, is lined with a layer known as the **endosteum** (end-OS-tee-um).

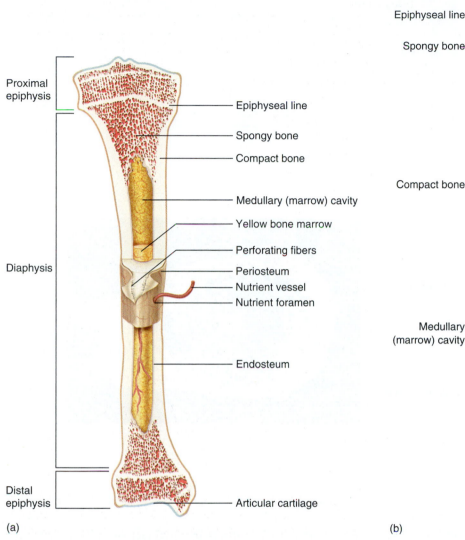

Proximal
epiphysis

Epiphyseal line

Spongy bone

Compact bone

Medullary (marrow) cavity

Yellow bone marrow

Perforating fibers

Periosteum

Nutrient vessel

Nutrient foramen

Diaphysis

Endosteum

Distal
epiphysis

Articular cartilage

(a)

Epiphyseal line

Spongy bone

Proximal
epiphysis

Compact bone

Diaphysis

Medullary
(marrow) cavity

Distal
epiphysis

(b)

 **FIGURE 6.3**

Long Bone, Longitudinal Section (a) Diagram; (b) photograph.

MICROSCOPIC STRUCTURE OF BONE

Ground Bone   Examine a prepared slide of ground bone and lo-cate the circular structures in the field of view. These modular units of bone are called **osteons.** Each osteon has a hole in the middle called a **central,** or **Haversian canal,** which houses blood vessels and nerves in the dense bone tissue. Around the central canal are rings of dark spots. These are the **lacunae** (singular, **lacuna**), which are spaces that the bone cells, or osteocytes, occupy in living tissue. In the prepared slide, the lacunae have filled with bone dust from the grinding process. The lacunae are connected to each other by thin tubes called **canaliculi** (CAN-uh-LIC-you-lie). The role of canaliculi is to provide a passageway through the dense bone mate-rial. Osteocytes are living and need to receive oxygen and nutrients and remove wastes. They do this by extending cellular strands through the canaliculi between osteocytes. The canaliculi connect to the **central canal.** The osteon has layers of dense mineral salts that form concentric rings between the lacunae. These are the **lamellae.** Locate these structures in figure 6.5.

DECALCIFIED BONE

Examine a slide of decalcified bone. In this preparation the bone salts have been dissolved away and the remaining thin section of bone has been sliced, mounted, and stained. In this way the perios-teum is visible on the surface of the bone, with the compact bone in the middle and the endosteum on the inside of the bone (fig. 6.6). There are three types of bone cells in osseous tissue. These are the

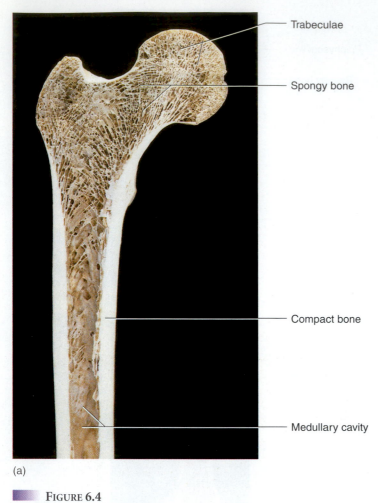

(a)

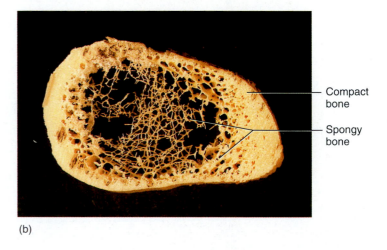

(b)

FIGURE 6.4

Cut Bone (a) Long section; (b) cross section.

osteoblasts, osteocytes, and **osteoclasts.** Osteoblasts are active bone-forming cells. They arise from **osteogenic cells**. They are frequently found in areas where additional bone is being formed (fig. 6.7). Examine a slide of decalcified bone and locate the osteoblasts.

Osteocytes (mature bone cells) are the most common cells in compact bone. In ground bone only the spaces where the osteocytes occur are visible (the lacunae), but osteocytes are visible in decalcified bone. Examine the slide of decalcified bone for osteocytes.

Osteoclasts are responsible for the remodeling of bone tissue and are found in areas where bone is being removed. They arise

from **stem cells**. Osteoclasts are multinucleate and secrete lysosomal enzymes that digest osseous material. Examine the inner portion of the same prepared slide and locate the osteoclasts. Compare them with figure 6.7.

Skeletal Anatomy of the Cat

If you are using cats, turn to section 1, "Skeletal Anatomy of the Cat," on page 410 of this lab manual.

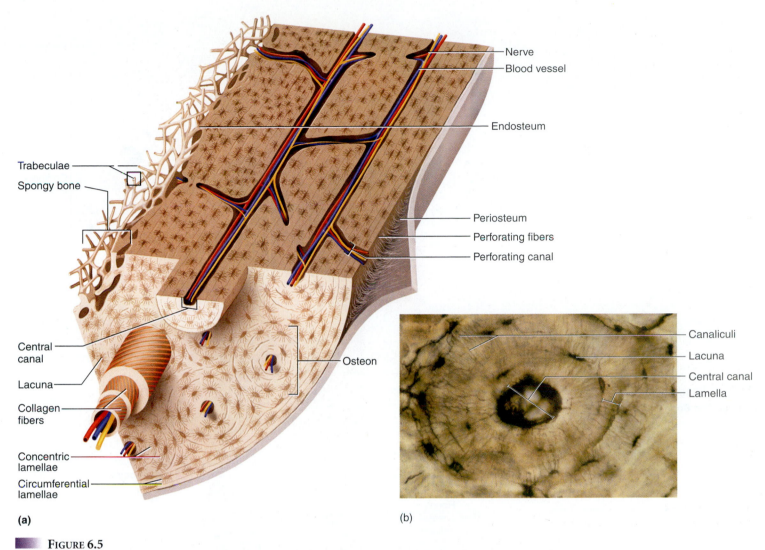

(a)

Trabeculae
Spongy bone

Central canal
Lacuna
Collagen fibers
Concentric lamellae
Circumferential lamellae

Nerve
Blood vessel
Endosteum
Periosteum
Perforating fibers
Perforating canal
Osteon

(b)

Canaliculi
Lacuna
Central canal
Lamella

FIGURE 6.5

Histology of Ground Bone (Osteon) (a) Diagram; (b) cross section (400×).

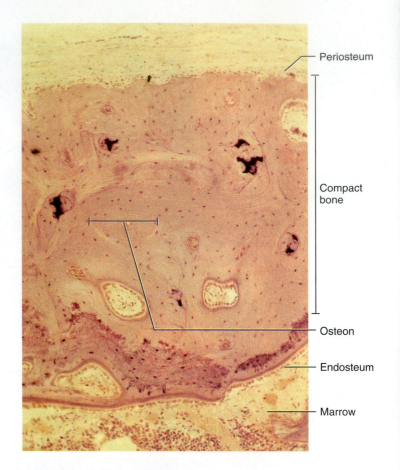

Periosteum

Compact bone

Osteon

Endosteum

Marrow

FIGURE 6.6
Decalcified Bone (100×)

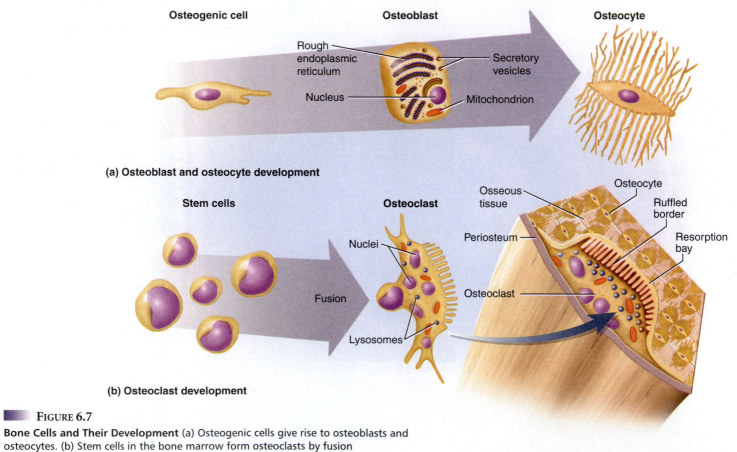

Osteogenic cell

Osteoblast

Osteocyte

Rough endoplasmic reticulum

Secretory vesicles

Nucleus

Mitochondrion

(a) Osteoblast and osteocyte development

Stem cells

Osteoclast

Osseous tissue

Osteocyte

Ruffled border

Resorption bay

Periosteum

Nuclei

Osteoclast

Fusion

Lysosomes

(b) Osteoclast development

FIGURE 6.7

Bone Cells and Their Development (a) Osteogenic cells give rise to osteoblasts and osteocytes. (b) Stem cells in the bone marrow form osteoclasts by fusion

Introduction to the Skeletal System

Name _____ Date _____

1. The hyoid bone belongs to the

   a. appendicular skeleton.   b. axial skeleton.   c. upper limb.   d. skull.

2. The clavicle belongs to the

   a. axial skeleton.   b. pectoral girdle.   c. pelvic girdle.   d. upper limb.

3. In the disease osteoporosis, there is a significant loss of cancellous bone. Explain how the loss of this specific bone material can weaken a bone.

4. Label the following illustration using the terms provided.

   central canal   osteon   lamella

   lacuna   canaliculi

   a. _____
   b. _____
   c. _____
   (the thin lines)
   d. _____
   e. _____
   (the dark spot)

5. The ends of a long bone are known as the _____.

6. A carpal bone is classified as a(n) _____ bone in terms of shape.

7. The radius is part of which skeletal division?

   a. axial   b. appendicular

8. The ribs are part of which skeletal division?

   a. axial   b. appendicular

9. What role do osteoblasts have in maintaining bone tissue?

10. The patella is part of which skeletal division?

    a. axial   b. appendicular

11. A young adult has how many bones, on average?

12. The inorganic portion of bone tissue is made of what mineral salt?

13. In terms of shape, what type are the upper bones of the cranium?

    a. long   b. short   c. flat   d. irregular

14. In terms of shape, what type of bone is the sternum?

    a. long   b. short   c. flat   d. irregular

15. In terms of shape, what type of bone are the metacarpals?

    a. long   b. short   c. flat   d. irregular

16. In terms of shape, what type of bone are the tarsals?

    a. long   b. short   c. flat   d. irregular

17. How does the shape of a long bone resist breaking when put under stress?

18. What is an osteon?

19. Synthetic bone material known as hydroxyapatite is often molded into the shape of bone. This procedure is done to replace bone that has been diseased, damaged, or surgically removed. Cells in the body remodel the synthetic material and produce new bone. What bone cells do you think are involved in this remodeling, and what role do these cells have?

(Continued next page)

20. Fill in the major bones or bony regions of
the body on the illustration using the
provided terms.

humerus

tarsals

scapula

metatarsals

femur

vertebral column

tibia

clavicle

hip bone

carpals

ribs

metacarpals

fibula

skull

phalanges (hand)

phalanges (foot)

patella

sternum

a. _____

b. _____

c. _____

d. _____

e. _____

f. _____

g. _____

h. _____

i. _____

j. _____

k. _____

l. _____

m. _____

n. _____

o. _____

p. _____

q. _____

r. _____

Appendicular Skeleton

Skeletal System

INTRODUCTION

The appendicular skeleton consists of 126 bones in four major groups. These are the pectoral girdle, upper extremity, pelvic girdle, and lower extremity. The **pectoral girdle** is composed of a scapula and clavicle on each side of the body. The **upper extremity** consists of the humerus, radius, ulna, carpals of the wrist, metacarpals of the palm, and phalanges of the fingers. The **pelvic girdle** (hip bones) each consists of three fused bones—the ilium, the ischium, and the pubis. The **lower extremity** consists of the femur, patella, tibia, fibula, tarsals of the ankle, metatarsals of the foot, and phalanges of the toes.

LEARNING OBJECTIVES

At the end of this exercise you should be able to
1. locate and name all the bones of the appendicular skeleton;
2. name the significant surface features of the major bones;
3. place a bone into one of the four major groups (pectoral girdle bones, upper extremity bones, pelvic girdle bones, lower extremity bones);
4. determine whether a selected bone is from the left side or right side of the body;
5. locate the major surface features of the scapula and hip bone;
6. name the individual carpal bones and tarsal bones in articulated hands and feet, respectively.

MATERIALS

Articulated skeleton

Disarticulated skeleton

Charts of the skeletal system

Plastic drinking straws (cut on a bias) or pipe cleaners for pointer tips

Foam pads of various sizes (to protect real bones from hard countertops)

PROCEDURE

Review the bones of the appendicular skeleton with the material in the lab and in laboratory exercise 6 (fig. 6.1). In this exercise you study the bones of the appendicular skeleton by examining an isolated (or disarticulated) bone in conjunction with the articulated skeleton. Hold the individual bone up to the skeleton to see how it is positioned in relation to the other bones of the body. When examining bones in the lab, do not use your pen or pencil to locate a structure. They leave marks on the bone. Use a cut drinking straw, a wooden applicator stick, or a pipe cleaner to point out structures. Your instructor may want you to place real bone material on foam pads to cushion the bone from the tabletop. Use your time in lab to find the material at hand. Once you are at home, draw or trace material from your manual and be able to name all of the parts.

Pectoral Girdle

The pectoral, or shoulder, girdle consists of the right and left **scapulae** (singular, **scapula**) and the right and left **clavicles.** The function of the pectoral girdle is to provide a movable yet stable support for the upper extremity.

SCAPULA

The scapula is commonly known as the shoulder blade and is found on the posterior, superior portion of the thorax. The scapula is roughly triangular and has three borders—a **superior border,** a **vertebral (medial) border,** and an **axillary (lateral) border**. On the superior border of the scapula is an indentation known as the **scapular notch,** which provides space for a nerve.

On the anterior surface of the scapula is a smooth, hollowed depression known as the **subscapular fossa.** The posterior surface of the scapula is divided by the **scapular spine** into a superior depression known as the **supraspinous fossa** and an inferior depression known as the **infraspinous fossa.** The spine of the scapula is a bony crest that runs from the vertebral border to the lateral edge of the scapula, where it expands to form the **acromion** (ah-CRO-me-on).

Another projection from the scapula is the **coracoid** (COR-uh-coyd) **process,** which projects anteriorly. The scapula also has two sharp angles known as the **inferior angle** and the **superior angle.** The features of the scapula are important because numerous muscles attach to them. The humerus moves in the **glenoid cavity (glenoid fossa),** a depression on the lateral edge of the scapula. Above the fossa is the supraglenoid tubercle, a process on which the biceps brachii muscle attaches. Just below the glenoid cavity is a bump known as the **infraglenoid tubercle,** which is an attachment point for the triceps brachii muscle.

You should also be able to distinguish a right scapula from a left one. This can be done if you recognize that the spine of the scapula is posterior, the inferior angle is less than 90°, and the glenoid cavity is lateral. Locate the features of the scapula on figure 7.1.

CLAVICLE

The **clavicle** is a small bone that serves as a strut between the scapula and the sternum. The blunt, vertically truncated end of the scapula is the **sternal end** and the horizontally flattened end is the **acromial**

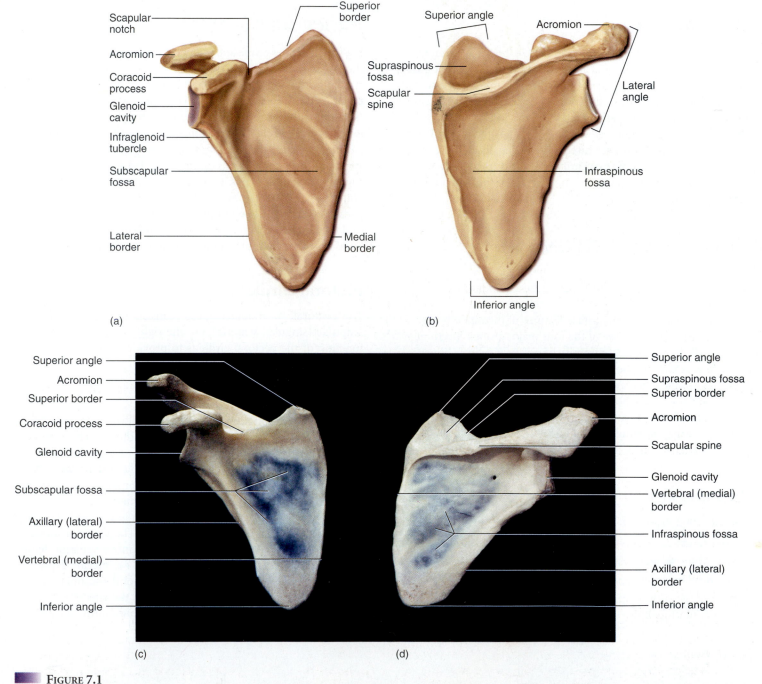

FIGURE 7.1

Right Scapula Diagram (a) anterior view; (b) posterior view. Photograph (c) anterior view; (d) posterior view.

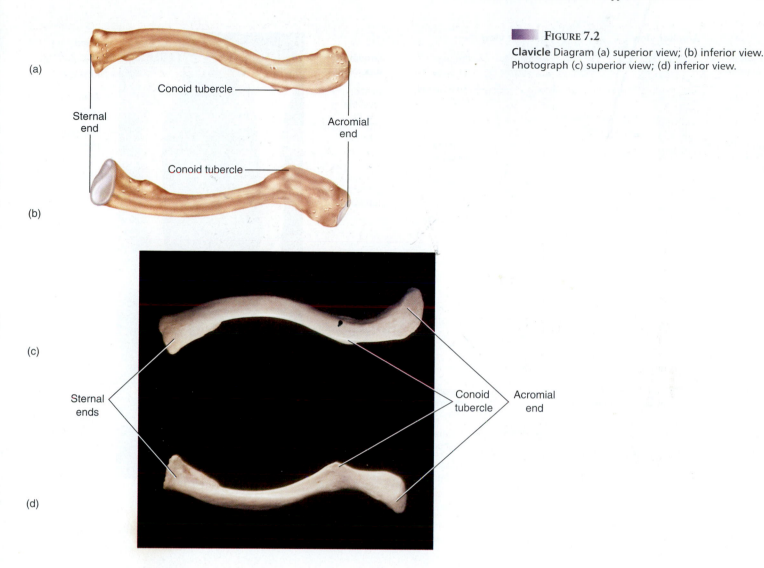

(a)

Conoid tubercle

Sternal
end

Acromial
end

Conoid tubercle

(b)

(c)

Sternal
ends

Conoid
tubercle

Acromial
end

(d)

FIGURE 7.2
Clavicle Diagram (a) superior view; (b) inferior view.
Photograph (c) superior view; (d) inferior view.

end. Note the **conoid (CON-oyd) tubercle** on the inferior surface of the clavicle. The clavicle is the most frequently fractured bone in the body. The arm transmits force to the clavicle, which is sandwiched between the scapula and the sternum. Examine the clavicles in the lab and compare them with figure 7.2.

Upper Extremity

HUMERUS

The **humerus** is a long bone that articulates with the scapula at its proximal end and with the radius and ulna at its distal end. The majority of the length of the humerus is the **diaphysis,** or **shaft,** and small holes penetrating the shaft are the **nutrient foramina.** Nutrient foramina are holes that occur in many bones and conduct blood vessels into these bones. The **head** of the humerus is a hemispheric structure that ends in a rim known as the **anatomical neck.** The anatomical neck is the site of the epiphyseal line. The humerus has two large processes—the **greater tubercle,** which is the largest and most lateral process, and the **lesser tubercle,** which is medial and smaller. These processes are attachment points for muscles

originating from the scapula. Between the tubercles is the **intertubercular,** or **bicipital, sulcus,** which is an elongated depression through which one of the biceps brachii tendons passes. Examine figure 7.3 and locate these features. Below the tubercles of the humerus is a constricted region known as the **surgical neck.** The surgical neck is so named because it is a frequent site of fracture. Locate the surgical neck of the humerus on your arm and name a muscle located nearby in the space provided.

Muscle near surgical neck _____.

A roughened area on the lateral surface of the humerus is the **deltoid tuberosity,** named for the attachment of the deltoid muscle.

At the distal end of the humerus are the regions of articulation with the bones of the forearm. On the lateral side is a round hemisphere known as the **capitulum** (ca-PIT-you-lum). This is where the head of the radius fits into the humerus. On the medial side is an hour glass–shaped structure called the **trochlea** (TROW-klee-uh). The trochlea is where the ulna attaches to the humerus. The capitulum and trochlea represent the **condyles** (KON-dials) of the humerus. To the sides of these condyles are the **epicondyles**

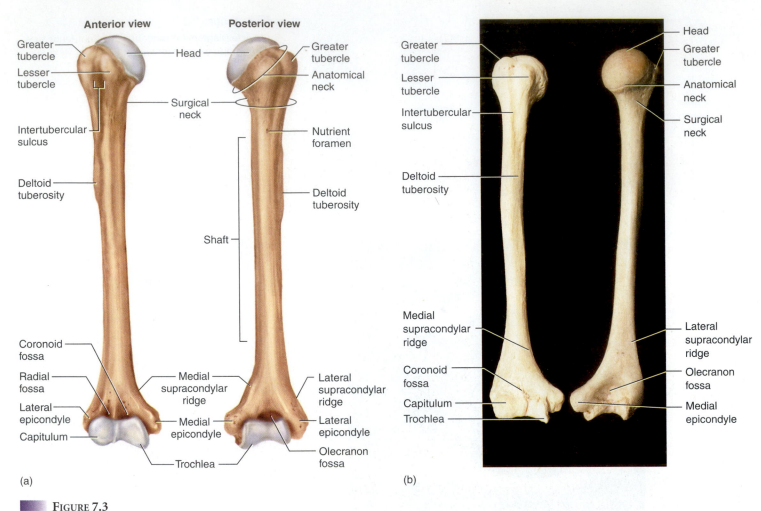

Anterior view   **Posterior view**

Greater tubercle — Head — Greater tubercle

Lesser tubercle — Anatomical neck

— Surgical neck

Intertubercular sulcus — Nutrient foramen

Deltoid tuberosity — Deltoid tuberosity

Shaft —

Coronoid fossa — Medial supracondylar ridge — Lateral supracondylar ridge

Radial fossa —

Lateral epicondyle — Medial epicondyle — Lateral epicondyle

Capitulum — Olecranon fossa

— Trochlea

(a)

Greater tubercle — Head — Greater tubercle

Lesser tubercle — Anatomical neck

Intertubercular sulcus — Surgical neck

Deltoid tuberosity —

Medial supracondylar ridge — Lateral supracondylar ridge

Coronoid fossa — Olecranon fossa

Capitulum — Medial epicondyle

Trochlea —

(b)

FIGURE 7.3

Right Humerus (a) Diagram; (b) photograph.

(EP-ee-KON-dials). The **medial epicondyle** of the humerus is a process commonly known as the "funny bone," and the **medial supracondylar ridge** forms a small wing of bone extending from the epicondyle to the distal shaft of the humerus. The **lateral epicondyle** also has a **lateral supracondylar ridge.** The epicondyles and supracondylar ridges are points of attachment for muscles that manipulate the forearm and hand.

At the distal end are two depressions in the humerus. The one on the anterior surface is the **coronoid fossa** and the one on the posterior surface is the **olecranon** (uh-LEC-ruh-non) **fossa.** These depressions accommodate the ulna as the forearm is flexed or extended. Locate these structures on specimens in the lab and on figure 7.3.

FOREARM

The radius and ulna make up the skeletal portion of the forearm. The **radius** is a lateral bone with a proximal **head,** resembling a wheel that articulates with the capitulum of the humerus. Distal to the head is the **tuberosity** of the radius, where the biceps brachii muscle inserts, and at the most distal and posterior end of the radius is the **styloid** (STY-loyd) **process.** On the distal end of the radius is a medial depression called the **ulnar notch.** This notch fits into the medial bone of the forearm, the ulna. Examine figure 7.4 for these structures and compare them with the bones in the lab.

The **ulna** has a U-shaped depression (some students remember this as "U for ulna") on the proximal portion, which is the **trochlear,** or **semilunar, notch** that articulates with the trochlea of the humerus. The most proximal portion of the ulna is the **olecranon,** commonly known as the elbow. The process distal to the trochlear notch is the **coronoid process,** which ensures a relatively tight fit with the humerus and provides an area for muscle attachment. On the proximal part of the ulna, where the head of the radius fits into the ulna, is a depression known as the **radial notch.** If you know that the trochlear notch is anterior and the radial notch is lateral, then you can determine if you are looking at a left or right ulna. In the following space, state what side of the body the bones of the forearm you are studying come from.

Side of the body: _____

The ulna has a *distal* **head,** unlike the radius (the head of the radius is proximal), and a **styloid process** on the distal, posterior side. Locate these structures in figure 7.4.

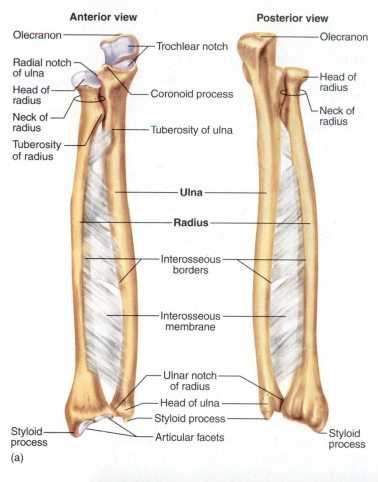

Anterior view

Olecranon

Radial notch of ulna

Head of radius

Neck of radius

Tuberosity of radius

Trochlear notch

Coronoid process

Tuberosity of ulna

Ulna

Radius

Interosseous borders

Interosseous membrane

Ulnar notch of radius

Head of ulna

Styloid process

Articular facets

Styloid process

Posterior view

Olecranon

Head of radius

Neck of radius

Styloid process

(a)

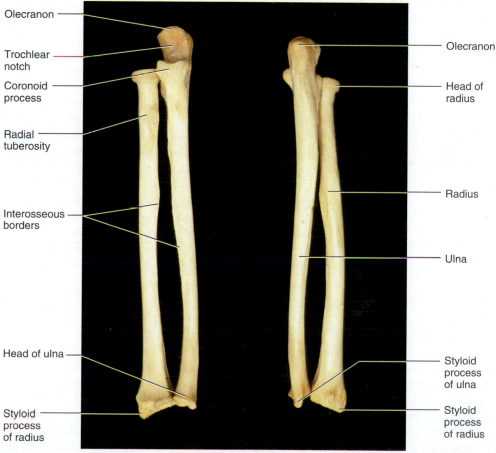

Olecranon

Trochlear notch

Coronoid process

Radial tuberosity

Interosseous borders

Head of ulna

Styloid process of radius

Olecranon

Head of radius

Radius

Ulna

Styloid process of ulna

Styloid process of radius

(b)

■ **FIGURE 7.4**

Right Radius and Ulna
(a) Diagram; (b) photograph.

83

HAND

Each hand consists of three groups of bones—the **carpals,** the **metacarpals,** and the **phalanges.** Over one-quarter of all of the bones in the body are in the hand. The hand is composed of eight carpal bones (fig. 7.5) that can be aligned in two rows. The proximal row from lateral to medial includes the **scaphoid** (SCAH-foyd) (navicular), **lunate, triquetrum** (tri-KWEE-trum) (triangular), and **pisiform** (PIE-sih-form). The distal row includes the **trapezium** (greater multangular), the **trapezoid** (lesser multangular), the **capitate,** and the **hamate.** On the hamate is a hooklike projection called the **hamulus** of the **hamate.** The carpal bones are named for their shape, with the scaphoid resembling a boat (*scaphos* = boat), the lunate named for the crescent moon, triquetrum (triangle), pisiform (pea-shaped), the trapezium (rhymes with "thumb" and named for a shape), the trapezoid (named for a shape), the capitate (headlike), and the hamate (*hamus* = hook). Another way to put the bones in order is to use a mnemonic device that uses the first letter of each carpal bone (*S* for scaphoid, *L* for lunate, etc.) in a sentence—"*Say Loudly To Pam, Time To Come Home.*" The carpal bones are regions of attachment for forearm muscles and for intrinsic muscles of the hand.

The **metacarpals** are long bones in the palm. Each metacarpal consists of a proximal **base,** a **body,** and a distal **head.** The metacarpals are labeled by either arabic (1–5) or roman numerals (I–V). Metacarpal I is proximal to the thumb and metacarpal V is proximal to the little finger. Metacarpals II through V have little movement, but metacarpal I allows for significant movement of the thumb. Locate metacarpal I on the skeletal material and on your own hand and note the substantial range of movement. This type of movement is covered in greater detail in laboratory exercise 10. Examine the carpals and metacarpals in the lab and in figure 7.5.

Distal to the metacarpals are the **phalanges.** These are long bones. Like the metacarpals, each phalanx has a **base, body,** and **head,** and the phalanges are classified as long bones. There are 14 phalanges in each hand. Each finger of the hand, except for the thumb, has 3 phalanges—a **proximal, middle,** and **distal phalanx.** The thumb is known as the **pollex** and has only a proximal and a distal phalanx. You should be able to identify each bone by noting the digit it comes from and whether it is proximal, middle, or distal. The tip of the thumb is the distal phalanx I, whereas the base of the little finger is the proximal phalanx V. Locate the various bones of the fingers in material in the lab and in figure 7.5.

Pelvic Girdle

The pelvic girdle consists of the two **hip bones** (**innominate bones**). Each adult hip bone results from the fusion of three bones—the ilium, the ischium, and the pubis. The hip bones are joined in the front by the **symphysis pubis,** or **pubic symphysis,** which is a fibrocartilage pad. The pelvic girdle (hip bones) along with the sacrum (part of the vertebral column) form the **pelvis.** The hip bones join together anteriorly by the pubic symphysis. The union of the pubic bones forms the **subpubic angle,** or **pubic arch.** This arch is greater than 90° in females and less than 90° in males. Changes in the shape of the pelvis occur due to the influence of hormones.

ILIUM

The most superior hip bone is the **ilium** (ILL-ee-um). If you feel the top edge of your hip, you are feeling the **iliac crest,** which is a long, crescent-shaped ridge of the ilium. The two processes that jut from the ilium in the front are the **anterior superior iliac spine** and the **anterior inferior iliac spine.** These are attachment points for some of the thigh flexor muscles. The posterior ilium has the **posterior superior iliac spine** and the **posterior inferior iliac spine.** Below the posterior inferior iliac spine is a large depression known as the **greater ischiadic (sciatic) notch.** The outer surface of the ilium is an attachment point for the gluteal muscles. The interior, shallow depression of the ilium is known as the **iliac fossa.** Locate these features in figure 7.6.

ISCHIUM AND PUBIS

The **ischium** (ISS-kee-um) is inferior to the ilium and is the part of the hip bone on which you sit. The **ischial spine** is a sharp projection on the ischium, and the **ischial tuberosity** is an attachment site for the hamstring muscles. Anterior to the ischial tuberosity is the **obturator** (OB-tur-aye-tur) **foramen,** a large hole underneath a cuplike depression known as the **acetabulum** (a region where all three hip bones are joined). The acetabulum is the socket into which the femur fits. The ischium connects to the pubis by an elongated portion of bone known as the **ischial ramus.** This ramus connects to the **inferior pubic ramus,** which is posterior to the symphysis pubis. The pubis is a U-shaped bone that connects inferiorly to the ischium and superiorly to the ilium. Examine these bones in lab.

If the two hip bones are articulated, you can see the upper basin of the pelvis, which is the **false,** or **greater, pelvis.** The false pelvis is medial to the iliac fossa. There is a rim of bone that separates the false pelvis from a deeper, smaller basin known as the **true pelvis.** This rim is known as the **pelvic brim** in an intact pelvis or the **arcuate line** on an isolated pelvic bone. The true pelvis is medial to the obturator foramen. Compare the material in lab to figure 7.6.

Examine the disarticulated hip bones in lab. Determine if you have a left or right bone as well as the sex of the individual. In a female pelvis the greater sciatic notch has a fairly broad angle, which approximates the arc inscribed by your outstretched thumb and index finger. In a male the greater sciatic notch is less than 90° and approximates the angle if you were to separate the index finger and the middle finger. As you examine the bones in the lab, look at figure 7.7 and locate the various terms listed on the illustration.

Lower Extremity

FEMUR

The **femur** is the longest and heaviest bone in the body and it is the only bone of the thigh. It articulates proximally with the hip and inferiorly with the tibia. The femur has a proximal, medial, spherical **head,** which inserts into the acetabulum of the hip bone. The **fovea capitis** is a depression in the head where a ligament attaches. Inferior to the head is an elongated, constricted **neck.** Inferior to the neck are two large processes called the **trochanters,** which are areas of muscle attachment. The larger, proximal process is the

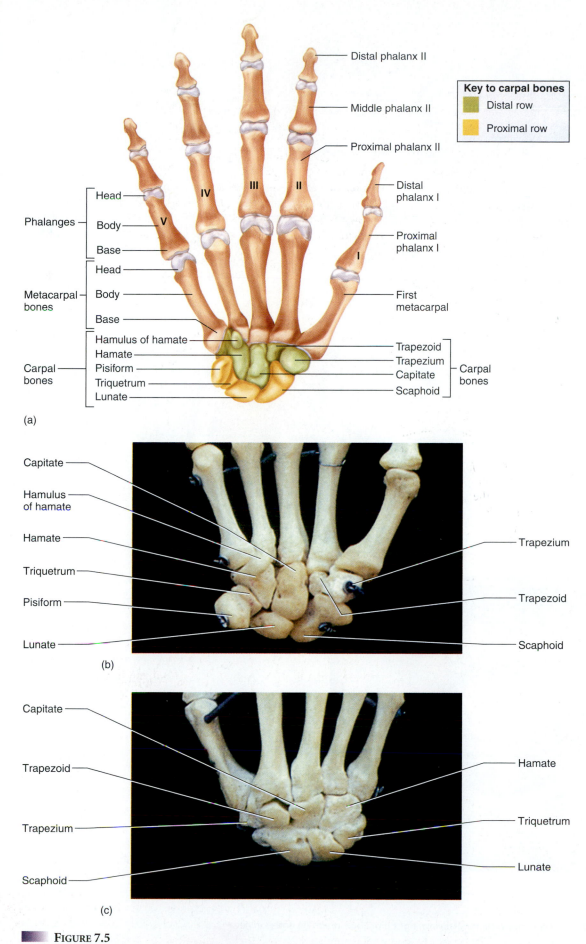

Distal phalanx II

Middle phalanx II

Key to carpal bones
Distal row
Proximal row

Proximal phalanx II

Distal phalanx I

Proximal phalanx I

First metacarpal

Phalanges
Head
Body
Base

Metacarpal bones
Head
Body
Base

Carpal bones
Hamulus of hamate
Hamate
Pisiform
Triquetrum
Lunate

Trapezoid
Trapezium
Capitate
Scaphoid
Carpal bones

(a)

Capitate

Hamulus of hamate

Hamate

Triquetrum

Pisiform

Lunate

Trapezium

Trapezoid

Scaphoid

(b)

Capitate

Trapezoid

Trapezium

Scaphoid

Hamate

Triquetrum

Lunate

(c)

FIGURE 7.5

Bones of the Right Hand Diagram (a) anterior view. Photograph (b) anterior view; (c) posterior view.

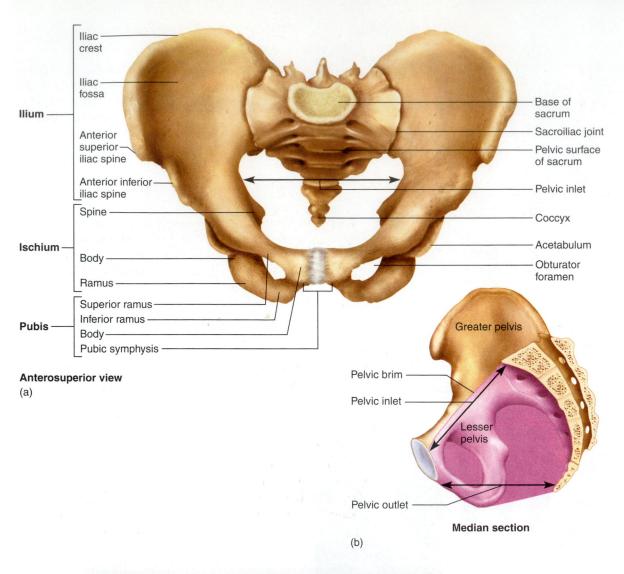

Ilium

Iliac crest

Iliac fossa

Anterior superior iliac spine

Anterior inferior iliac spine

Base of sacrum

Sacroiliac joint

Pelvic surface of sacrum

Pelvic inlet

Ischium

Spine

Body

Ramus

Coccyx

Acetabulum

Obturator foramen

Pubis

Superior ramus

Inferior ramus

Body

Pubic symphysis

Anterosuperior view
(a)

Greater pelvis

Pelvic brim

Pelvic inlet

Lesser pelvis

Pelvic outlet

Median section

(b)

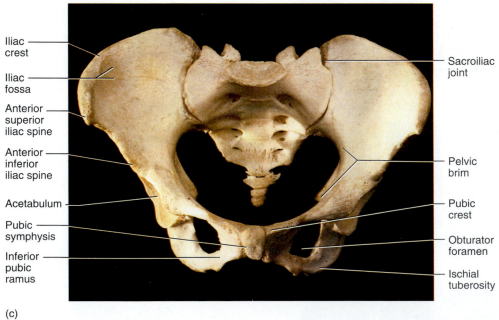

Iliac crest

Iliac fossa

Anterior superior iliac spine

Anterior inferior iliac spine

Acetabulum

Pubic symphysis

Inferior pubic ramus

Sacroiliac joint

Pelvic brim

Pubic crest

Obturator foramen

Ischial tuberosity

(c)

FIGURE 7.6

Bones of the Pelvic Girdle Diagram (a) anterior view; (b) medial view, right hip. Photograph (c) anterior view.

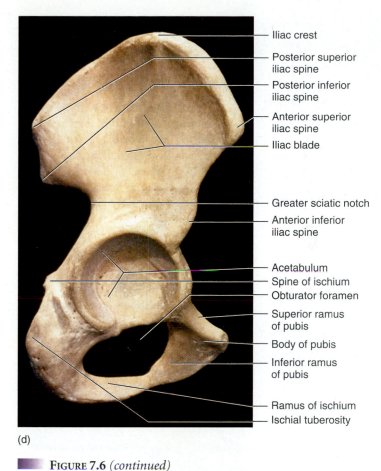

Iliac crest
Posterior superior iliac spine
Posterior inferior iliac spine
Anterior superior iliac spine
Iliac blade

Greater sciatic notch
Anterior inferior iliac spine

Acetabulum
Spine of ischium
Obturator foramen
Superior ramus of pubis
Body of pubis
Inferior ramus of pubis

Ramus of ischium
Ischial tuberosity

(d)

FIGURE 7.6 *(continued)*
Bones of the Pelvic Girdle Diagram (d) lateral view, right hip.

greater trochanter and the smaller, distal process is the **lesser trochanter.** Locate these structures in figure 7.8. You should also find the **intertrochanteric crest,** which is a posterior ridge between the two trochanters and the anterior **intertrochanteric line.** In the following space, determine which is broader—the intertrochanteric crest or intertrochanteric line.

Broadest surface feature _____

The **shaft** of the femur is curved and bows anteriorly with the posterior shaft marked by the **linea aspera** (meaning rough line). At the proximal and lateral portion of the linea aspera is a roughened area called the **gluteal tuberosity,** an attachment site for the gluteus maximus muscle. The distal portion of the femur consists of **medial** and **lateral condyles.** These two smooth processes articulate with the condyles of the tibia. To the sides of the condyles are bulges known as the **epicondyles.** Locate the **lateral epicondyle** and the **medial epicondyle** on the femur. Also note the location of the **adductor tubercle,** a small, triangular process proximal to the medial epicondyle. This is one of the attachment points for one of the adductor muscles.

PATELLA

The **patella,** commonly known as the kneecap, is a bone formed in the tendon of the quadriceps femoris muscles on the anterior thigh. The patella is a specific type of bone known as a **sesamoid bone**. Sesamoid bones develop in the tendons. The patella ossifies between ages 3 and 6. When the leg is flexed, as when kneeling, the

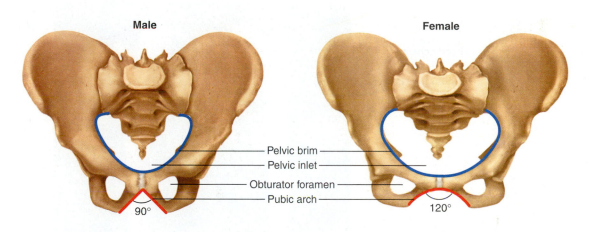

Male

Female

Pelvic brim
Pelvic inlet
Obturator foramen
Pubic arch

90°

120°

FIGURE 7.7
Comparison of the Male and Female Pelvic Girdles The pelvic brim is the margin highlighted in blue; the pubic arch is highlighted in red.

FIGURE 7.8

Right Femur (a) Diagram; (b) photograph.

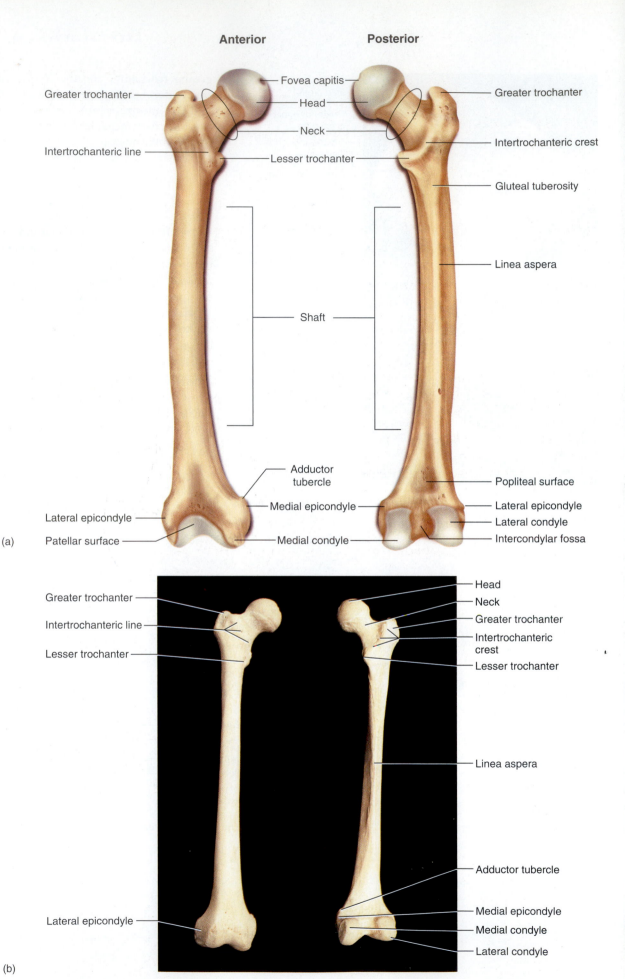

Anterior

Posterior

Greater trochanter

Fovea capitis

Head

Neck

Greater trochanter

Intertrochanteric line

Intertrochanteric crest

Lesser trochanter

Gluteal tuberosity

Linea aspera

Shaft

Adductor tubercle

Popliteal surface

Medial epicondyle

Lateral epicondyle

Lateral epicondyle

Lateral condyle

Patellar surface

Medial condyle

Intercondylar fossa

(a)

Greater trochanter

Head

Neck

Intertrochanteric line

Greater trochanter

Lesser trochanter

Intertrochanteric crest

Lesser trochanter

Linea aspera

Adductor tubercle

Lateral epicondyle

Medial epicondyle

Medial condyle

Lateral condyle

(b)

Anterior surface   **Posterior surface**

Base of patella ——

Apex of patella ——

—— Articular facets

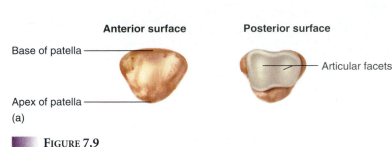

(a)

Base of patella ——

Apex of patella ——

—— Base

—— Articular facets

—— Apex

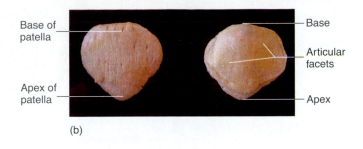

(b)

FIGURE 7.9

Patella, Anterior and Posterior Views (a) Diagram; (b) photograph.

patella protects the ligaments of the knee joint. Examine the patella in the lab and compare it with figure 7.9.

LEG

The **tibia** is the largest bone of the leg. It is the weight-bearing bone, inferior to the femur. The tibia is roughly triangular in cross section and is medial to the fibula. The **tibial condyles** are separated by the **intercondylar eminence** and they articulate with the condyles of the femur. There is a process on the proximal, anterior surface of the tibia known as the **tibial tuberosity.** This is the attachment point for the patellar ligament. At the distal end of the tibia is an extension of bone known as the **medial malleolus** (MAH-lee-OH-lus). This process is one-half of the ankle joint that articulates with the talus of the foot. Along the length of the tibia is the **anterior crest,** a ridge that runs from superior to inferior. This is the shin that is bruised if you run into something (such as a coffee table) in the middle of the night. Locate the structures of the tibia in the lab and in figure 7.10.

The **fibula** is smaller than the tibia (a way to remember that it is smaller than the tibia is this mnemonic: "tell a little *fib*"). The fibula is lateral to the tibia and is not a weight-bearing bone but is an attachment point for muscles. The fibula has a proximal **head** and a distal process called the **lateral malleolus.** The lateral malleolus is the other half of the ankle that forms a joint with the talus. Examine the bones in the lab and compare the fibula with figure 7.10.

FOOT

The feet have 26 bones each and, if you add up all of the bones of the feet along with the hands, you have 106 bones, or more than 50% of all of the bones in the body!

There are seven **tarsal bones** of the foot, including the **talus,** which articulates with the tibia and fibula. Directly inferior to the talus is the **calcaneus** (cal-CAY-nee-us), which is commonly known as the heel bone. The bone anterior to both the talus and the calcaneus is the **navicular.** The foot has three **cuneiform** (cue-NEE-ih-form) **bones,** which are wedge-shaped bones (*cuneiform* = wedge-shaped). These are the **medial,** or **first, cuneiform;** the **intermediate,** or **second, cuneiform;** and the **lateral,** or **third, cuneiform.** Locate these on specimens in the lab and in figure 7.11. The lateral cuneiform bone is not the most lateral bone of the foot. The cuboid is lateral to the lateral cuneiform. One way to remember the tarsal bones is with the mnemonic "Children that never march in line cry." The beginning letter of each word is the same as a tarsal bone.

The pattern of the **metatarsals** is similar to that of the metacarpals of the hand. The first metatarsal is under the largest digit, the **hallux** (big toe), and the fifth metatarsal is under the smallest digit. The pattern of the phalanges of the foot is the same as in the hand, with toes two to five each having a **proximal, middle,** and **distal phalanx** and the big toe having only a proximal and a distal phalanx. Examine the material in the lab and compare the metatarsals and phalanges with figure 7.11.

When you have finished locating the major features of the bones of the appendicular skeleton, test yourself and your lab partner by pointing to various structures and seeing how accurate you are at identifying the structures.

STUDY HINTS

Use your time in lab to *find* the material at hand. Once you are at home, draw or trace material from your laboratory manual and be able to name all of the parts.

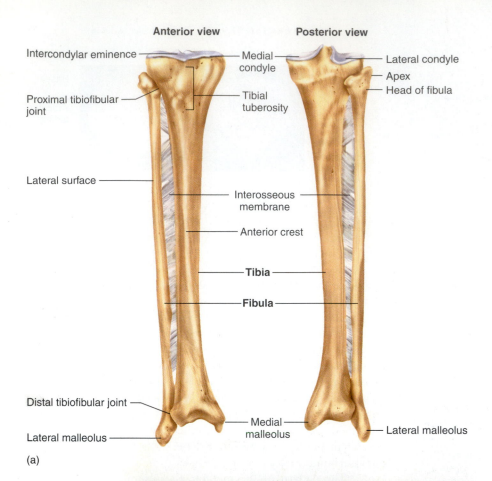

Anterior view   **Posterior view**

Intercondylar eminence

Medial condyle

Proximal tibiofibular joint

Tibial tuberosity

Lateral surface

Interosseous membrane

Anterior crest

Tibia

Fibula

Lateral condyle

Apex

Head of fibula

Distal tibiofibular joint

Medial malleolus

Lateral malleolus

Lateral malleolus

(a)

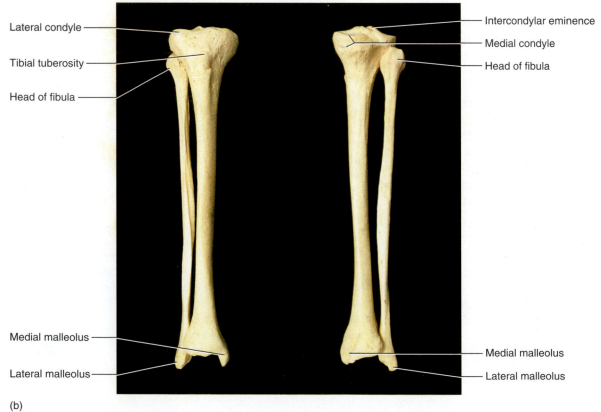

Lateral condyle

Tibial tuberosity

Head of fibula

Intercondylar eminence

Medial condyle

Head of fibula

Medial malleolus

Lateral malleolus

Medial malleolus

Lateral malleolus

(b)

FIGURE 7.10

Right Tibia and Fibula (a) Diagram; (b) photograph.

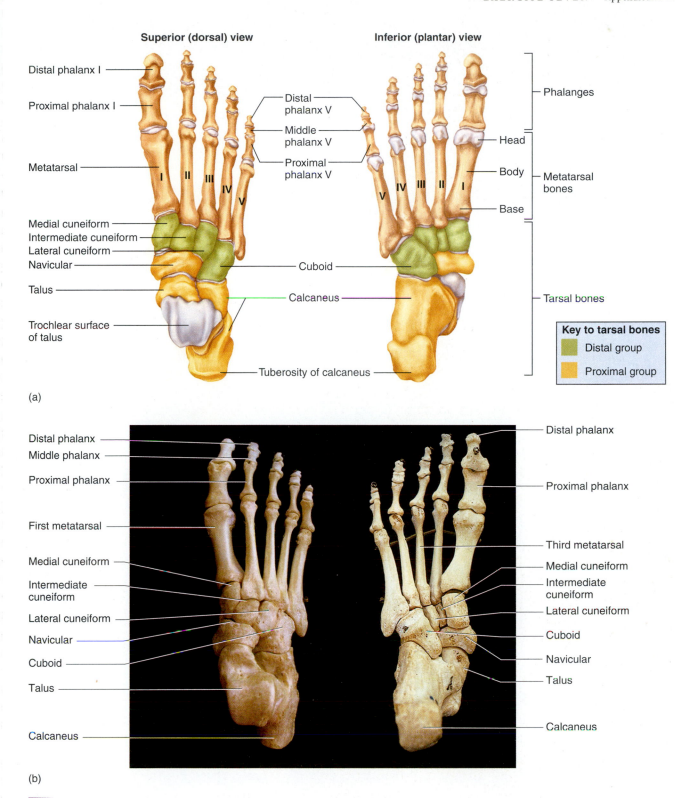

Superior (dorsal) view

Distal phalanx I

Proximal phalanx I

Metatarsal

I   II   III   IV   V

Distal phalanx V

Middle phalanx V

Proximal phalanx V

Inferior (plantar) view

Phalanges

Head

Body

Base

Metatarsal bones

V   IV   III   II   I

Medial cuneiform
Intermediate cuneiform
Lateral cuneiform
Navicular

Talus

Trochlear surface of talus

Cuboid

Calcaneus

Tuberosity of calcaneus

Tarsal bones

Key to tarsal bones

Distal group

Proximal group

(a)

Distal phalanx
Middle phalanx
Proximal phalanx

First metatarsal

Medial cuneiform

Intermediate cuneiform

Lateral cuneiform

Navicular

Cuboid

Talus

Calcaneus

Distal phalanx

Proximal phalanx

Third metatarsal
Medial cuneiform
Intermediate cuneiform
Lateral cuneiform

Cuboid

Navicular

Talus

Calcaneus

(b)

FIGURE 7.11

Bones of the Right Foot (a) Diagram; (b) photograph.

Name _____ Date _____

1. Name the anterior depression on the scapula.

2. The humerus fits into what specific part of the scapula?

3. Match the bone with the region it comes from.

   hallux   hip bone

   ilium   hand

   clavicle   upper extremity

   scaphoid   pectoral girdle

   radius   foot

4. What specific part of the clavicle attaches to the scapula?

5. Frequently, the clavicle is broken when the arms are extended to brace a fall. Explain how hitting the ground with your hands can fracture the clavicle.

6. The proximal epiphyseal line on the humerus has what other name?

7. The medial and lateral condyles of the humerus have specific names. What are they?

8. Name the depression in the ulna into which the humerus inserts.

9. Name the bony process that extends distally from the head of the ulna.

10. How does the head of the ulna differ in position from the head of the radius?

11. Each metacarpal bone consists of three major regions. What are they?

12. Name the carpal bone at the base of the thumb.

13. What is the name of the joint at the anterior junction of the pubic bones?

14. Name the notch found just inferior to the posterior inferior iliac spine.

15. Explain the difference between the false pelvis and the true pelvis.

16. Name the roughened line that runs along the length of the posterior femur.

17. Name the sesamoid bone that forms in the quadriceps femoris tendon.

18. Name the weight-bearing bone of the leg.

19. What is the name of the process on the distal portion of the tibia?

20. What is the function of the fibula?

21. What is another name for the heel bone?

22. What is the name of the bone of the foot that joins with the tibia and fibula?

23. What is another name for the big toe?

24. How many bones are in the ankle versus the number of bones in the wrist?

25. Label the parts of the scapula in the following illustration using the terms provided.

medial border   acromion
scapular spine   infraspinous fossa
supraspinous fossa   inferior angle
coracoid process   lateral border

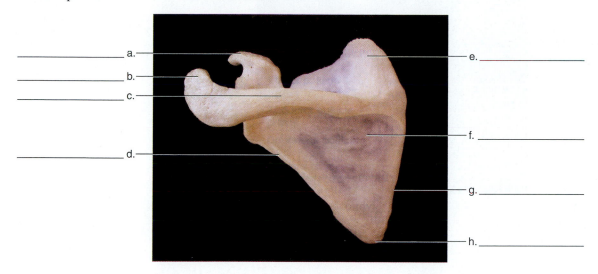

26. Label the following illustration of the foot using the terms provided.

calcaneus   medial cuneiform
cuboid   intermediate cuneiform
talus   lateral cuneiform
navicular

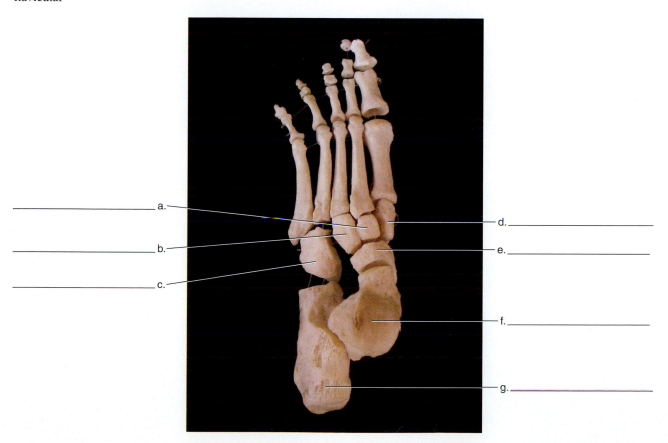

27. Label the following illustration of the hand using the terms provided.

lunate   trapezoid
hamate   capitate
pisiform

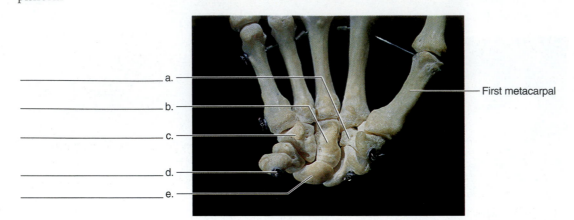

a.

b.

c.

d.

e.

First metacarpal

Axial Skeleton: Vertebrae, Ribs, Sternum, Hyoid

Skeletal System

INTRODUCTION

The axial skeleton consists of 80 bones, including the skull, hyoid, vertebrae (singular, *vertebra*), sternum, and ribs. In this exercise you examine all parts of the axial skeleton except the skull, which will be covered in laboratory exercise 9. There are 33 individual vertebrae, which form the vertebral column. Some fuse into larger structures, such as the sacrum. The hyoid is a small, floating bone between the floor of the mouth and the upper anterior neck; it assists the tongue in swallowing and the larynx in speaking.

The thoracic cage consists of the 12 pairs of ribs and the sternum, which protect the lungs and heart yet provide for flexibility during breathing.

LEARNING OBJECTIVES

At the end of this exercise you should be able to
1. locate and name all the bones of the vertebral column on an articulated skeleton;
2. name the significant surface features of individual vertebrae;
3. describe the differences among cervical, thoracic, and lumbar vertebrae and be able to name and distinguish between the atlas and the axis;
4. locate the major features of the hyoid bone;
5. distinguish a left rib from a right rib;
6. list the anatomical features of a rib;
7. demonstrate the location of the three major regions of the sternum.

MATERIALS

Articulated skeleton or plastic casts of a skeleton

Disarticulated skeleton or plastic casts of bones

Charts of the skeletal system

Plastic drinking straws cut at an angle or pipe cleaners for pointer tips

Foam pads of various sizes (to protect bone from hard countertops)

A cardboard box or piece of wood about 1 foot square and 3 inches deep

PROCEDURE

Review the bones of the axial skeleton in the lab and in laboratory exercise 6 (fig. 6.1). In this exercise you study the bones of the axial skeleton by examining both the disarticulated bones and the articulated skeleton. Hold each bone up to the skeleton to see how it is positioned in relation to the other bones of the body. When examining bones in the lab, do not use your pen or pencil to locate a structure. Use a cut drinking straw or a pipe cleaner to point out structures. Your instructor may want you to place real bone material on foam pads to cushion the bone from the tabletop.

Vertebrae

OVERVIEW OF THE VERTEBRAE

The **vertebral column** of humans is significantly different from that of all other mammals in that humans are bipedal. The lower vertebrae are larger than the upper vertebrae. This is due to the increase in weight on the lower vertebrae. Between the vertebrae are fibrocartilaginous pads known as **intervertebral discs.** Obtain a vertebra and examine it for the features represented in a typical vertebra (fig. 8.1). Place the vertebra in front of you, so that you can see through the large hole known as the **vertebral foramen,** where the spinal cord is located. Note the large **body** of the vertebra, which supports the weight of the vertebral column and is in contact with the intervertebral discs. Posterior to the body is the **vertebral,** or **neural, arch,** which consists of two pedicles (PED-ik-uls) and two laminae (LAM-in-aye). The **pedicles** are the parts of the arch that extend from the body of the vertebra to the two lateral projections called the **transverse processes.** Each **lamina** (LAM-in-uh) is a broad, flat structure between the transverse process and the dorsal **spinous process.**

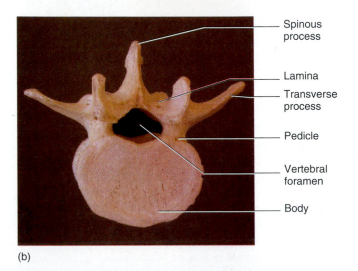

(a)

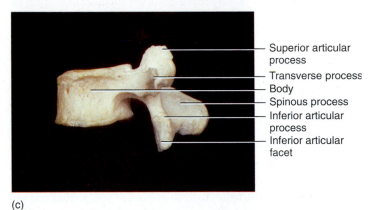

(b)

Superior articular
process
Transverse process
Body
Spinous process
Inferior articular
process
Inferior articular
facet

(c)

FIGURE 8.1

Features of a Typical Vertebra (a) Diagram, superior view. Photograph
(b) superior view; (c) lateral view.

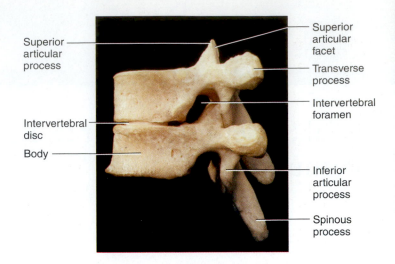

FIGURE 8.2
Articulated Vertebrae, Lateral View

If you rotate the vertebra as illustrated in figure 8.1c, you should be able to see the vertebral body in lateral view with the **superior articular process** and the **superior articular facet.** The superior articular facet is a flat surface that articulates with the vertebra above. There is also an **inferior articular process** and an **inferior articular facet** that articulates with the vertebra below. If you put two adjacent vertebrae together and view them from the side, you can see the **intervertebral foramen,** a hole where the spinal nerve is located. This is illustrated in figure 8.2.

SPINAL CURVATURES

The spine has four curvatures, which alternate from superior to inferior. The **cervical** (SERVE-ik-ul) **curvature** is convex (bowed forward) when seen from anatomical position. The **thoracic** (thor-AH-sic) **curvature** is concave, while the **lumbar curvature** is convex and the **sacral (pelvic) curvature** is concave. These curvatures allow for balance in an upright posture and provide room for the lungs and abdominal organs. Locate them in figure 8.3.

CERVICAL VERTEBRAE

There are seven **cervical vertebrae** (C1–C7), and these can be distinguished from all other vertebrae in that each vertebra has three foramina. The **vertebral foramen** is the largest opening, and the two **transverse foramina** are found only in cervical vertebrae. The transverse foramina house the vertebral arteries and vertebral veins. Some of the cervical vertebrae (C2–C6) have **bifid** (BY-fid) **spinous processes** (*bifid* = split in two), and the bodies of the cervical vertebrae are less massive than inferior vertebrae. Examine the isolated cervical vertebrae in the lab and compare them with figure 8.4. Three cervical vertebrae are unique enough to warrant special attention. The first cervical vertebra (C1) is known as the **atlas** and is the only cervical vertebra without a body. Atlas was a titan in Greek mythology who held up the world. The atlas joins with the head and allows you to nod your head to indicate "yes." The second cervical vertebra (C2) is the **axis,** and it has a process called the **dens,** or **odontoid process,** which runs superiorly through the atlas. The odontoid process allows the atlas to move on the axis and allows you to rotate your head to indicate "no." The seventh cervical vertebra (C7) is known as the **vertebra prominens.** It has a spinous process that projects sharply in a posterior direction and can be palpated (felt) as a significant bump at the junction of the neck and the back. Look at an atlas, an axis, and a vertebra prominens in the lab and compare them with figure 8.5.

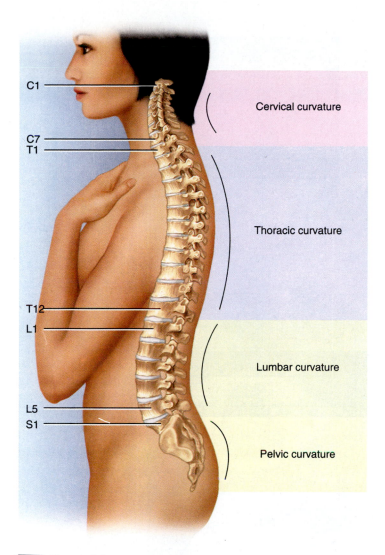

FIGURE 8.3
Spinal Curvatures, Lateral View

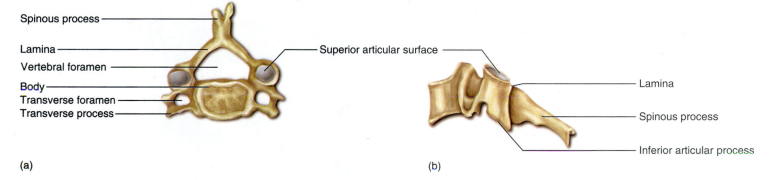

FIGURE 8.4
Cervical Vertebra Diagram (a) superior view; (b) lateral view. *(continued)*

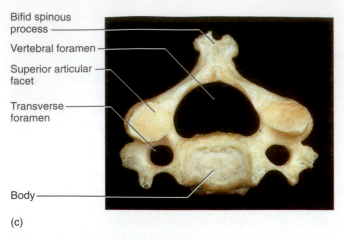

Bifid spinous process
Vertebral foramen
Superior articular facet
Transverse foramen
Body

(c)

Superior articular process
Transverse foramen
Spinous process
Body

(d)

■ FIGURE 8.4 (continued)
Cervical Vertebra photograph (c) superior view; (d) lateral view.

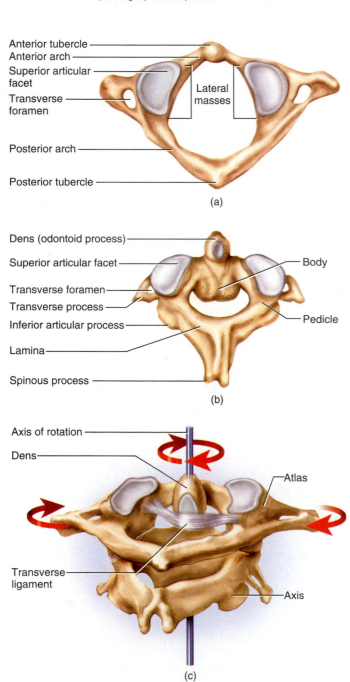

Anterior tubercle
Anterior arch
Superior articular facet
Transverse foramen
Lateral masses
Posterior arch
Posterior tubercle

(a)

Dens (odontoid process)
Superior articular facet
Transverse foramen
Transverse process
Inferior articular process
Lamina
Spinous process
Body
Pedicle

(b)

Axis of rotation
Dens
Atlas
Transverse ligament
Axis

(c)

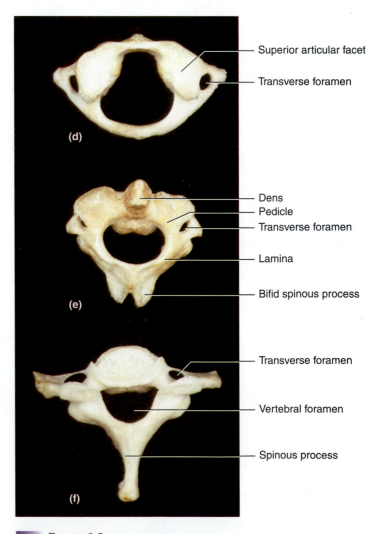

Superior articular facet
Transverse foramen

(d)

Dens
Pedicle
Transverse foramen
Lamina
Bifid spinous process

(e)

Transverse foramen
Vertebral foramen
Spinous process

(f)

■ FIGURE 8.5
Atlas, Axis, and Vertebra Prominens (a) Atlas, superior view; (b) axis, superoposterior view; (c) atlas and axis joined. Photographs of three vertebrae, superior view (d) atlas; (e) axis; (f) vertebra prominens.

THORACIC VERTEBRAE

There are 12 thoracic vertebrae. The **thoracic vertebrae** are distinguished from all other vertebrae by markings on their lateral posterior bodies, which are attachment points for ribs. In some areas, a rib attaches to just 1 vertebra, leaving a mark on the vertebral body known as a **rib facet.** In other areas, the head of a rib attaches to 2 vertebrae. In this case, each vertebra has a smooth attachment point known as a **demifacet.** The head of the rib spans both of the vertebrae and articulates with the demifacet of the superior and inferior vertebrae.

The thoracic vertebrae also have longer spinous processes than the cervical vertebrae, and the spinous processes of the thoracic vertebrae tend to angle in a more inferior direction. Thoracic vertebrae do not have transverse foramina. Examine figure 8.6 and note the characteristics of the vertebrae as seen in the lab.

LUMBAR VERTEBRAE

There are five lumbar vertebrae. The **lumbar vertebrae** are distinguished from the other vertebrae by having neither transverse foramina nor rib facets. The spinous processes of the lumbar vertebrae tend to be more horizontal than the thoracic vertebrae, and the bodies of the lumbar vertebrae are larger, as they carry more weight than the thoracic vertebrae, yet the twelfth thoracic and the first lumbar vertebrae are remarkably similar. The last thoracic vertebra has rib facets on it, and the first lumbar does not. Look at the lumbar vertebrae in the lab and compare them with figure 8.7.

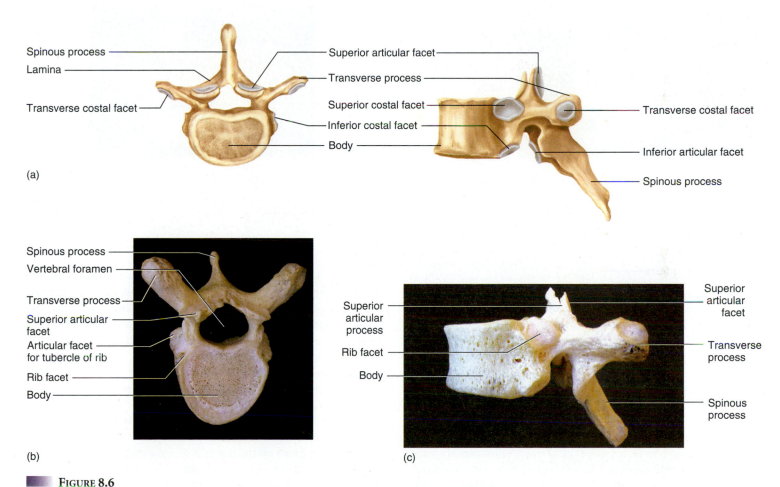

FIGURE 8.6

Thoracic Vertebrae Diagram (a) superior and lateral views. Photograph (b) superior view; (c) lateral view.

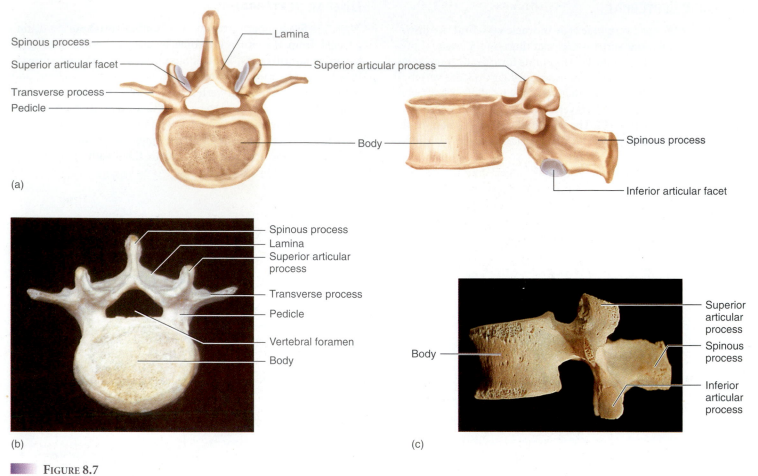

■ **FIGURE 8.7**

Lumbar Vertebrae Diagram (a) superior and lateral views. Photograph (b) superior view; (c) lateral view.

SACRUM

The **sacrum** is a large, wedge-shaped bone composed of five fused vertebrae. The lines of fusion are called **transverse lines,** and these may be seen on both the anterior and posterior sides. Notice how the sacrum is shaped like a shallow, triangular bowl. If you place the sacrum in front of you, as you would a cereal bowl, the shallow depression is the **anterior surface.** The two rows of holes you see are the **anterior sacral foramina.** On the posterior surface are the **posterior sacral foramina.** Compare a sacrum from your lab with figure 8.8.

The **sacral promontory** is a rim on the anterior superior part of the sacrum, and on each side of the sacral promontory is an **ala,** an expanded, winglike region. The roughened areas, lateral to these, are the **articular,** or **auricular** (ear-shaped), **surfaces** of the sacrum; each joins with the ilium to form the **sacroiliac** (SACK-ro-ILL-ee-ak) **joint.** As additional force is applied to the sacrum, it wedges itself into the ilium. If you examine the posterior surface of

the sacrum, you will notice the posterior sacral foramina, the **median** and **lateral sacral crests,** and the superior and inferior openings of the **sacral canal.** As with the other vertebrae, the **superior articular processes** and the **superior articular facets** join with the next most superior vertebra (the fifth lumbar vertebra). The opening at the inferior part of the sacrum is the **sacral hiatus** (a gap). These features can be seen in figure 8.8.

The sacrum is wedge-shaped, which allows for the upper body to carry greater weight than if the sacrum were box-shaped, as it is in many quadrupeds. Hold a smooth block of wood or a box between your hands, as illustrated in figure 8.9a. Have your lab partner push down on the box and see how much force is required for the box to slip through your hands. This would be similar to the forces that would act on a rectangular sacrum. Now turn the box or block of wood so that it forms a wedge in your hands, as illustrated in figure 8.9b. Have your lab partner push down on the box. Is it easier or harder to dislodge the "sacrum" in this way?

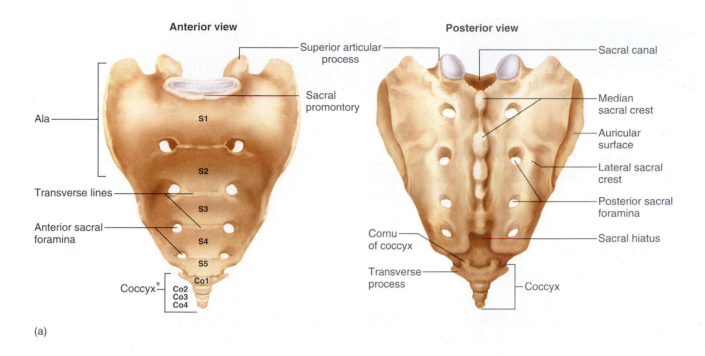

(a)

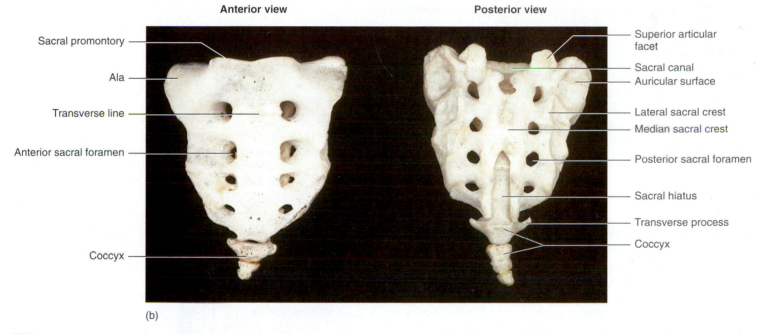

(b)

FIGURE 8.8

Sacrum and Coccyx (a) Diagram; (b) photograph.

(a)   (b)

FIGURE 8.9

Forces Acting on the Sacrum (a) Rectangular sacrum; (b) wedge-shaped sacrum.

COCCYX

The **coccyx** is the terminal portion of the vertebral column, and it usually consists of four fused vertebrae. The coccyx may be fused with the sacrum in some older females who have had several children and in individuals who have fallen backwards and landed on a hard surface. Examine the coccyx in the lab and compare it with figure 8.8.

Hyoid

The **hyoid** is a floating bone (it has no direct bony attachments) found at the junction of the floor of the mouth and the neck. The hyoid is anchored by muscles from the anterior, posterior, and inferior directions and aids tongue movement and swallowing. Locate the central **body** of the hyoid, the **greater horn,** or **cornua,** (COR-new-uh) and the **lesser horn,** or **cornu,** (singular, COR-new) on the material in the lab and in figure 8.10.

Ribs

There are 12 pairs of **ribs** in the human; along with the sternum, they make up the **thoracic cage.** Each rib has a **head** that articulates with the body of one or more vertebrae. On the head is a **facet,** which is the site of articulation. A constricted region near the head is the **neck** of the rib. Locate these features on figure 8.11. The process near the neck on ribs 1–10 is the **tubercle** of the rib, which articulates with the transverse process of the vertebra. Examine the articulated skeleton from the back and notice that the **shaft** of the ribs bends at about the level of the inferior angle of the scapula. This bend can be seen on ribs 2–10 and is known as the **angle** of the rib, or the **costal angle.** Some ribs have a truncated **sternal end** that attaches to a costal cartilage prior to joining the sternum. The superior edge of the rib is more rounded than the inferior edge. The

depression that runs along the inferior edge of each rib is known as the **costal groove.** With the costal groove in an inferior position and the blunt sternal end toward the midline, determine whether you are looking at a left rib or a right rib.

The first 7 pairs of ribs are **true ribs.** A true rib is one that attaches to the sternum by its own cartilage. Ribs 8–12 are **false ribs,** because they do not attach to the sternum by their own cartilage. Ribs 8–10 attach to the sternum by way of the cartilage of rib 7. Ribs 11 and 12 do not attach to the sternum at all and are known as

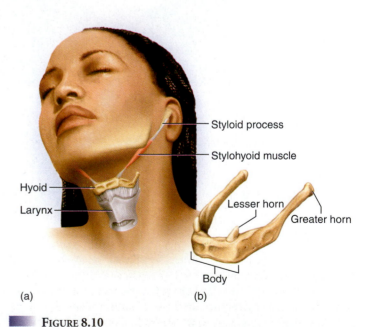

(a)   (b)

FIGURE 8.10

Hyoid Bone, Anterior View (a) In situ; (b) details of hyoid.

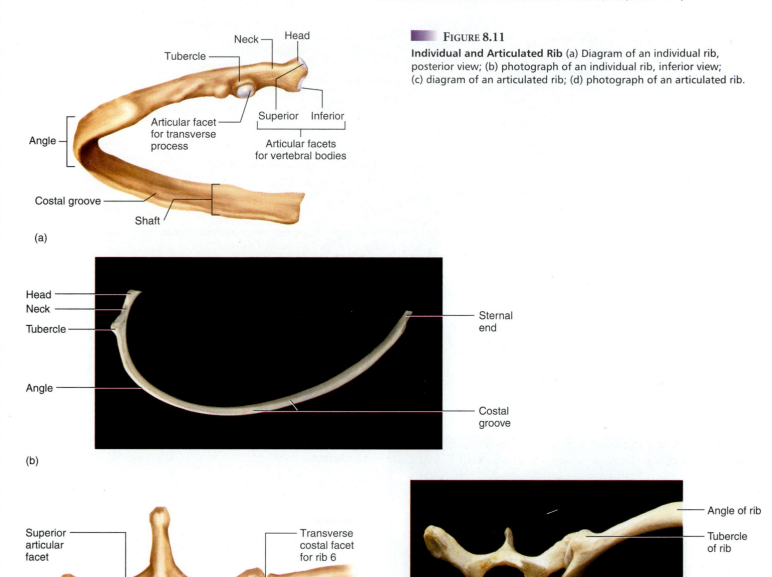

FIGURE 8.11

Individual and Articulated Rib (a) Diagram of an individual rib, posterior view; (b) photograph of an individual rib, inferior view; (c) diagram of an articulated rib; (d) photograph of an articulated rib.

(a) Labels: Neck, Head, Tubercle, Angle, Articular facet for transverse process, Superior, Inferior, Articular facets for vertebral bodies, Costal groove, Shaft

(b) Labels: Head, Neck, Tubercle, Angle, Sternal end, Costal groove

(c) Labels: Superior articular facet, Transverse costal facet for rib 6, Tubercle, Superior costal facet for rib 6, Neck, Head, Rib 6, T6

(d) Labels: Angle of rib, Tubercle of rib, Neck of rib, Head of rib, Demifacet, Body

floating ribs (a specific type of false rib). Examine the articulated skeleton in the lab and compare it with figure 8.12.

Sternum

The **sternum** is composed of three fused bones. The superior segment is the **manubrium** (ma-NOO-bree-um) and the depression at the top of the manubrium is a **jugular,** or median **suprasternal, notch.** The lateral indentations on the manubrium are sites of articulation with the clavicles known as **clavicular notches.** The main portion of the sternum is the **body,** or **gladiolus,** and between the

body and the manubrium is the **sternal angle.** This is a landmark for finding the second rib when using a stethoscope to listen to heart sounds. Locate the **costal notches** on the body of the sternum. These are where the cartilages of the ribs attach.

The narrow, bladelike part that is the most inferior segment of the sternum is the **xiphoid** (ZYE-foyd) **process.** Care must be taken when performing CPR (cardiopulmonary resuscitation) so that pressure is applied to the body of the sternum and not to the xiphoid process. If the force is applied to the xiphoid process, it could fracture and penetrate the liver. Examine the structures of the sternum on lab specimens and in figure 8.12.

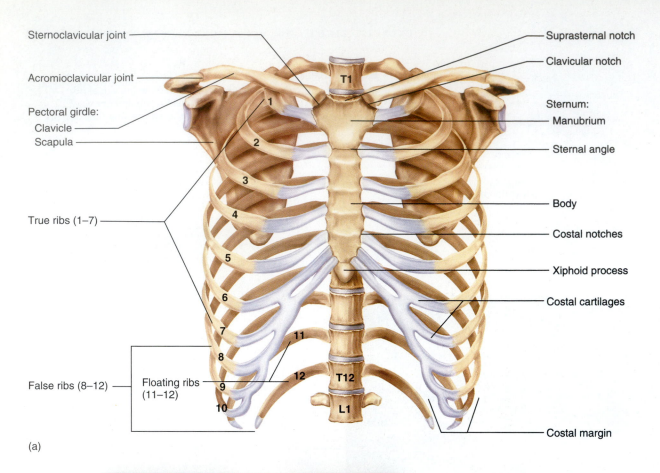

Sternoclavicular joint

Acromioclavicular joint

Pectoral girdle:
Clavicle
Scapula

True ribs (1–7)

False ribs (8–12) Floating ribs (11–12)

Suprasternal notch

Clavicular notch

Sternum:
Manubrium

Sternal angle

Body

Costal notches

Xiphoid process

Costal cartilages

Costal margin

T1
1
2
3
4
5
6
7
8
9
10
11
12
T12
L1

(a)

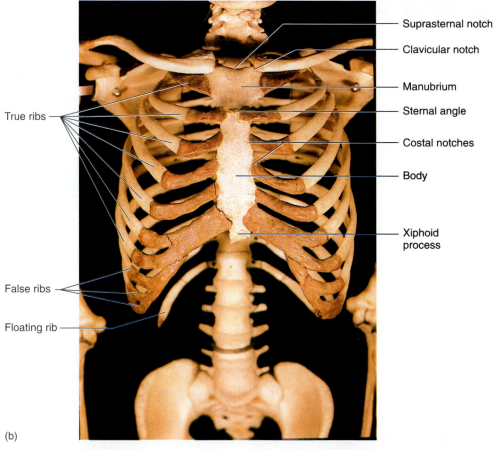

Suprasternal notch

Clavicular notch

Manubrium

Sternal angle

Costal notches

Body

Xiphoid process

True ribs

False ribs

Floating rib

(b)

FIGURE 8.12

Thoracic Cage (a) Diagram; (b) photograph.

Axial Skeleton: Vertebrae, Ribs, Sternum, Hyoid

Name _____ Date _____

1. Label the following illustration, using the terms provided.

a. _____

b. _____

c. _____

d. _____

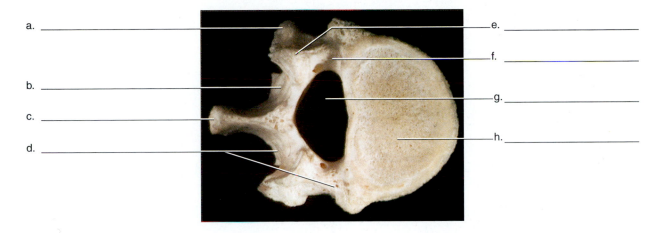

e. _____

f. _____

g. _____

h. _____

vertebral foramen   pedicle

vertebral body   spinous process

superior articular process   transverse process

lamina   vertebral arch

2. What are the names of the fibrocartilage pads found between adjacent bodies of the vertebrae?

3. What structures make up the vertebral arch?

4. The superior articular facet of a vertebra articulates with what specific structure?

5. Which spinal curvature is the most superior one?

6. On what vertebra would you find the odontoid process?

7. Which vertebra has no body?

8. What two features do cervical vertebrae have that no other vertebrae have?

9. Match the cervical vertebrae with their numeric names.

    axis   C1

    vertebra prominens   C2

    atlas   C7

10. Determine from what part of the spinal column the vertebra in the following illustration comes.

11. How many lumbar vertebrae are in the human body?

12. Distinguish between the posterior sacral foramina and the sacral canal.

13. Name the horizontal lines that result from the fusion of the sacral vertebrae.

14. What is the joint between the sacrum and the hip bones called?

15. Match the vertebrae from a region with the features or structures found there.

Vertebrae	Features or Structures
_____ cervical	a. ala
_____ thoracic	b. vertebrae with the largest vertebral bodies
_____ lumbar	c. vertebrae with rib facets
_____ sacral	d. typically four fused vertebrae
_____ coccyx	e. transverse foramina

16. Name the major features of the hyoid.

17. Does the hyoid have any solid bony attachments?

18. A rib that attaches to the sternum by the cartilage of rib 7 has what name?

19. Is the angle of the rib on the anterior or posterior side of the body?

20. Which ribs (by number) are the floating ribs?

21. What part of a rib articulates with the transverse process of a vertebra?

22. The most inferior portion of the sternum is the
    a. body.   b. manubrium.   c. angle.   d. xiphoid.

23. The superior portion of the sternum has what name?

Axial Skeleton: Skull

Skeletal System

INTRODUCTION

The bones of the skull can be divided into three groups: bones of the cranial vault, bones of the face, and bones of the middle ear. The skull is the most complex region of the skeleton and not only houses the brain but contains a significant number of sense organs as well. In this exercise you learn the details of the skull, particularly those of the cranial vault and the face. You will study the bones of the middle ear in this laboratory manual in laboratory exercise 18, "Introduction to Sensory Receptors." Be careful if you handle real bones, as they are fragile and can easily be broken.

LEARNING OBJECTIVES

At the end of this exercise you should be able to
1. list all the bones of the cranial vault and the major features of those bones;
2. list all the bones of the face and the major features of those bones;
3. name all the bones that occur singly or in pairs in the skull;
4. locate the major foramina of the skull;
5. find the specific bony markings of representative bones;
6. name the major sutures of the skull;
7. list all the fontanels of the fetal skull.

MATERIALS

Disarticulated skull, if available

Articulated skulls with the calvaria cut

Foam pads to cushion skulls from the desktop

Pipe cleaners

PROCEDURE

Overview of the Cranial Vault

Familiarize yourself with the bones of the skull in the **cranial vault** and the **face.** The skull bones in the cranial vault are listed next with a number, in parentheses, after the name of the bone indicating whether the bone occurs singly or as a pair. Take a skull back to your lab table and locate the bones listed in figure 9.1.

Frontal (1)   Parietal (2)

Occipital (1)   Temporal (2)

Sphenoid (1)   Ethmoid (1)

You can remember the bones of the cranium with the mnemonic "of pets." Each letter represents a cranial bone (*o* for occipital, *f* for frontal, etc.)

Overview of the Face

The bones of the face are listed next. Locate these bones on a skull in the lab, as shown in figures 9.1 and 9.2.

Maxilla (2)   Nasal (2)

Mandible (1)   Palatine (2)

Vomer (1)   Zygomatic (2)

Lacrimal (2)   Inferior nasal concha (plural, *conchae*) (2)

The facial bones can be remembered by this mnemonic device: "Manny makes naturally insane zigzags pleasing Vera Lynn." The first names (Manny and Vera) represent the single bones of the face (mandible and vomer) and the other bones are paired.

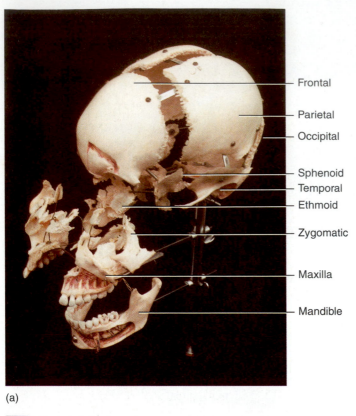

(a)

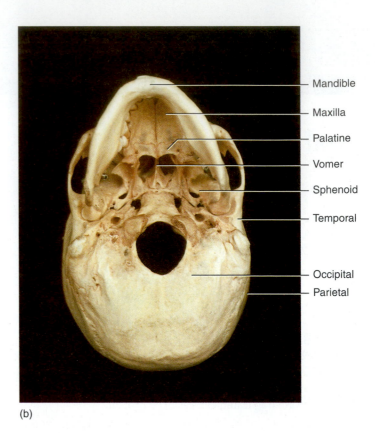

(b)

FIGURE 9.1
Overview of the Skull (a) Superolateral view; (b) inferior view.

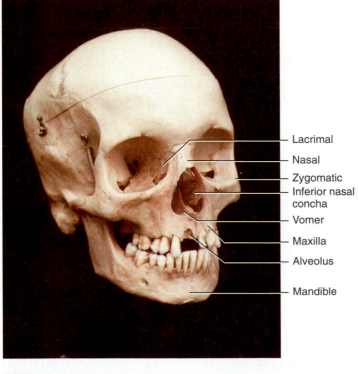

FIGURE 9.2
Bones of the Face

Be very careful handling skulls. As you locate the various bones on a skull in the lab, make sure you do not poke pencils, pens, or fingers into the delicate regions of the orbits (eye sockets) or nasal cavity. Do not break the delicate structures in these regions. Once you have found the major bones of the skull, use the following descriptions and illustrations to find the specific structures of the skull. The skull is described from a series of views, and you should locate the anatomical features listed for each of those views.

ANTERIOR VIEW

The large bone that makes up the forehead is the **frontal bone.** It is a single bone that makes up the superior portion of the **orbits** and has two ridges above the eyes called **supraorbital ridges,** or **margin,** (eyebrows). There is a hole in each of these ridges called the **supraorbital foramen** that allows for nerves and arteries to reach the face. Most of what you see inferior to the frontal bone are facial bones. The bony part of the nose is composed of the paired **nasal bones,** which join with the cartilage that forms the tip of the nose. Posterior to the nasal bones are thin strips of the upper maxillae, bones that hold the upper teeth; posterior to the **maxilla** are the thin **lacrimal bones,** which contain the **nasolacrimal ducts,** into which tears drain from the eyes to the nose. Locate these bones in figure 9.3. Posterior to the lacrimal bones is the **ethmoid bone,** which is very delicate and frequently broken on skulls that have been mishandled. Posterior to the ethmoid is the **sphenoid bone,** which forms the posterior wall of the orbit and contains not only the **optic canal** (a passageway for the optic nerve) but also the **superior orbital fissure** and the **inferior orbital fissure.** The bones on the lateral side of the orbit are the **zygomatic bones,** commonly known as the cheekbones.

The major bone that makes up the floor of the orbit and below is the maxilla. The two maxillae have sockets called **alveoli** (singular, **alveolus**) which contain the teeth and extensions of bone between each pair of sockets called **alveolar processes.** The **infraorbital foramen** is a small hole in the maxilla below the eye; it is a passageway for nerves and blood vessels. The most inferior bone of the face is the **mandible,** commonly known as the jaw. The mandible also has alveoli and alveolar processes (fig. 9.4). The mandible begins as two bones in utero and fuses at the midline of the chin. This fusion results in the **mental symphysis** (review fig. 9.3). Lateral to the mental symphysis are the **mental foramina,** which conduct nerves and blood vessels to the tissue anterior to the jaw.

SUPERIOR VIEW

Locate the major suture lines of the skull from this view, as seen in figure 9.5. The frontal bone is separated from the pair of parietal bones by the **coronal suture.** The parietal bones are separated from one another by the **sagittal suture.** The parietal bones are separated from the occipital bone by the **lambdoid suture.** The lambdoid suture is named after the Greek letter lambda (**λ**), which is **Y**-shaped. There may be small bones that occur between the occipital bone and the parietal bones (or between other skull bones), and these are known as **sutural,** or **Wormian, bones.**

LATERAL VIEW

The coronal and lambdoid sutures can also be seen from the lateral view along with the **squamous suture,** which separates the **temporal bone** from the **parietal bone.** Locate the cranial bones from a lateral view. These are the **frontal, parietal, occipital, temporal, sphenoid,** and **ethmoid bones.** You may be able to see a bump at the back of the occipital bone from this angle. This is known as the **external occipital protuberance.** The hole on the side of the head where the ear attaches is the **external acoustic (auditory) meatus.** The large process posterior and inferior to this opening is the **mastoid process.** A long, thin spine medial to the mastoid process is the **styloid process,** which is an attachment point for muscles. The sphenoid bone can be seen from this view just adjacent to the temporal bone. Anterior to the sphenoid bone is the zygomatic bone forming the lateral wall of the orbit. The zygomatic bone has a **temporal process** that connects with the temporal bone. The temporal bone has a **zygomatic process** and both of these processes make up the **zygomatic arch.** The inner wall of the orbit is made up of the ethmoid bone, the lacrimal bone, the maxilla, and the nasal bone. Examine these structures in figure 9.6.

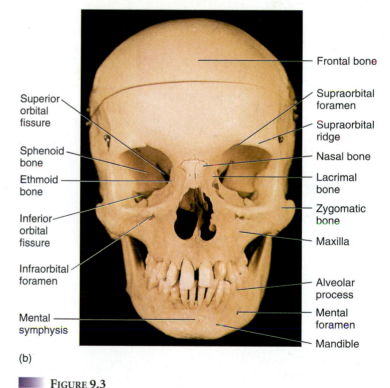

Frontal bone — Glabella — Coronal suture — Squamous suture — Sphenoid bone — Lacrimal bone — Nasal bone — Middle nasal concha — Infraorbital foramen — Vomer — Mandible — Mental protuberance

Supraorbital foramen — Parietal bone — Supraorbital margin — Temporal bone — Ethmoid bone — Zygomatic bone — Inferior nasal concha — Maxilla — Mental foramen

(a)

Superior orbital fissure — Sphenoid bone — Ethmoid bone — Inferior orbital fissure — Infraorbital foramen — Mental symphysis

Frontal bone — Supraorbital foramen — Supraorbital ridge — Nasal bone — Lacrimal bone — Zygomatic bone — Maxilla — Alveolar process — Mental foramen — Mandible

(b)

FIGURE 9.3

Skull, Anterior View (a) Diagram; (b) photograph.

The mandible has a **condylar process** with a terminal **mandibular condyle** articulating with the temporal bone; a **coronoid process,** medial to the zygomatic arch; and a **mandibular notch,** a depression between the condyloid process and coronoid process. The vertical section of the mandible is the **mandibular**

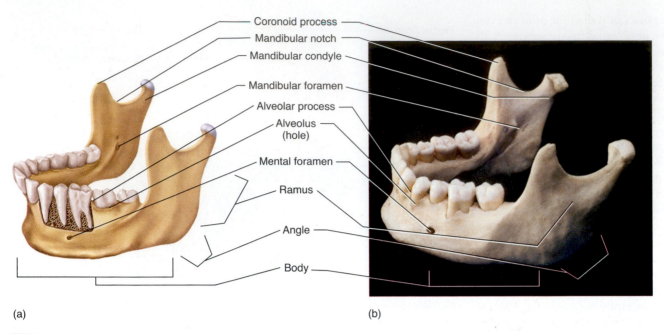

(a)   (b)

FIGURE 9.4
Mandible, Lateral View (a) Diagram; (b) photograph.

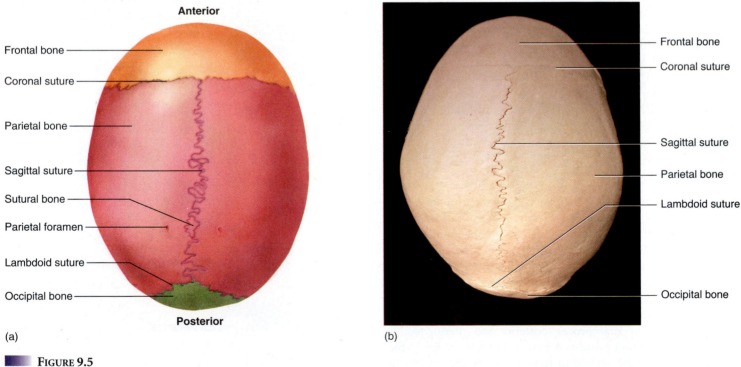

(a)   (b)

FIGURE 9.5
Skull, Superior View (a) Diagram; (b) photograph.

ramus (*ramus* = branch), and the horizontal portion of the mandible is the **body.** The **angle** of the mandible is at the posterior part of the bone at the junction of the body and the ramus. On the inside of each ramus of the mandible is the **mandibular foramen,** a conduit for a nerve. The parts of the mandible can be identified in figures 9.4 and 9.6.

INFERIOR VIEW

Place the skull in front of you with the mandible and the top of the skull removed (fig. 9.7). The largest hole in the skull, the **foramen magnum,** should be close to you. The foramen magnum is in the occipital bone and is the dividing line between the brain and the

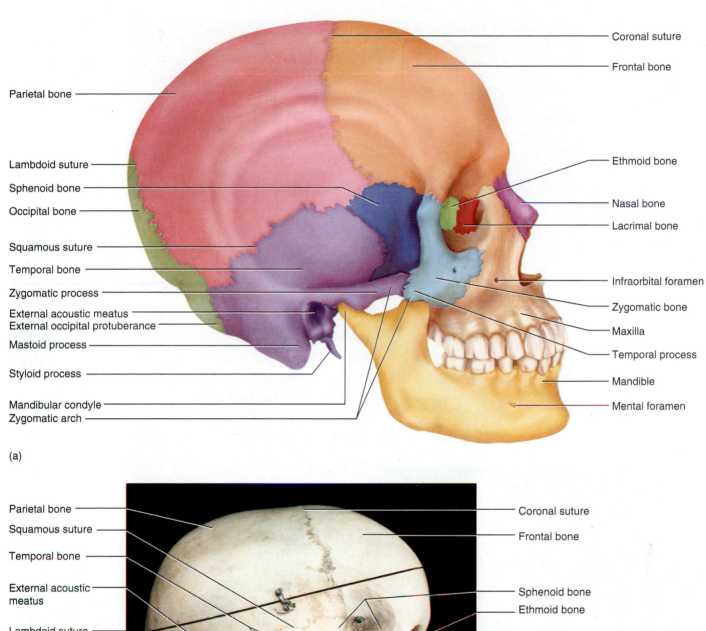

Coronal suture

Frontal bone

Parietal bone

Ethmoid bone

Lambdoid suture

Sphenoid bone

Occipital bone

Nasal bone

Lacrimal bone

Squamous suture

Temporal bone

Infraorbital foramen

Zygomatic process

Zygomatic bone

External acoustic meatus
External occipital protuberance

Maxilla

Mastoid process

Temporal process

Styloid process

Mandible

Mandibular condyle
Zygomatic arch

Mental foramen

(a)

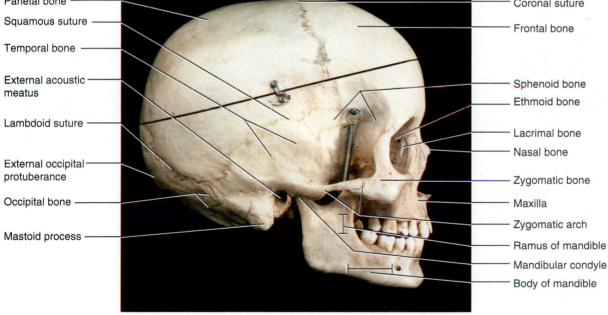

Parietal bone

Squamous suture

Temporal bone

External acoustic
meatus

Lambdoid suture

External occipital
protuberance

Occipital bone

Mastoid process

Coronal suture

Frontal bone

Sphenoid bone

Ethmoid bone

Lacrimal bone

Nasal bone

Zygomatic bone

Maxilla

Zygomatic arch

Ramus of mandible

Mandibular condyle

Body of mandible

(b)

FIGURE 9.6
Skull, Lateral View (a) Diagram; (b) photograph.

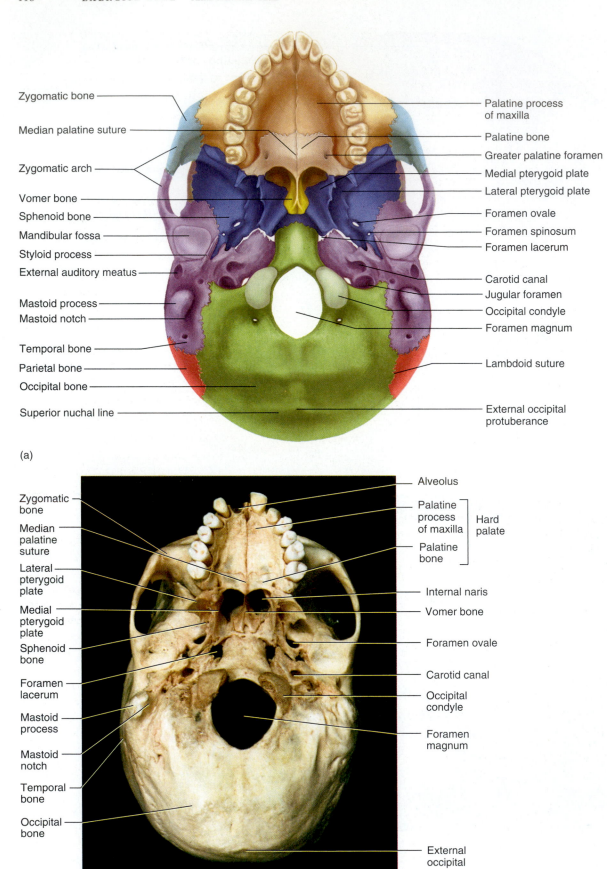

Zygomatic bone

Median palatine suture

Zygomatic arch

Vomer bone

Sphenoid bone

Mandibular fossa

Styloid process

External auditory meatus

Mastoid process

Mastoid notch

Temporal bone

Parietal bone

Occipital bone

Superior nuchal line

Palatine process of maxilla

Palatine bone

Greater palatine foramen

Medial pterygoid plate

Lateral pterygoid plate

Foramen ovale

Foramen spinosum

Foramen lacerum

Carotid canal

Jugular foramen

Occipital condyle

Foramen magnum

Lambdoid suture

External occipital protuberance

(a)

Zygomatic bone

Median palatine suture

Lateral pterygoid plate

Medial pterygoid plate

Sphenoid bone

Foramen lacerum

Mastoid process

Mastoid notch

Temporal bone

Occipital bone

Alveolus

Palatine process of maxilla

Palatine bone

Hard palate

Internal naris

Vomer bone

Foramen ovale

Carotid canal

Occipital condyle

Foramen magnum

External occipital protuberance

(b)

FIGURE 9.7

Skull, Inferior View (a) Diagram; (b) photograph.

spinal cord. Lateral to the foramen magnum are the **occipital condyles,** processes that articulate with the superior articular facets of the first cervical vertebra. A small bump at the posterior part of the occipital bone is the external occipital protuberance. Two small openings are found near the foramen magnum: the **hypoglossal canals,** which allow the passage of the hypoglossal nerve. At the junction of the occipital bone and the temporal bone is the **jugular foramen,** a hole through which the jugular vein passes. If you carefully insert a pipe cleaner into the jugular foramen and turn the skull over, you will see that the foramen leads to the posterior part of the skull.

You can see the mastoid process of the temporal bone and a depression just medial to the process known as the **mastoid notch.** Find the styloid processes in your specimen, though they may be hard to locate, since they are frequently broken in lab specimens. The styloid process is an attachment point for the muscles that move the tongue, larynx, and hyoid. Medial to the styloid process is the **carotid canal,** through which passes the internal carotid artery to the brain. If you carefully insert a pipe cleaner into the carotid canal of a skull, you will notice that the canal bends at about a 90° angle; if you turn the skull over, you will note that the opening occurs in the middle of the skull. You cannot do this with most plastic casts of skulls. At the junction of the temporal bone and the sphenoid bone is the **foramen lacerum,** which is adjacent to the carotid canal. The temporal bone also has a **mandibular fossa,** which is the articulation site of the mandible. The zygomatic process of the temporal bone can also be seen from this view.

From the inferior view the sphenoid bone can be seen as a bone that runs from one side of the skull to the other. Two pairs of flattened shelves can also be seen, the **lateral pterygoid plates** and the **medial pterygoid plates** (*pterygoid* = winglike). These are attachments for muscles that extend from the sphenoid to the mandible. Just posterior to the pterygoid plate is the **foramen ovale,** a hole that conducts one of the branches of the trigeminal nerve to the mandible.

In the midline of the skull, and sometimes looking like a part of the sphenoid, is the **vomer.** The two large holes on each side of the vomer are the **internal nares.** Connected to the vomer and forming part of the **hard palate** is the **palatine bone.** The palatine bones are **L**-shaped bones with a horizontal plate and a vertical plate. The horizontal plates normally join together at the **median palatine suture.** If this suture does not fuse completely at birth, an individual has a cleft palate. The anterior portion of the hard palate is made of the **palatine processes of the maxillae.**

The major openings of the skull are presented in table 9.1. Locate the openings and note the number of structures that pass through these holes.

INTERIOR OF THE CRANIAL VAULT

With the superior portion of the skull removed, you can see the **cranial cavity** divided into three major regions. These are the **anterior cranial fossa,** a depression anterior to the lesser wings of the sphenoid; the **middle cranial fossae,** which lie between the **lesser wings of the sphenoid** and the petrous portion of the temporal bone; and the **posterior cranial fossa,** which is posterior to the petrous portion of the temporal bone. Examine figure 9.8 for a view of the interior of the cranium.

TABLE 9.1
Openings of the Skull

Opening	Function or Structure in Opening
Carotid canal	Internal carotid artery, nerves
External acoustic (auditory) meatus	Opening for sound transmission
Foramen lacerum	Internal carotid artery
Foramen magnum	Spinal cord, vertebral arteries, accessory nerves
Foramen ovale	Mandibular branch of trigeminal nerve
Foramen rotundum	Maxillary branch of trigeminal nerve
Foramen spinosum	Meningeal blood vessels
Hypoglossal canal	Hypoglossal nerve
Inferior orbital fissure	Maxillary branch of trigeminal nerve
Infraorbital foramen	Infraorbital nerve and artery for the face
Internal acoustic (auditory) meatus	Vestibulocochlear nerve and facial nerve
Jugular foramen	Internal jugular vein, vagus, and other nerves
Mandibular foramen	Mandibular branch of trigeminal nerve and blood vessels
Mental foramen	Mental nerve and blood vessels
Optic canal	Optic nerve
Stylomastoid foramen	Facial nerve exits skull
Superior orbital fissure	Nerves to the eye and face
Supraorbital foramen	Supraorbital nerve and artery for the face

Beginning with the anterior cranial fossa, you should find the centrally located **ethmoid bone.** A sharp ridge known as the **crista galli** is part of the ethmoid and projects from the main portion of this bone. The small, horizontal plate of bone with numerous holes lateral to the crista galli is the **cribriform plate.** It is also part of the ethmoid bone. The holes in this plate, called **cribriform (olfactory) foramina,** transmit the sense of smell from the nose to the brain. If the skull was cut close to the orbit, you can see the **frontal sinus** as a hollow space in the anterior portion of the frontal bone.

The dividing line between the anterior and middle cranial fossae is the sphenoid bone. Locate the lesser wings of the sphenoid and the **sella turcica** (turk's saddle) just posterior to it. The **greater wings of the sphenoid** are more inferior than the lesser wings, and each one contains the **foramen rotundum,** which takes a branch of the trigeminal nerve to the maxilla. The **foramen ovale** can be seen from this view as well.

Posterior to the sphenoid bone is the temporal bone, having a flattened lateral section known as the **squamous portion** and a heavier mass of bone known as the petrous portion. The **petrous (rocklike) portion** divides the middle and posterior cranial fossae. The petrous portion also has a hole in the posterior surface, which is the **internal acoustic (auditory) meatus.** This is a passageway for the nerves that come from the inner ear. The posterior cranial fossa is located dorsal to the petrous portion of the temporal bone. It contains the foramen magnum and the jugular foramina. Most of this fossa is formed by the occipital bone.

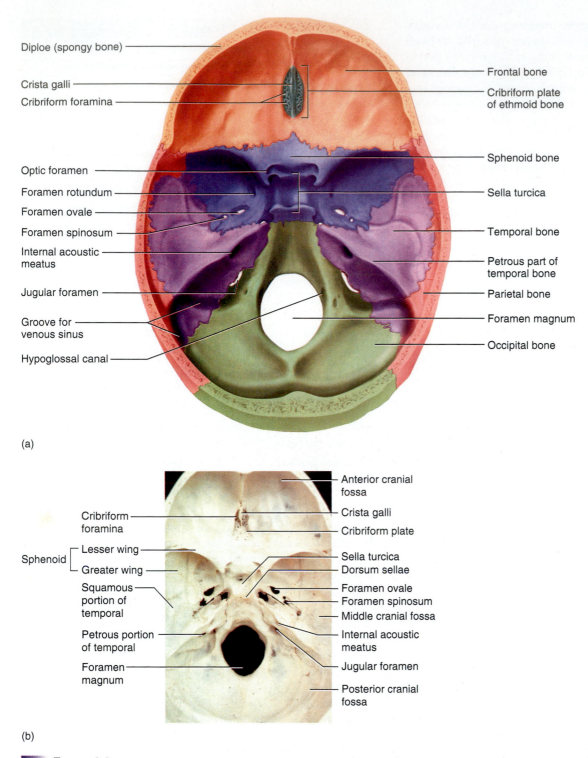

Diploe (spongy bone)

Crista galli

Cribriform foramina

Frontal bone

Cribriform plate of ethmoid bone

Sphenoid bone

Optic foramen

Foramen rotundum

Foramen ovale

Foramen spinosum

Internal acoustic meatus

Jugular foramen

Groove for venous sinus

Hypoglossal canal

Sella turcica

Temporal bone

Petrous part of temporal bone

Parietal bone

Foramen magnum

Occipital bone

(a)

Anterior cranial fossa

Cribriform foramina

Crista galli

Cribriform plate

Sphenoid {
Lesser wing
Greater wing

Sella turcica

Dorsum sellae

Foramen ovale

Squamous portion of temporal

Foramen spinosum

Middle cranial fossa

Petrous portion of temporal

Internal acoustic meatus

Foramen magnum

Jugular foramen

Posterior cranial fossa

(b)

FIGURE 9.8

Interior of the Cranium (a) Diagram; (b) photograph.

MIDSAGITTAL SECTION OF THE SKULL

Examine the structures that can be seen in a midsagittal section. Be extremely careful with the specimen, since many of the internal structures are fragile. Use figure 9.9 as a guide to the midsagittal section of the skull, whether or not one is present in the lab. Locate the **nasal septum,** which is composed of the **vomer,** the **perpendicular plate of the ethmoid bone,** and the **nasal cartilage** (absent in skull preparations). If the nasal septum is removed, you should be able to see the **superior nasal concha** and the **middle nasal concha** of the ethmoid bone. Below these is the **inferior nasal concha** (review fig. 9.2), which is a separate bone. Look for the junction between the palatine bone and the palatine process of the

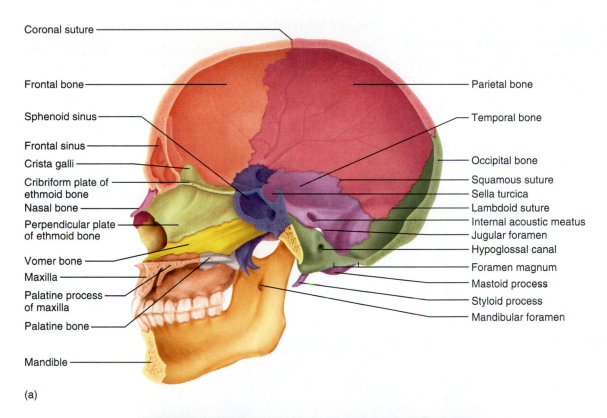

Coronal suture

Frontal bone

Sphenoid sinus

Frontal sinus

Crista galli

Cribriform plate of ethmoid bone

Nasal bone

Perpendicular plate of ethmoid bone

Vomer bone

Maxilla

Palatine process of maxilla

Palatine bone

Mandible

Parietal bone

Temporal bone

Occipital bone

Squamous suture

Sella turcica

Lambdoid suture

Internal acoustic meatus

Jugular foramen

Hypoglossal canal

Foramen magnum

Mastoid process

Styloid process

Mandibular foramen

(a)

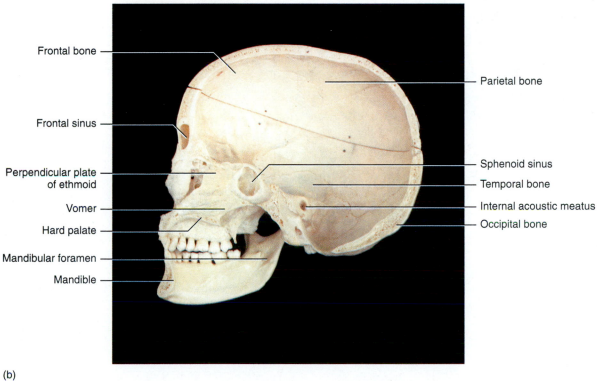

Frontal bone

Frontal sinus

Perpendicular plate of ethmoid

Vomer

Hard palate

Mandibular foramen

Mandible

Parietal bone

Sphenoid sinus

Temporal bone

Internal acoustic meatus

Occipital bone

(b)

FIGURE 9.9

Skull, Midsagittal Section (a) Diagram; (b) photograph.

maxilla. These two bony plates make up the hard palate. If the mandible is present, locate the **mandibular foramen,** which is on the inner aspect of the mandible and transmits branches of the trigeminal nerve and blood vessels to the mandible.

SINUSES

There are numerous sinuses and air cells in the skull. These sinuses provide resonance to the voice and give shape to the skull while decreasing its weight. The **paranasal sinuses** are located around the region of the nose and are named for the bones in which they are found. They include the **frontal sinus,** the **maxillary sinus,** the **ethmoid sinus** (ethmoid air cells), and the **sphenoid sinus.** These sinuses may fill with fluid when a person has a cold and harbor bacteria in secondary infections. Locate the sinuses in skulls in the lab and compare them with figure 9.10.

SELECT INDIVIDUAL BONES OF THE SKULL

Ethmoid   The **ethmoid bone** is located in the middle of the skull. The **perpendicular plate** of the ethmoid can be seen in midsagittal view or from the anterior view through the external nares (nostrils). The **orbital plate** is the part of the ethmoid that lines the medial wall of the orbit. The **middle nasal conchae** can be seen from the nasal cavity as well, but the **superior nasal conchae** are best seen by looking at an inferior view of the skull through the internal nares or in a midsagittal view with the nasal septum removed. Examine isolated ethmoid bones in the lab and find the **crista galli, cribriform plate,** and other structures, as shown in figure 9.11.

Sphenoid   The **sphenoid bone** is seen in figure 9.12. The sella turcica has a small depression in it called the **hypophyseal fossa,** in which the pituitary gland sits. The posterior, raised part of the sella

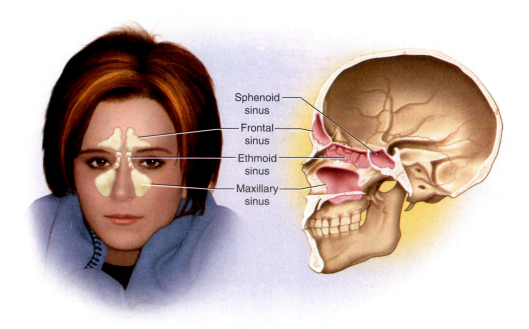

Sphenoid sinus

Frontal sinus

Ethmoid sinus

Maxillary sinus

FIGURE 9.10

Paranasal Sinuses of the Skull

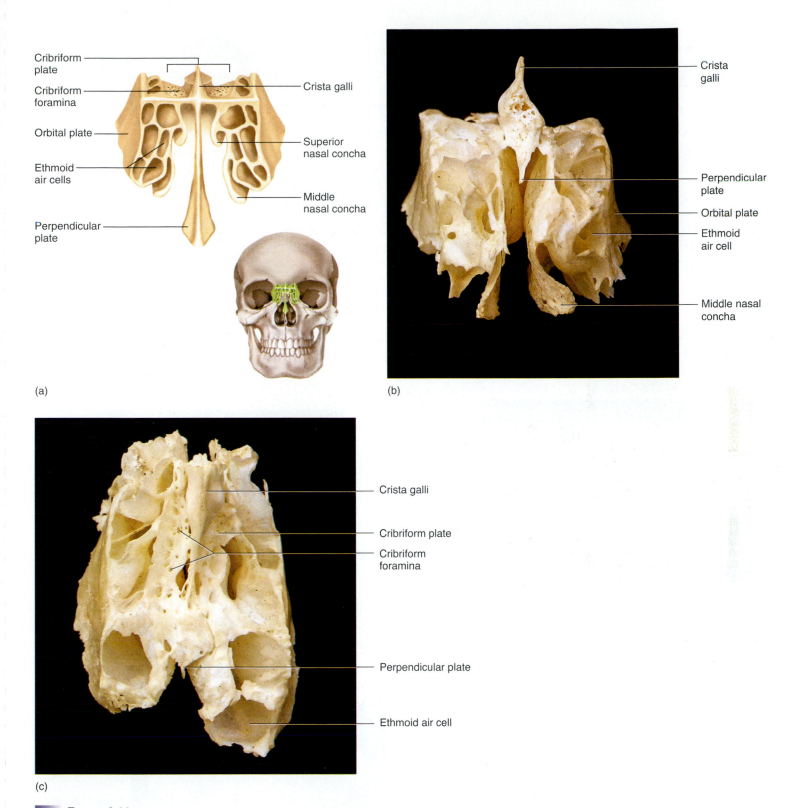

Cribriform plate

Cribriform foramina

Orbital plate

Ethmoid air cells

Perpendicular plate

Crista galli

Superior nasal concha

Middle nasal concha

(a)

Crista galli

Perpendicular plate

Orbital plate

Ethmoid air cell

Middle nasal concha

(b)

Crista galli

Cribriform plate

Cribriform foramina

Perpendicular plate

Ethmoid air cell

(c)

FIGURE 9.11

Ethmoid Bone Diagram (a) anterior view. Photograph (b) anterior view; (c) superior view.

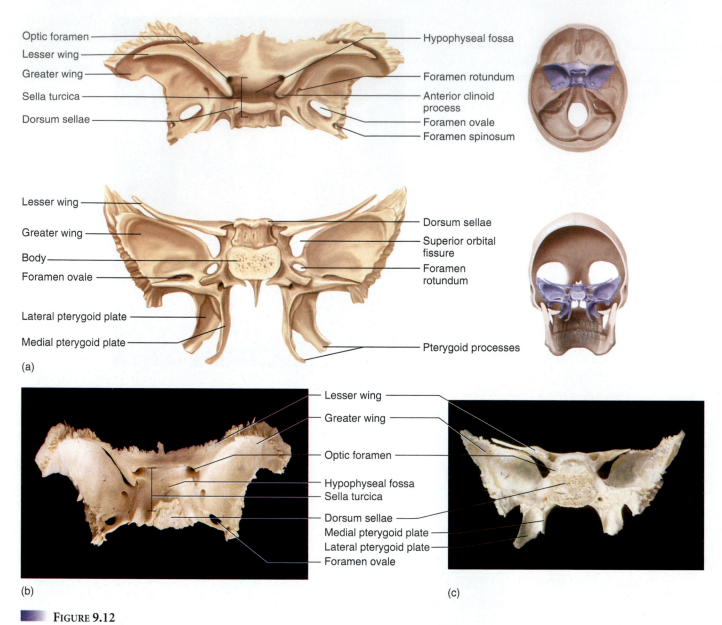

Optic foramen
Lesser wing
Greater wing
Sella turcica
Dorsum sellae

Hypophyseal fossa
Foramen rotundum
Anterior clinoid process
Foramen ovale
Foramen spinosum

Lesser wing
Greater wing
Body
Foramen ovale
Lateral pterygoid plate
Medial pterygoid plate
(a)

Dorsum sellae
Superior orbital fissure
Foramen rotundum
Pterygoid processes

Lesser wing
Greater wing
Optic foramen
Hypophyseal fossa
Sella turcica
Dorsum sellae
Medial pterygoid plate
Lateral pterygoid plate
Foramen ovale

(b)   (c)

FIGURE 9.12

Sphenoid Bone Diagram (a) superior and posterior views. Photograph (b) superior view; (c) posterior view.

turcica is known as the **dorsum sellae.** Examine an isolated sphenoid bone in the lab and locate the **greater wings,** the **lesser wings,** the **medial** and **lateral pterygoid plates,** the **sella turcica,** the **hypophyseal fossa,** the **dorsum sellae,** and other features, as seen in figure 9.12.

Temporal   There are two **temporal bones** in the cranium. Each has a flat, superior **squamous portion,** which forms part of the cranial vault, and a medial part called the **petrous portion.** The petrous portion contains the ear ossicles and the opening of the **internal acoustic (auditory) meatus,** as seen in the medial view of the temporal bone in figure 9.13. Examine an isolated temporal bone in lab and locate these features, as well as the **zygomatic process,** which articulates with the zygomatic bone, and the **mastoid process,** which can be palpated (felt) as a bump posterior to the ear.

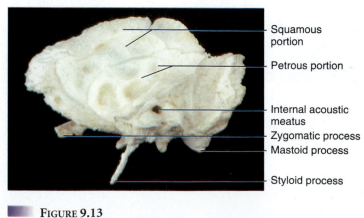

Squamous portion
Petrous portion
Internal acoustic meatus
Zygomatic process
Mastoid process
Styloid process

FIGURE 9.13
Right Temporal Bone, Medial View

FONTANELS

The fontanels are the "soft spots" of an infant's skull. There are four types of fontanels, and they allow for the passage of the skull through the birth canal by enabling the bones of the cranium to slide over one another in a process called molding. After birth the fontanels allow for further expansion of the skull. The **anterior (frontal) fontanel** is an area between the frontal bone and the parietal bones. The **posterior (occipital) fontanel** is between the oc- cipital bone and the parietal bones. The **anterolateral (sphenoid) fontanels** are paired structures on each side of the skull and are located superior to the sphenoid bone, and the **posterolateral (mastoid) fontanels** are also paired structures posterior to the temporal bone. Most fontanels fuse typically before 1 year of age, though the anterior fontanel may fuse as late as age 2. Locate these structures in figure 9.14 and on the material available in the lab.

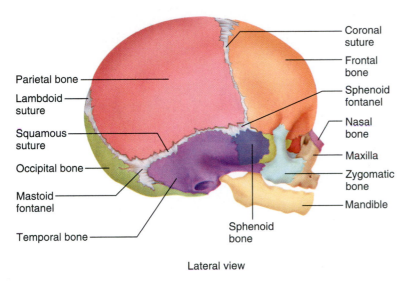

Lateral view

(a)

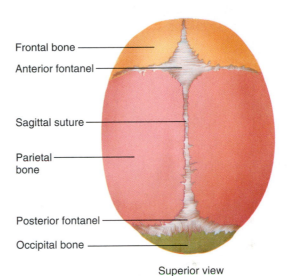

Superior view

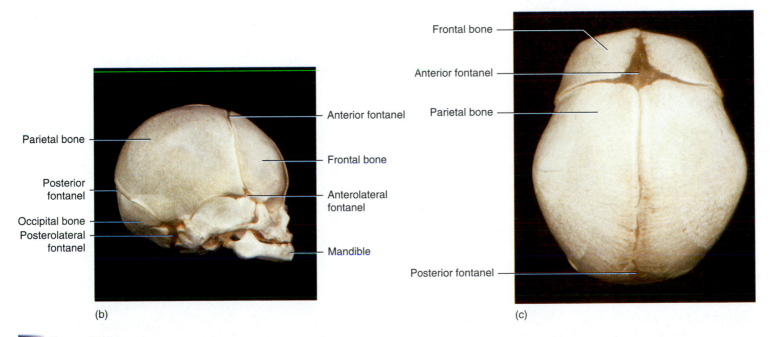

(b)

(c)

FIGURE 9.14

Fetal Skull and Fontanels (a) Diagram of lateral and superior views. Photograph (b) lateral view; (c) superior view.

Name _____ Date _____

1. The eyebrows are superfical to what bone?

2. What is the common name for the zygomatic bone?

3. What is the name of the process directly posterior to the earlobe?

4. The hard palate is made up of what bones?

5. What are the names of the major paranasal sinuses?

6. The mandible fits into what part of the temporal bone to form the jaw joint?

7. What bone is found just posterior to the ethmoid bone?

8. The sella turcica is part of what bone?

9. What occupies the depression in the sella turcica?

10. What are the names of the bones that surround the opening of the nose?

11. The upper teeth are held in place by what bones?

12. In what bone would you find the foramen magnum?

13. What are the two bony structures that make up the nasal septum?

14. The sagittal suture separates the _____ from the _____ .

    a. sphenoid, ethmoid   b. left parietal, right parietal   c. frontal, parietal   d. parietals, occipital

15. Which bone is *not* located in the orbit?

    a. maxilla   b. zygomatic   c. ethmoid   d. sphenoid   e. temporal

16. Which bone is *not* a paired bone of the skull?

    a. zygomatic   b. temporal   c. lacrimal   d. vomer

17. Label the following illustration using the terms provided.

maxilla	palatine bone	occipital condyle
vomer	temporal bone	internal naris
mastoid process	foramen magnum	jugular foramen
carotid canal	occipital bone	zygomatic bone

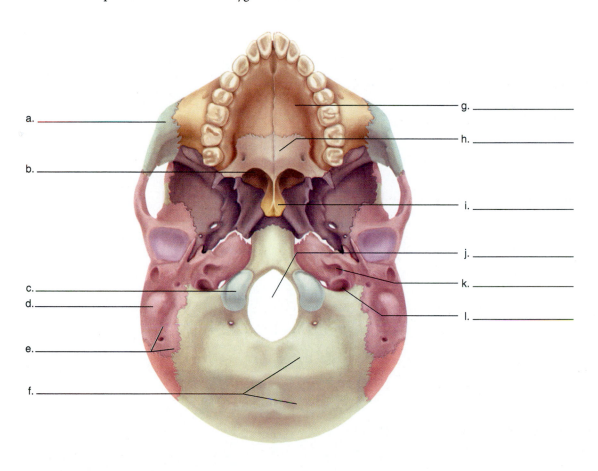

a. _____

b. _____

c. _____

d. _____

e. _____

f. _____

g. _____

h. _____

i. _____

j. _____

k. _____

l. _____

18. Label the following illustration using the terms provided.

mental foramen

coronoid process

ramus

angle

body

condylar process

mandibular notch

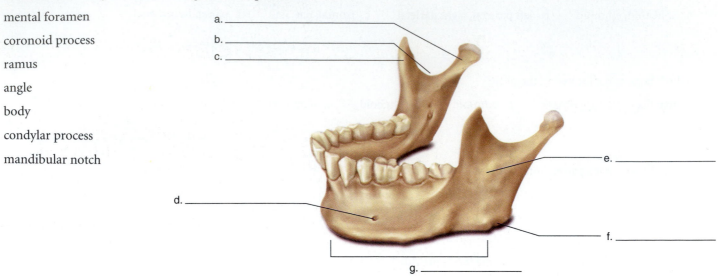

a. _____

b. _____

c. _____

d. _____

e. _____

f. _____

g. _____

19. Why would a fractured maxilla create more problems, in terms of bone healing, than a fractured femur?

20. Name all of the fontanels in the fetal skull.

21. Which one of the fontanels is the most dorsal?

Articulations

Skeletal System

INTRODUCTION

The study of the joints between bones is called **arthrology.** Joints are known as **articulations.** They are important as points of interplay between the skeletal system and the muscular system and because of the significant trauma and disease that occur in them. Diseases such as arthritis affect millions of people worldwide. The development of artificial joints is continually advancing to replace specific joints of the body that can no longer provide either a comfortable range of movement or the stability required for body support.

In this exercise you study the structure and function of various types of joints in the body. Joints are classified according to their physical composition. As you study the joints in the lab, use the joints in your body as references and mimic the actions of specific articulations.

LEARNING OBJECTIVES

At the end of this exercise you should be able to
1. distinguish among bony, fibrous, cartilaginous, and synovial joints;
2. explain the structure of the knee, hip, jaw, elbow, and shoulder;
3. discuss the nature of a synovial joint;
4. locate the fibrous capsule, synovial membrane, synovial fluid, and articular cartilage in a dissected synovial joint;
5. list five types of synovial joints;
6. describe actions at joints, such as flexion, extension, medial rotation, and lateral rotation.

MATERIALS

Mammal joint with intact synovial capsule
Dissection tray with scalpel or razor blades, blunt probe, and
 protective gloves

Waste container
Model or chart of joints, including those of the shoulder, elbow,
 knee, hip, and jaw
Articulated skeleton

PROCEDURE

Types of Joints

Joints are bony, fibrous, cartilaginous, or synovial (table 10.1).

BONY JOINTS

In **bony joints,** or **synostoses** (singular, **synostosis**), two separate bones fuse and become immovable joints. Developing bones in young individuals fuse and become synostoses. As a person approaches the age of 35 or so, some of the sutures of the skull begin to fuse from the region closest to the brain toward the superficial surface of the skull. This fusion leads to the complete union of two bones. The two frontal bones, present as separate bones in the fetal skeleton, fuse together as a synostosis and form the single frontal bone.

FIBROUS JOINTS

Fibrous joints are those with two bones held together by connective tissue fibers; they allow for little or no movement. If the bones are bound closely together, the joint does not allow for movement between the bones. One example of this type of joint is called a **suture;** it occurs between adjacent bones in the cranium. In these joints the bones are tightly held together by dense fibrous connective tissue. Examine a skull in the lab and locate the sutures represented in figure 10.1. These sutures may be **serrated, lap,** or **plane sutures.**

Another type of fibrous joint is a **gomphosis** (plural, **gomphoses**), which is represented by teeth in the alveoli of the maxilla and the mandible. This type of joint is like a peg in a socket.

TABLE 10.1
Classification of Joints

Bony Joints—Joints That Result from the Fusion of Two Bones	
Synostosis	Fusion of parietal bones in older individuals

Fibrous Joints—Joints Held Together by Collagenous Fibers	
Suture	Frontal and parietal bones
Gomphosis	Teeth and mandible
Syndesmosis	Distal tibia and distal fibula

Cartilaginous Joints (Synchondroses)—Joints Held Together by Cartilage	
Epiphyseal plate	Humeral head and shaft
Costal cartilages	Ribs and sternum
Symphysis	Joint between two vertebral bodies

Synovial Joints—Joints Enclosed by Synovial Capsule	
Plane	Between carpal bones
Hinge	Humerus and ulna
Pivot	Atlas and axis, radius and ulna
Condylar	Radius and scaphoid, atlas and occipital
Saddle	Trapezium and first metacarpal
Ball-and-socket	Acetabulum and femur

A gomphosis connects the bone of the jaw to the tooth by fibrous connective tissue called **periodontal ligaments.** Examine a jaw or complete skull in the lab and locate a gomphosis. Compare your specimen with figure 10.2.

A **syndesmosis** (plural, **syndesmoses**) is a fibrous joint where the fibrous connective tissue is longer than in sutures or gomphoses, allowing for limited movement. An example of a syndesmosis is the connection between interosseus membranes of the radius and ulna or the tibia and fibula. Examine an articulated skeleton in the lab and compare it with figure 10.3.

CARTILAGINOUS JOINTS

Cartilaginous joints have cartilage between the bones and may be either immovable or slightly movable. The articulation is known as a **synchondrosis** (plural, **synchondroses**). If the cartilage is thin between the bones, the joint is immovable (a synarthrosis). An example of this is the **epiphyseal plate** (fig. 10.4). This joint eventually fuses to form a single bone. Another kind of synchondrosis is the **costal cartilages** between the true ribs and the sternum. The cartilage in this joint is longer than one in an epiphyseal plate and the joint is more movable (amphiarthrotic), allowing the ribs limited movement. Examine these joints in lab and compare them with figure 10.5.

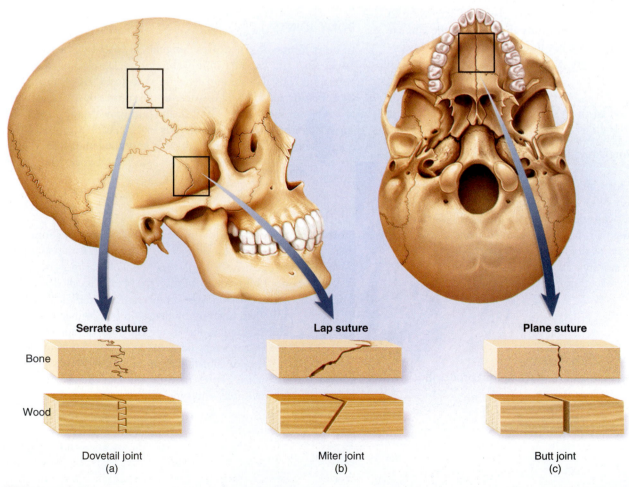

Serrate suture   **Lap suture**   **Plane suture**

Bone

Wood

Dovetail joint   Miter joint   Butt joint
(a)   (b)   (c)

FIGURE 10.1

Sutures (top) seen in position in the skull; (middle) a close up of the joint in bone and (bottom) analogous wood joints.

FIGURE 10.2
Gomphosis

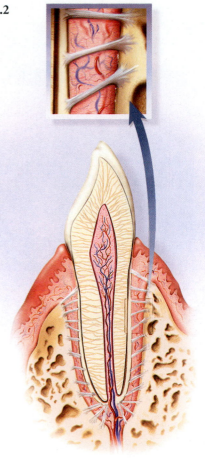

FIGURE 10.3
Syndesmosis

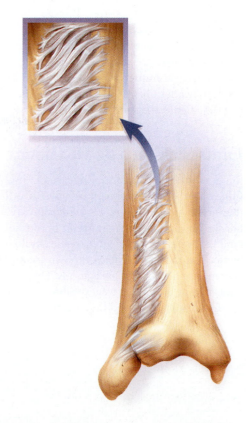

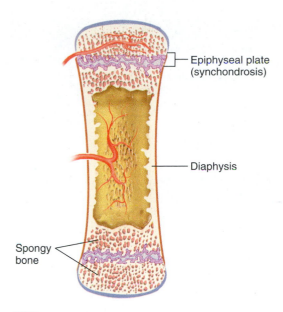

FIGURE 10.4
Cartilaginous Joint Epiphyseal plate.

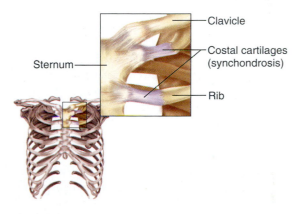

FIGURE 10.5
Synchondrosis of an Amphiarthrosis

A **symphysis** is a fibrocartilaginous pad found between the pubic bones (for example, the pubic symphysis) and the intervertebral discs. These joints allow for limited movement. The physical stress in a symphysis is greater than in other cartilaginous joints, such as the costal cartilages, and the joint reflects this with the presence of fibrocartilage. Fibrocartilage endures much greater stress than hyaline cartilage. Locate a symphysis in lab and compare it with figure 10.6.

SYNOVIAL JOINTS

Of all the types of joints, **synovial joints** have the most complex structure, including a joint capsule, an inner membrane, and synovial fluid, and they are the most movable. The joint capsule is made of an outer **fibrous capsule** and an inner **synovial membrane.** The synovial membrane secretes **synovial fluid,** a lubricating liquid that reduces friction inside the joint. The space inside the joint is called the **synovial cavity,** and each bone of the joint ends

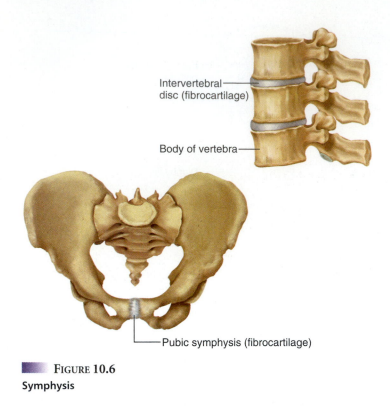

FIGURE 10.6
Symphysis

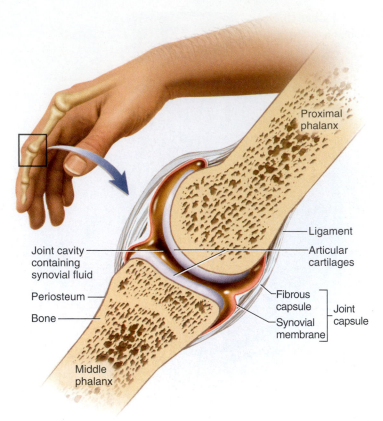

FIGURE 10.7
Synovial Joint Structure

in a hyaline cartilage cap called the **articular cartilage.** There are other structures in synovial joints that are characteristic of specific joints, and these are discussed later in this exercise. The shape of bones determines the type of movement at the articulation. In general, the more movable a joint is, the less stable it is. The stability of a joint depends on the number and types of ligaments, tendons, and muscles and on the the way the bones fit together. Compare the features of the joint in figure 10.7 with models or charts in the lab. Determine whether the knee joint is more stable than the hip joint based on the bony fit.

Dissection of a Synovial Joint   To understand the structure of a synovial joint, examine a fresh or recently thawed mammal joint. Wear protective gloves as you dissect the joint, and wash your hands thoroughly with soap and water after the dissection. Place the joint in front of you on a dissecting tray and cut into the joint capsule with a scalpel or razor blade. Note the tough, white material that surrounds the joint. This is the joint capsule, and it may be fused with ligaments that bind the bones of the joint together. Notice the synovial fluid, which is a slippery substance that provides a slick feel to the inside of the capsule, reducing friction between the bones. Once you have cut into the joint, examine the articular cartilage on the ends of the bones. *Carefully* cut into this cartilage with a scalpel or razor blade and notice how the material chips away from the bone. When you have finished with the dissection, make sure to rinse off the dissection equipment and dispose of the joint in the appropriate animal waste container.

Modified Synovial Structures   Bursae and tendon sheaths are modified synovial structures. **Bursae** (singular, **bursa**) are small synovial sacs between tendons and bones or other structures. The bursae cushion the tendons as they pass over the other structures. **Tendon sheaths** are modified synovial structures encircling tendons. There are numerous tendon sheaths in the palm of the hand. Tendon sheaths lubricate the tendons as they slide past one another. Examine figure 10.8 for these structures.

Joints Classified by Movement

Joints are classified as immovable, semimovable, or freely movable. Immovable joints are known as **synarthrotic joints.** The bones in these joints are tightly bound by connective tissue. Synarthrotic joints may be fibrous or cartilaginous. In fibrous joints, collagenous fibers bind the bones. In cartilaginous joints, hyaline cartilage joins the two bones. Semimovable joints are known as **amphiarthrotic joints,** and they may be fibrous or cartilaginous. The cartilage of amphiarthrotic joints may be hyaline cartilage or fibrocartilage. Freely movable joints are **diarthrotic joints** and they are always synovial joints.

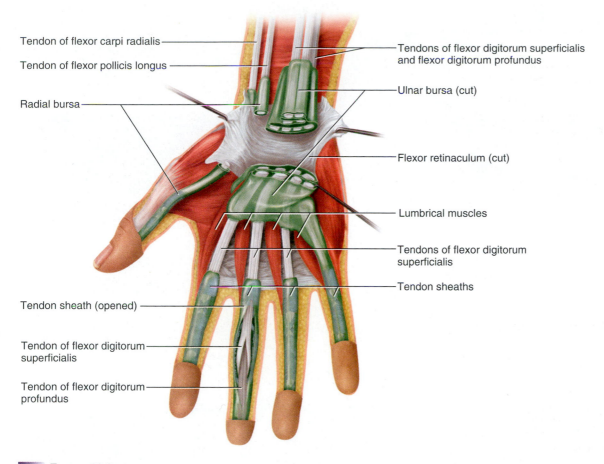

Tendon of flexor carpi radialis

Tendon of flexor pollicis longus

Radial bursa

Tendon sheath (opened)

Tendon of flexor digitorum superficialis

Tendon of flexor digitorum profundus

Tendons of flexor digitorum superficialis and flexor digitorum profundus

Ulnar bursa (cut)

Flexor retinaculum (cut)

Lumbrical muscles

Tendons of flexor digitorum superficialis

Tendon sheaths

FIGURE 10.8
Modified Synovial Structures

SYNOVIAL JOINTS CLASSIFIED BY MOVEMENT

Synovial joints are classified according to the type of movement they allow between the articulating bones. The joints are listed here in the general order of least movable to most movable. Examine the joints on an articulated skeleton in lab and compare them with figure 10.9.

1. **Plane joints** allow for movement between two flat surfaces, such as between the superior and inferior facets of adjacent vertebrae or intertarsal or intercarpal joints.
2. **Hinge joints** allow for angular movement, such as in the elbow, in the knee, or between the phalanges of the fingers. You can increase or decrease the angle of the two bones with this joint.
3. **Pivot joints** allow for rotational movement between the atlas and the axis when moving the head to indicate "no." They are also located at the proximal radius and ulna.
4. **Condylar joints** allow significant movement in two planes, such as at the base of the fingers (between the metacarpals and phalanges). Your fingers can easily move anterior to

posterior or medial to lateral, yet they do not move well at 45° angles to these planes. Condylar joints consist of a convex surface paired with a concave surface. The junction between the radius and scaphoid bone and between the atlas and occipital bone are good examples of condylar joints.

5. **Saddle joints** have two concave surfaces that articulate with one another. An example of a saddle joint is between the trapezium and the first metacarpal of the thumb. This provides for more movement in the thumb than the condylar joint of the wrist.
6. **Ball-and-socket joints** consist of a spherical head in a round concavity, such as in the shoulder and the hip There is extensive movement in these joints, yet they are less stable due to the freedom of movement they afford.

Monaxial joints move in only one plane. Hinge, pivot, and plane joints are monaxial joints. **Biaxial joints** move in two planes. Condylar and saddle joints are biaxial joints. **Multiaxial joints** move in many planes. Ball-and-socket joints are multiaxial joints

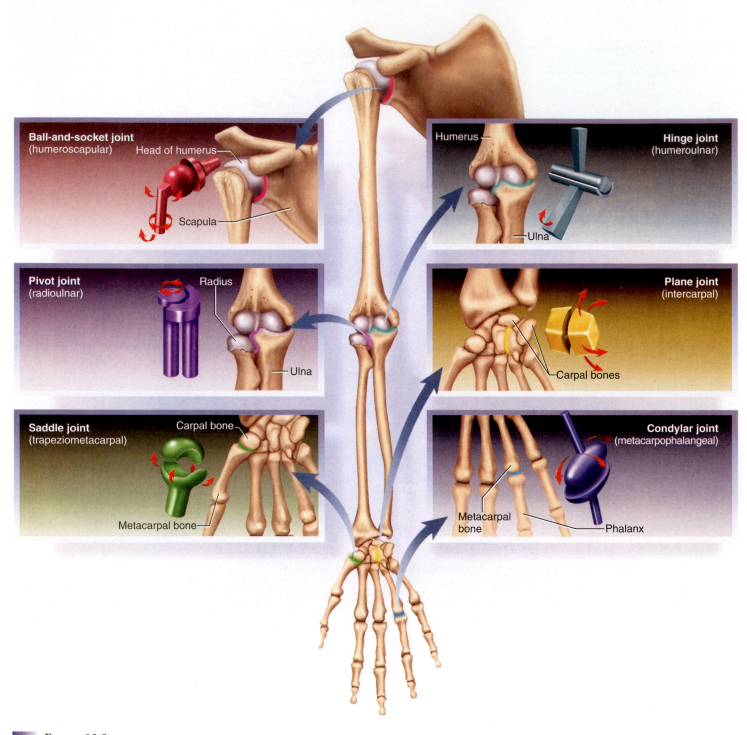

Ball-and-socket joint
(humeroscapular)
Head of humerus
Scapula

Hinge joint
(humeroulnar)
Humerus
Ulna

Pivot joint
(radioulnar)
Radius
Ulna

Plane joint
(intercarpal)
Carpal bones

Saddle joint
(trapeziometacarpal)
Carpal bone
Metacarpal bone

Condylar joint
(metacarpophalangeal)
Metacarpal bone
Phalanx

FIGURE 10.9

Synovial Joints Six different synovial joints are unique due to their bony fit.

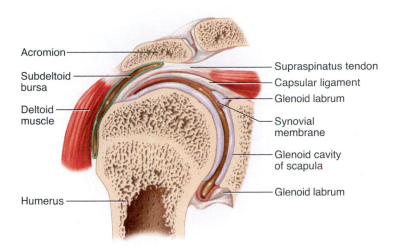

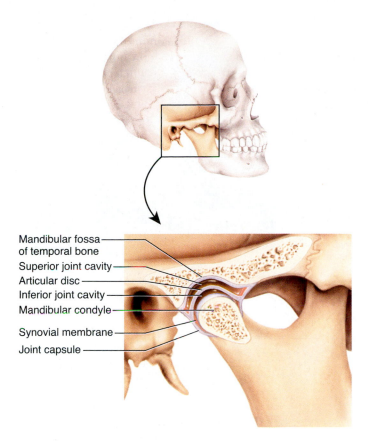

Mandibular fossa of temporal bone
Superior joint cavity
Articular disc
Inferior joint cavity
Mandibular condyle
Synovial membrane
Joint capsule

FIGURE 10.10
Temporomandibular Joint (TMJ)

Acromion
Subdeltoid bursa
Deltoid muscle
Humerus
Supraspinatus tendon
Capsular ligament
Glenoid labrum
Synovial membrane
Glenoid cavity of scapula
Glenoid labrum

FIGURE 10.11
Glenohumeral Joint

Specific Joints of the Body

JAW JOINT

The **temporomandibular joint (TMJ)** is the only diarthrotic joint of the skull (fig. 10.10). This joint has an **articular disc,** a pad of fibrocartilage that provides a cushion between the condylar process of the mandible and the temporal bone. Numerous ligaments strengthen this joint. Examine a skull with the mandible attached for the nature of the temporomandibular joint.

SHOULDER JOINT

The **humeral** (shoulder) **joint** is known as the **glenohumeral joint** because the glenoid cavity articulates with the head of the humerus.

The shallow glenoid cavity is deepened by the **glenoid labrum** (*labrum* = lip), a cartilaginous ring that surrounds the cavity. Numerous bursae, ligaments, a tough joint capsule, and the **rotator cuff muscles** also stabilize the joint. Compare figure 10.11 with a model or an actual joint in the lab.

ELBOW JOINT

The elbow joint consists of both the humeroulnar joint and the humeroradial joint. The humeroulnar joint provides the hinge mechanism of the elbow and the humeroradial joint allows for rotational movement at the joint. The **radial collateral ligament** and the **ulnar collateral ligament** strengthen the bony fit of the joint, along with tendons from the biceps brachii muscle and triceps brachii muscle. The **annular ligament** wraps around the head of the radius and is subject to dislocation when the forearm is pulled abruptly. Examine articulated skeletons or models in lab and compare them with figure 10.12

HIP JOINT

The hip joint is known as the **acetabulofemoral joint.** As with the shoulder joint, a labrum—in this case, the **acetabular labrum**—deepens the hip socket. Numerous ligaments, including the iliofemoral,

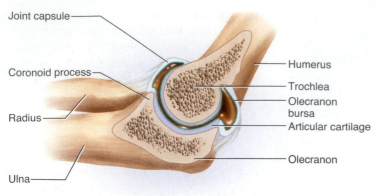

FIGURE 10.12
Elbow Joint

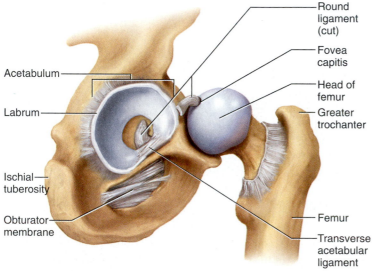

FIGURE 10.13
Acetabulofemoral Joint

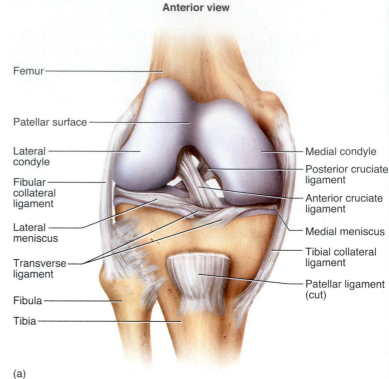

(a)

(b)

FIGURE 10.14
Tibiofemoral Joint

pubofemoral, and ischiofemoral ligaments bind the femur to the hip bone. The **round ligament** is a band of dense connective tissue that attaches the acetabulum to the **fovea capitis** of the femur. Compare the models, charts, or specimens in the lab with figure 10.13.

TIBIOFEMORAL JOINT

The **tibiofemoral joint**, or knee joint, is the largest, most complex joint of the body. It consists of several major ligaments, including the **tibial (medial) collateral ligament**, the **fibular (lateral) collateral ligament**, the **anterior cruciate ligament**, and the **posterior cruciate ligament**. Another important structure is the **patellar tendon**, which runs from the quadriceps femoris muscle to the tibial tuberosity. The **patellar ligament** is a specific connection between the patella and the tibial tuberosity. The other major structures of the tibiofemoral joint are the **medial** and **lateral menisci** (singular, **meniscus**), wedge-shaped pads of fibrocartilage that provide a cushion between the femur and tibia. The knee is primarily a hinge joint with a little lateral movement allowed. Compare specimens in the lab with figure 10.14.

Levers

There are three parts to a lever system (fig. 10.15). These are the fulcrum, the resistance, and the effort. The **fulcrum** is the point of movement or rotation of the lever. The **resistance** is the part of the lever that is to be moved, and the **effort** is the part to which an action is applied to move the lever. There are three classes of levers:

1. A **first class lever** is one in which the effort is on one side of the fulcrum and the resistance is on the other side. A classic example of this is a see-saw and a common example in humans is the junction between the head and atlas. Examine an articulated skeleton in lab and identify the fulcrum, the resistance, and the effort for this articulation.
2. A **second class lever** has the fulcrum at one end, the resistance in the middle, and the effort at the other end. A

wheelbarrow is a good representative of a second class lever. The wheel is the fulcrum, the load in the basin of the wheelbarrow is the resistance, and the effort is your arms pulling up on the handles of the wheelbarrow. A muscle called the digastric opens the jaw while the temporalis muscle resists this motion. Locate the hinge of the jaw in lab and determine the parts of the lever system.

3. A **third class lever** has the fulcrum at one end, the effort in the middle, and the resistance at the other end. Opening a trap door from the top is an example of a third class lever. The hinge on the door represents the fulcrum, the handle in the middle of the door is the effort, and the weight of the door is the resistance. Most of the joints of the body involve third class levers and a common example is the elbow joint. Examine this joint in an articulated skeleton and identify the fulcrum, the effort, and the resistance.

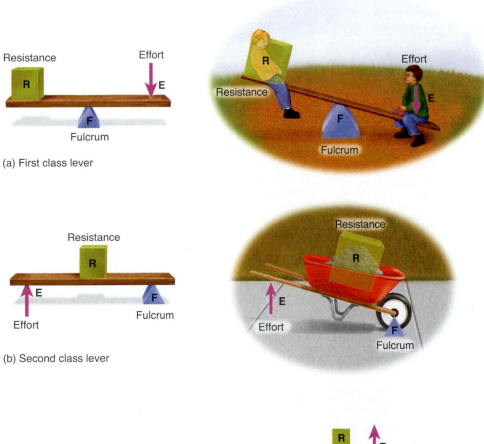

(a) First class lever

(b) Second class lever

(c) Third class lever

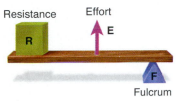
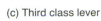
FIGURE 10.15
Levers

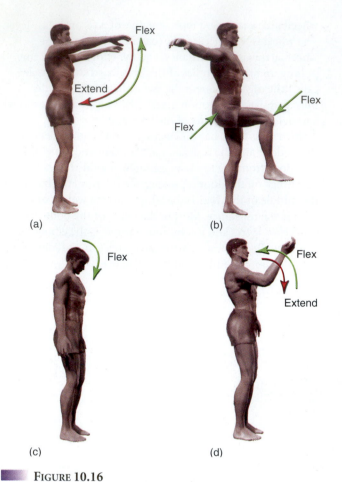

(a)   (b)

(c)   (d)

FIGURE 10.16

Flexion and Extension of Selected Joints. (a) Flexion and extension of the arm; (b) flexion of the thigh and leg; (c) flexion of the head; (d) flexion and extension of the forearm.

Movement at Joints

Many kinds of movements occur at joints. These movements, controlled by muscles, are called actions. Actions can decrease a joint angle, increase a joint angle, and cause rotation at a joint, among other movements. The specific types of action at a joint are as follows.

Flexion is a decrease in the joint angle from anatomic position. If you bend your elbow, you are flexing your forearm. Flexion of the thigh is in the anterior direction, yet flexion of the leg is in the *posterior* direction. Bending forward at the waist is flexion of the vertebral column. Looking at your toes is flexion of the head. Examine figure 10.16 for examples of flexion.

Extension is a return to anatomic position of a part of the body that was flexed. If you are looking at your toes and lift your head back to anatomic position, you are extending your head. If you straighten your knee after it is bent (flexed), you are extending the leg. Examine figure 10.16 for examples of extension. Extension of the part of the body beyond anatomic position is known as **hyperextension.** When you are about to roll a bowling ball and your arm reaches the very back of the arc, you are hyperextending your arm.

Abduction (*abduct* = to take away) is movement of the limbs in the coronal plane away from the body. Abducting the fingers is spreading the fingers apart in the frontal plane. Abduction is taking away a part of the body in a lateral direction, as seen in figure 10.17.

Adduction is the return of the part of the body to anatomic position after abduction. When your fingers are straight and together, they are adducted. In doing "jumping jacks," you are abducting and adducting in series. Think of adduction as "adding" a limb back to the body, as seen in figure 10.17.

Rotation is the circular movement of a part of the body, as seen in figure 10.18. **Lateral rotation** moves the anterior surface of

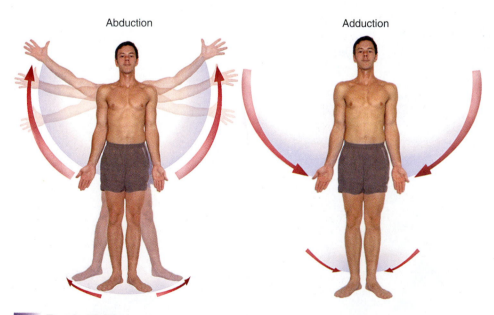

FIGURE 10.17

Abduction and Adduction of the Arm and Thighs

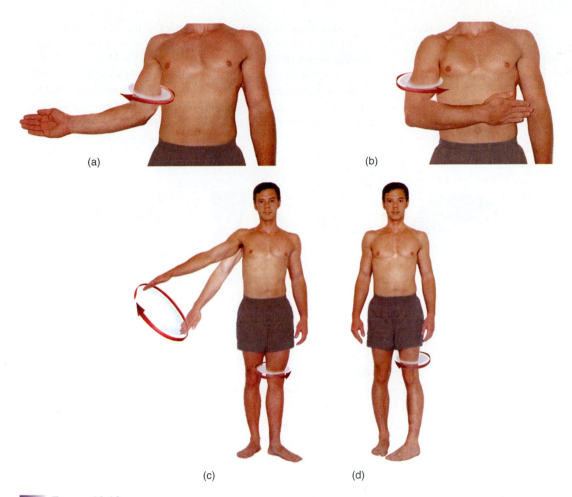

FIGURE 10.18

Rotation of Joints. (a) Lateral rotation of the arm; (b) medial rotation of the arm; (c) circumduction of the arm and lateral rotation of the thigh; (d) medial rotation of the thigh.

the limb toward the lateral side of the body. **Medial rotation** turns the anterior surface of the limb toward the midline.

Supination is lateral rotation of the hand. The hands are supinated when the body is in anatomic position.

Pronation is medial rotation of the hands. When you turn your palms posteriorly from anatomic position, you are pronating your hands.

Circumduction is the movement of a muscle in a conical shape, with the point of the cone being proximal. Examine figure 10.18 for an example of circumduction.

Elevation is movement in a superior direction. Elevation occurs when you shrug your shoulders or close your mouth.

Depression is the opposite of elevation. It is movement in the inferior direction. Opening your mouth is depression of the mandible. Elevation and depression are seen in figure 10.19.

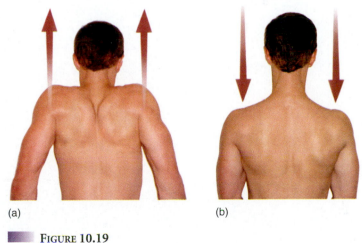

FIGURE 10.19

Elevation and Depression of the Shoulders (a) Elevation; (b) depression.

Protraction is a horizontal movement in the anterior direction, as in jutting the chin forward.

Retraction is the reverse of protraction. A jaw that moves from anterior to posterior is retracted. These two actions are illustrated in figure 10.20.

Inversion is movement of the feet—turning the soles of the feet medially so they face each other. Sometimes inversion of the foot is known as *supinating* the foot.

Eversion means turning the soles of the feet laterally. This is also known as *pronating* the foot. Inversion and eversion can be seen in figure 10.21.

Fixing a muscle prevents motion in either direction. This is done with opposing muscles contracting simultaneously. The muscle that has the main force on a joint is called the **prime mover,** or **agonist.** Muscles that assist with the prime mover are **synergists,** while those that oppose the muscle are **antagonists.**

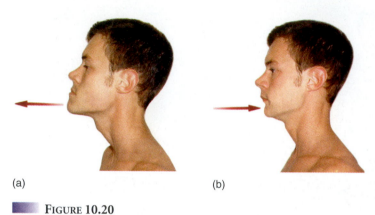

(a)   (b)

FIGURE 10.20

Protraction and Retraction of the Mandible (a) Protraction;
(b) retraction.

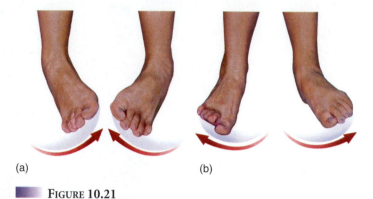

(a)   (b)

FIGURE 10.21

Inversion and Eversion of the Feet (a) Inversion; (b) eversion.

Name _____ Date _____

1. The study of articulations, or joints, is known as _____.

2. What kind of joint (based on joint structure) is one in which the bones are held together by collagenous fibers?

3. When two bones fuse together into a single bone, this union is called a _____.

4. The teeth are held into the jaw by what kind of joint?

5. What kind of joint is found between the distal tibia and fibula?

6. Bones held together by cartilage (cartilaginous joints) are also known as _____ joints.

7. The epiphyseal plate is a cartilaginous joint. In terms of movement, it is also called a _____ .

8. Label the following illustration using the terms provided.

joint capsule

bone

synovial cavity

synovial membrane

articular cartilage

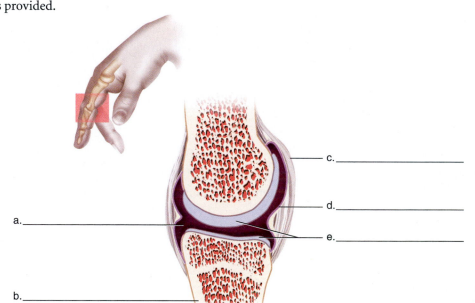

c._____

d._____

e._____

a._____

b._____

9. What is the name of a joint that is held together by a joint capsule?

10. Synovial fluid is secreted by what structure?

11. How much movement occurs in a suture between two bones of the skull?

12. Rank the following joints in terms of least movable to most movable, with 1 being the least movable and 5 being the most movable.

    plane   saddle   suture   syndesmosis   ball-and-socket

13. Which one of the following joints has the greatest range of movement?

    a. gomphosis   b. suture   c. synchondrosis   d. hinge

14. In which of these joints would you find a meniscus?

    a. cartilaginous   b. fibrous   c. synovial

15. What is the function of the meniscus in the knee?

16. Match the joint in the left column with the type of joint in the right column.

    _____ acetabulofemoral   a. hinge
    _____ radiocarpal   b. ball-and-socket
    _____ tibiofemoral   c. plane
    _____ intercarpal   d. condylar

17. What is the function of the labrum in the glenohumeral joint?

18. The joint between the trapezium and the first metacarpal is what kind of joint?

19. The joint between the femur and the tibia is what type of joint?

20 What kind of joint is found at the wrist (between the radius and the carpal bones)?

Introduction to the Study of Muscles and Muscles of the Shoulder and Upper Extremity

Muscular System

INTRODUCTION

Laboratory exercises 11 to 14 focus on skeletal muscles. There are over 600 skeletal muscles in the human body. The muscles are grouped according to their location, with most muscles occurring as pairs. In laboratory exercises 11 to 14 you will learn about some of the major muscles of the body. The origin, insertion, action, and innervation are listed for about 100 muscles in this lab manual. Your instructor may wish to customize the list, so that you learn specific muscles or specific things about each muscle. The study of muscles can be both fun and challenging. Begin your study of muscles early and do not wait to learn them until the night before the lab exam!

Skeletal muscle is voluntary muscle. It contracts when consciously stimulated by specific nerves. In this exercise you are introduced to things that concern skeletal muscles and you study the major muscles of the shoulder and upper limb. Some of these muscles originate on the scapula and move or stabilize the humerus. Others originate on the humerus and insert on the forearm.

The muscles of the forearm and hand are very important, particularly in terms of rehabilitating limbs after surgery to relieve carpal tunnel syndrome or after trauma to the limb. You should learn the origins and insertions as you learn the muscles of the forearm, since many of these muscles look quite similar. By knowing the origins and insertions of a muscle you will not easily mistake it for another muscle.

LEARNING OBJECTIVES

At the end of this exercise you should be able to

1. locate the muscles of the shoulder and upper extremity on a torso model, chart, cadaver (if available);
2. list the origin, insertion, and action of each muscle presented;
3. describe what nerve controls (innervates) each muscle;
4. list what muscles function as synergists or antagonists to the prime mover;

5. name all the muscles that have an action on a joint, such as all the muscles that laterally rotate the arm or muscles that flex the hand;
6. reproduce the actions of each listed muscle by moving the appropriate limbs to demonstrate the actions.

MATERIALS

Human torso model and upper extremity model

Human muscle charts

Articulated skeleton

Cadaver (if available)

PROCEDURE

The study of human musculature is seen as a daunting task by some students, while other students recall the muscle section of the course as their favorite part of the study of human anatomy. Studying muscles is more satisfying if attainable goals are set. This is best accomplished by choosing small numbers of muscles to study at a sitting and using flash cards as study aids. It is vital that you know the bones and bony markings before studying muscles. Look over laboratory exercise 7, "Appendicular Skeleton," if you need to review the skeleton. You may also want to make a quick sketch of bones and how muscles attach to them to help you visualize points of attachment.

A thorough study of muscles involves knowing not only the name of the muscle but also its origin, insertion, action, and innervation. The muscles are listed in this exercise by their name followed by their origin. The **origin** is the attachment point of the muscle that does not move during muscular contraction. The muscles also are listed with their **insertion,** which is the attachment that moves during contraction. Each muscle has an **action,** which is the effect the muscle has on a part of the body (such as *flexion* of the arm). Finally, the **innervation** of a muscle is the specific nerve that controls it. Nerves are important in that, if the motor portion of a nerve is damaged, the muscle innervated by that nerve cannot function.

Origins and Insertions

Examine an articulated skeleton as you study the muscles. Try to visualize the muscle as it attaches to the origin and the insertion. As you study the muscles locate them on your body and try to determine which part is the origin and which is the insertion. It also helps to think of the body in anatomic position as you look at the origins and the insertions.

Actions

To fully understand muscles you must know how they move joints and associate the actions with the body in anatomic position. When you are learning the action of a muscle, imagine a piece of string tied between the origin and the insertion of the muscle. Think of how the bones move as you pull the insertion closer to the origin. Mimic the action of the muscle as you learn it. When you perform the action of a particular muscle, place your fingers on the muscle. You should be able to feel the muscle tighten as you perform the action.

Muscle Nomenclature

Another aid to learning muscles is to understand how they are named. Muscles are named by a number of criteria:

Location: Some muscles are named by their location—such as the *tibialis* anterior, the *frontalis,* or the *temporalis.*

Size: The adductor *magnus* is the largest of the adductor muscles. The gluteus *minimus* is the smallest of the gluteal muscles.

Shape: The *deltoid* muscle is shaped like a delta, or triangle.

Orientation: The *transversus* abdominis muscle has fibers that are oriented horizontally, or in a transverse direction.

Origin and insertion: The *infraspinatus* muscle originates on the infraspinous fossa of the scapula, while the flexor *hallucis* longus muscle inserts on the hallux, or big toe.

Number of heads: The *triceps* brachii muscle has three heads.

Action: The *extensor* digitorum is a muscle that extends the fingers.

Length: Muscles with the term *longus* or *brevis* are named for the overall length of the muscle.

Overview of the Muscles

Most people are familiar with the larger skeletal muscles, such as the pectoralis major, gluteus maximus, deltoid, and trapezius. Examine the charts and models in lab and locate as many of the most common muscles of the body as you can on the charts or models and in figure 11.1. You should later refer to this figure in the subsequent exercises if you need an overview of the muscles. You should also review the muscle nomenclature in this exercise and actions at joints as described in exercise 10. Muscles can have numerous actions at a joint. In addition to moving joints, muscles can also fix a joint.

Examination of Muscles

Look at the charts, models, and cadaver (if available) in the lab and locate the muscles presented in this section. Review the bones of the hand, as well as the bones of the arm and forearm, in laboratory exercise 7, "Appendicular Skeleton," to locate precisely the bony origins and insertions. Refer to the following descriptions as you examine the muscles. The specific details of the muscles are provided in table 11.1.

FIGURE 11.1

Overview of the Muscles of the Body. (a) Anterior view.

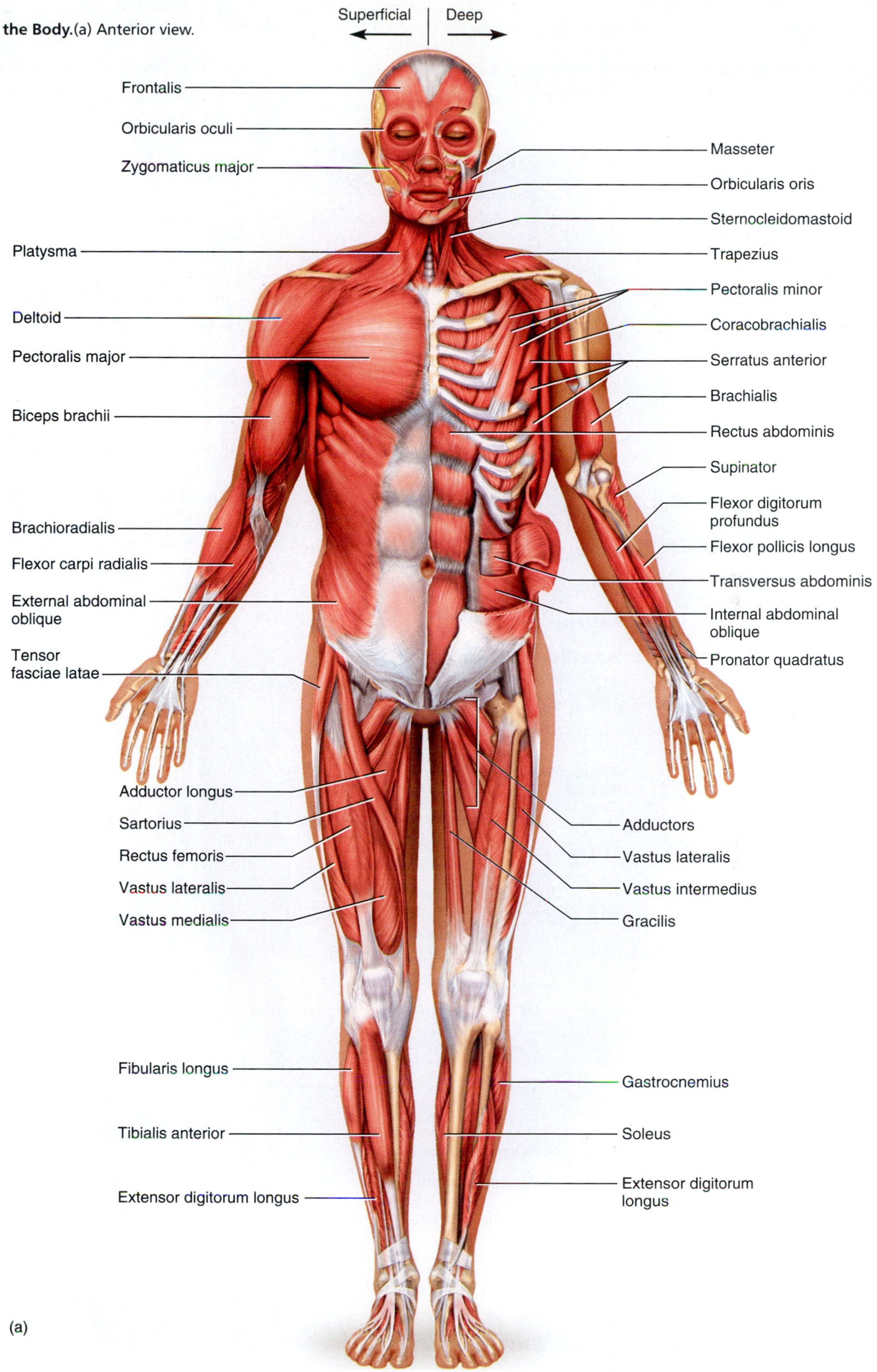

Superficial | Deep

Frontalis

Orbicularis oculi

Zygomaticus major

Masseter

Orbicularis oris

Sternocleidomastoid

Platysma

Trapezius

Pectoralis minor

Deltoid

Coracobrachialis

Pectoralis major

Serratus anterior

Brachialis

Biceps brachii

Rectus abdominis

Supinator

Flexor digitorum profundus

Brachioradialis

Flexor pollicis longus

Flexor carpi radialis

Transversus abdominis

External abdominal oblique

Internal abdominal oblique

Tensor fasciae latae

Pronator quadratus

Adductor longus

Adductors

Sartorius

Vastus lateralis

Rectus femoris

Vastus intermedius

Vastus lateralis

Gracilis

Vastus medialis

Fibularis longus

Gastrocnemius

Tibialis anterior

Soleus

Extensor digitorum longus

Extensor digitorum longus

Extensor digitorum longus

(a)

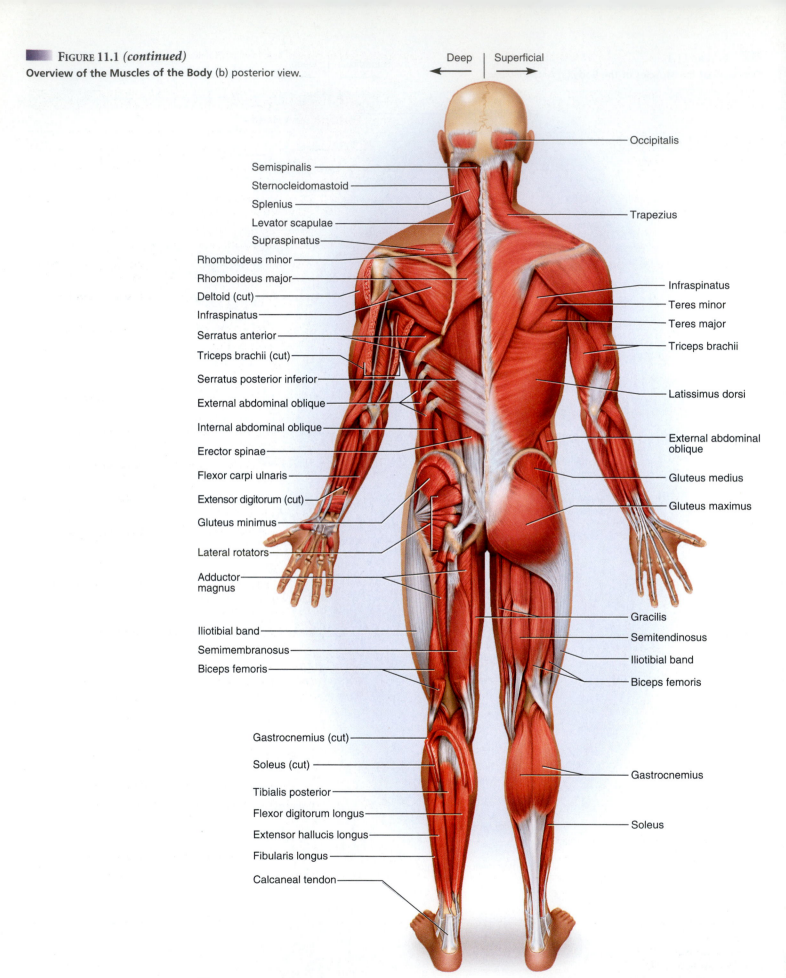

Deep | Superficial

Occipitalis

Semispinalis

Sternocleidomastoid

Splenius

Levator scapulae

Supraspinatus

Rhomboideus minor

Rhomboideus major

Deltoid (cut)

Infraspinatus

Serratus anterior

Triceps brachii (cut)

Serratus posterior inferior

External abdominal oblique

Internal abdominal oblique

Erector spinae

Flexor carpi ulnaris

Extensor digitorum (cut)

Gluteus minimus

Lateral rotators

Adductor magnus

Iliotibial band

Semimembranosus

Biceps femoris

Gastrocnemius (cut)

Soleus (cut)

Tibialis posterior

Flexor digitorum longus

Extensor hallucis longus

Fibularis longus

Calcaneal tendon

Trapezius

Infraspinatus

Teres minor

Teres major

Triceps brachii

Latissimus dorsi

External abdominal oblique

Gluteus medius

Gluteus maximus

Gracilis

Semitendinosus

Iliotibial band

Biceps femoris

Gastrocnemius

Soleus

(b)

TABLE 11.1

Muscles of the Shoulder, Arm, Forearm, and Hand

Muscles of the Shoulder and Arm

Name	Origin	Insertion	Action	Innervation
Superficial Muscles of the Shoulder				
Trapezius	Posterior occipital bone, ligamentum nuchae, C7–T12	Clavicle, acromion process, spine of scapula	Extends and abducts head, rotates and adducts scapula, fixes scapula	Accessory nerve (XI), spinal nerves C2–4
Deltoid	Clavicle, acromion process, spine of scapula	Deltoid tuberosity of the humerus	Abducts arm; flexes, extends, medially and laterally rotates arm	Axillary nerve
Pectoralis major	Clavicle, sternum, cartilages of ribs 1–7	Crest of greater tubercle of humerus	Flexes, adducts, and medially rotates arm	Medial and lateral pectoral nerves
Pectoralis minor	Ribs 3–5	Coracoid process of scapula	Depresses glenoid cavity, raises ribs 3–5	Medial pectoral nerve
Latissimus dorsi	T7–12, L1–5, S1–5, crest of ilium, ribs 10–12	Intertubercular groove of humerus	Extends, adducts, and medially rotates arm; draws shoulder inferiorly	Thoracodorsal nerve
Deep Muscles of the Shoulder				
Supraspinatus	Supraspinous fossa	Greater tubercle of humerus	Abducts arm, helps stabilize shoulder joint	Suprascapular nerve
Infraspinatus	Infraspinous fossa	Greater tubercle of humerus	Laterally rotates arm, stabilizes shoulder joint	Suprascapular nerve
Subscapularis	Subscapular fossa	Lesser tubercle of humerus	Medially rotates arm, stabilizes shoulder joint	Subscapular nerve
Teres minor	Lateral border of scapula	Greater tubercle of humerus	Laterally rotates and adducts arm, stabilizes shoulder joint	Axillary nerve
Teres major	Inferior angle of scapula	Crest of lesser tubercle of humerus	Extends, adducts, and medially rotates arm	Subscapular nerve
Muscles of the Arm				
Biceps brachii	Long head: superior margin of glenoid fossa Short head: coracoid process of scapula	Radial tuberosity	Flexes arm, flexes forearm, supinates hand	Musculocutaneous nerve
Triceps brachii	Infraglenoid tuberosity of scapula, lateral and posterior surface of humerus	Olecranon process of ulna	Extends and adducts arm, extends forearm	Radial nerve
Coracobrachialis	Coracoid process of scapula	Midmedial shaft of humerus	Flexes and adducts arm	Musculocutaneous nerve
Brachialis	Anterior, distal surface of humerus	Coronoid process of ulna	Flexes forearm	Musculocutaneous nerve
Brachioradialis	Lateral supracondylar ridge of humerus	Styloid process of radius	Flexes forearm	Radial nerve
Supinator	Lateral epicondyle of humerus, anterior ulna	Proximal radius	Supinates hand	Radial nerve
Pronator quadratus	Distal part of anterior ulna	Distal radius	Pronates hand	Median nerve
Pronator teres	Medial epicondyle of humerus, coronoid process of ulna	Lateral, middle shaft of radius	Pronates hand, flexes forearm	Median nerve
Palmaris longus	Medial epicondyle of humerus	Palmar aponeurosis	Flexes hand	Median nerve
Flexor carpi radialis	Medial epicondyle of humerus	Second and third metacarpals	Flexes and abducts hand	Median nerve
Flexor carpi ulnaris	Medial epicondyle of humerus, olecranon process and dorsal border of ulna	Pisiform, hamate, fifth metacarpal	Flexes and adducts hand	Ulnar nerve

C = cervical vertebrae
T = thoracic vertebrae
L = lumbar vertebrae
S = sacral vertebrae

(continued)

TABLE 11.1
Muscles of the Shoulder, Arm, Forearm, and Hand (continued)

Name	Origin	Insertion	Action	Innervation
Muscles of the Arm (continued)				
Flexor digitorum superficialis	Medial epicondyle of humerus, proximal ulna, proximal radius	Middle phalanges of second through fifth digits	Flexes proximal and middle phalanges, flexes hand	Median nerve
Flexor digitorum profundus	Anterior, proximal surface of ulna; interosseous membrane	Distal phalanges of second through fifth digits	Flexes phalanges, flexes hand	Median and ulnar nerves
Flexor pollicis longus	Anterior portion of radius and interosseous membrane	Distal phalanx of pollex (thumb)	Flexes thumb	Median nerve
Flexor pollicis brevis	Trapezium	Proximal phalanx of first digit	Flexes thumb	Median and ulnar nerves
Abductor pollicis brevis	Scaphoid and trapezium	Proximal phalanx of first digit	Abducts thumb	Median nerve
Opponens pollicis	Trapezium	First metacarpal	Opposes thumb	Median nerve
Abductor digiti minimi	Pisiform	Proximal phalanx of fifth digit	Abducts fifth digit	Ulnar nerve
Flexor digiti minimi brevis	Hamulus of hamate	Proximal phalanx of fifth digit	Flexes fifth digit	Ulnar nerve
Opponens digiti miniml	Hamulus of hamate	Fifth metacarpal	Opposes fifth digit	Ulnar nerve
Extensor carpi radialis longus	Lateral supracondylar ridge of humerus	Second metacarpal	Extends and abducts hand	Radial nerve
Extensor carpi radialis brevis	Lateral epicondyle of humerus	Third metacarpal	Extends and abducts hand	Radial nerve
Extensor carpi ulnaris	Lateral epicondyle of humerus, proximal ulna	Fifth metacarpal	Extends and adducts hand	Radial nerve
Extensor digitorum	Lateral epicondyle of humerus	Middle and distal phalanges of second through fifth digits	Extends phalanges, extends hand	Radial nerve
Abductor pollicis longus	Posterior radius and ulna, interosseous membrane	First metacarpal	Abducts thumb	Radial nerve
Extensor pollicis longus and brevis	Posterior radius and ulna, interosseous membrane	Proximal and distal phalanges of pollex (thumb)	Extends thumb	Radial nerve

MUSCLES OF THE SHOULDER

Locate the diamond-shaped **trapezius** (tra-PEE-zee-us) muscle, which has an origin in the midline of the vertebral column and head and inserts laterally. If the scapula is fixed, the head moves, yet, if the vertebral column and head are fixed, then the trapezius moves the scapula. Figure 11.2 illustrates the trapezius.

Another superficial muscle of the posterior surface is the **latissimus dorsi** (la-TISS-ih-muss DOR-sye) muscle, which arises from the vertebrae by way of a broad, flat **lumbodorsal fascia.** The latissimus dorsi originates on the back, but the insertion is on the anterior aspect of the humerus. It is a powerful extensor of the arm and is called the swimmer's muscle (fig. 11.2).

The **deltoid** is located on top of the shoulder and abducts the arm. Place your hand on your shoulder, abduct your arm, and feel the deltoid flex. It also flexes, extends, medially rotates, and laterally rotates the arm. This diverse action occurs because different parts of the deltoid can contract independently. Locate the deltoid on material in the lab and compare it with figures 11.2 and 11.3. It is a

fan-shaped muscle whose insertion partially covers the insertion of the **pectoralis** (PEC-tur-AL-is) **major.** The pectoralis major has fibers that run horizontally across the chest region, and it is a superficial muscle of the chest. Deep to the pectoralis major muscle is the **pectoralis minor** muscle, whose fibers run in a more vertical direction, as seen in figure 11.3.

Deep to the deltoid and the diamond-shaped trapezius muscle are muscles that originate on the scapula proper. The **supraspinatus** (SOO-pra-spy-NAY-tus) is named for its origin on the supraspinous fossa, and it is a synergist to the deltoid. The **infraspinatus** (IN-fra-spy-NAY-tus) is a muscle originating on the infraspinous fossa. It laterally rotates the arm. The **subscapularis** (sub-SCAP-you-LAR-is) is named for its origin on the subscapular fossa. The insertion of the subscapularis is on the anterior surface of the humerus; therefore, when it contracts, it medially rotates the arm. The subscapularis is located on the anterior surface of the scapula, between the scapula and the ribs, and cannot be seen in a posterior view. Locate these muscles in the lab and compare them with figure 11.4.

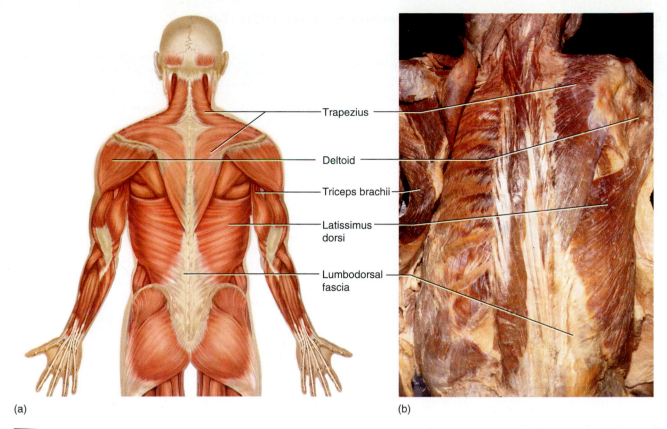

Trapezius

Deltoid

Triceps brachii

Latissimus dorsi

Lumbodorsal fascia

(a)

(b)

FIGURE 11.2
Superficial Back Muscles, Posterior View (a) Diagram; (b) photograph.

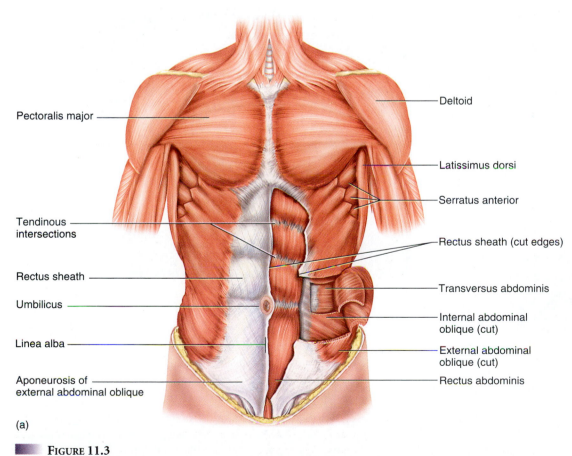

Deltoid

Pectoralis major

Latissimus dorsi

Serratus anterior

Tendinous intersections

Rectus sheath (cut edges)

Rectus sheath

Transversus abdominis

Umbilicus

Internal abdominal oblique (cut)

Linea alba

External abdominal oblique (cut)

Aponeurosis of external abdominal oblique

Rectus abdominis

(a)

FIGURE 11.3
Thoracic Muscles, Anterior View Diagram (a) superficial muscles.

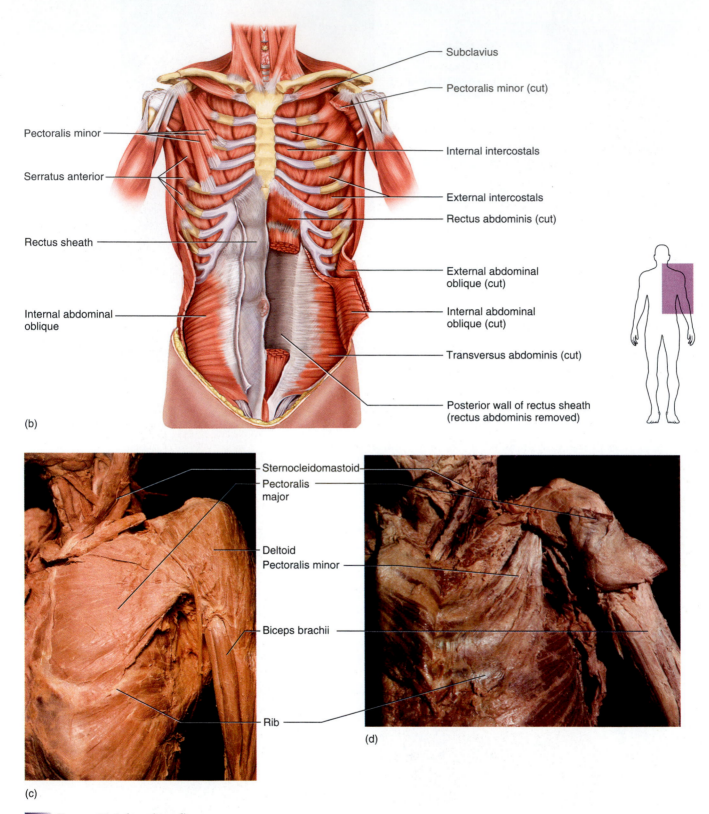

(b)

Subclavius

Pectoralis minor (cut)

Pectoralis minor

Internal intercostals

Serratus anterior

External intercostals

Rectus abdominis (cut)

Rectus sheath

External abdominal oblique (cut)

Internal abdominal oblique (cut)

Internal abdominal oblique

Transversus abdominis (cut)

Posterior wall of rectus sheath (rectus abdominis removed)

Sternocleidomastoid

Pectoralis major

Deltoid

Pectoralis minor

Biceps brachii

Rib

(c)

(d)

FIGURE 11.3 *(continued)*

Thoracic Muscles, Anterior View Diagram (b) deep muscles. Photograph (c) superficial muscles; (d) deep muscles.

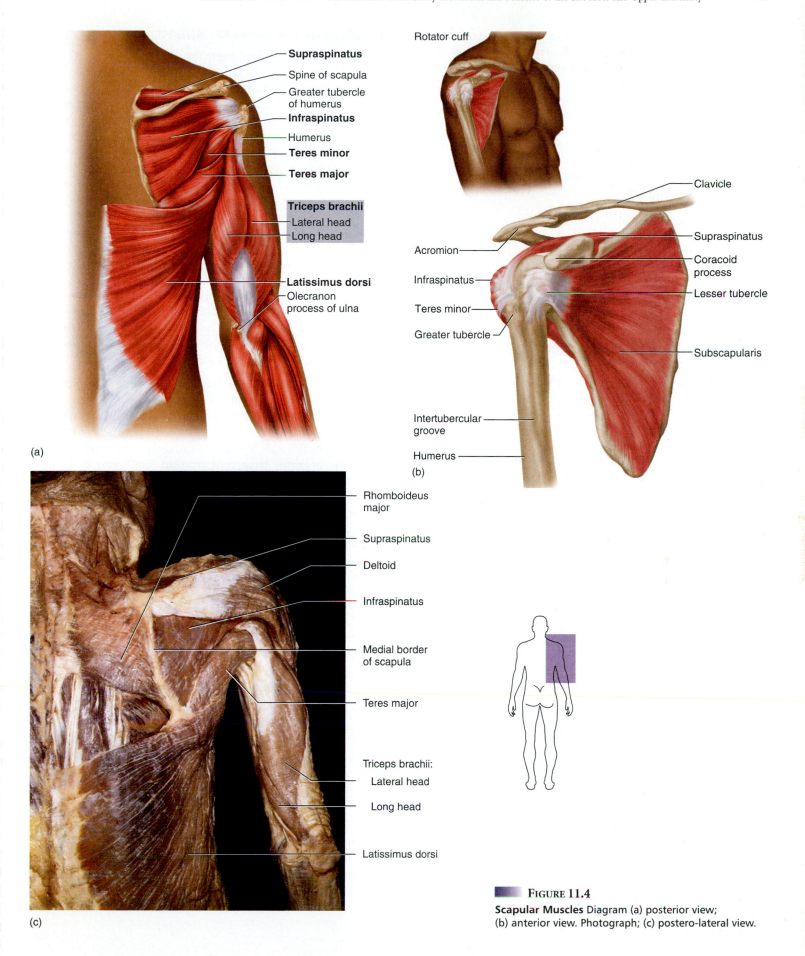

Supraspinatus
Spine of scapula
Greater tubercle
of humerus
Infraspinatus
Humerus
Teres minor
Teres major
Triceps brachii
Lateral head
Long head
Latissimus dorsi
Olecranon
process of ulna

(a)

Rotator cuff

Clavicle
Acromion
Supraspinatus
Infraspinatus
Coracoid
process
Teres minor
Lesser tubercle
Greater tubercle
Subscapularis
Intertubercular
groove
Humerus

(b)

Rhomboideus
major
Supraspinatus
Deltoid
Infraspinatus
Medial border
of scapula
Teres major
Triceps brachii:
Lateral head
Long head
Latissimus dorsi

(c)

FIGURE 11.4

Scapular Muscles Diagram (a) posterior view;
(b) anterior view. Photograph; (c) postero-lateral view.

The scapula gives rise to two other muscles, the teres minor and the teres major. The **teres** (TERR-eez) **minor** appears like a slip of the infraspinatus, while the **teres major** crosses to the medial side of the humerus (fig. 11.4).

The **rotator** (musculotendinous—MUS-que-lo-TEND-inus) **cuff** muscles stabilize the shoulder joint by holding the head of the humerus into the glenoid fossa. Muscles composing the cuff are the supraspinatus, infraspinatus, subscapularis, and teres minor. You can remember these muscles because the humerus "SITS" in the musculotendinous cuff. A rotator cuff injury occurs if there is damage to any of these muscles.

MUSCLES OF THE ARM

The **biceps brachii** (BY-seps BRAY-kee-eye) muscle is a two-headed muscle of the arm. It neither originates nor inserts on the humerus, yet it has an action on the arm. Flexion of the forearm occurs primarily from contraction of this muscle. The biceps brachii has a long tendon that originates on the scapula and runs between the intertubercular groove of the humerus and a short tendon that originates on the coracoid process. The insertion of the biceps brachii is on the radius.

Alongside the short head of the biceps brachii is a small slip of muscle known as the **coracobrachialis** (COR-uh-co-BRAY-kee-AL-iss). This muscle is a short, diagonal muscle of the arm. Underneath the biceps brachii on the anterior surface of the arm is the **brachialis** (BRAY-kee-AL-iss) muscle. The brachialis crosses the anterior surface of the elbow joint and flexes the forearm. The **triceps** (TRY-ceps) **brachii** is the only major muscle on the posterior surface of the humerus. The triceps *brachii* extends the arm and forearm and is an antagonist to the biceps brachii. It is seen in figure 11.5. Examine the material in lab and compare these muscles to figure 11.5 and with descriptions found in table 11.1.

MUSCLES OF THE FOREARM AND HAND

The muscles of the forearm, due to their similar appearance, provide greater challenges than do the muscles of the shoulder and arm. As you study these muscles, it is important that you locate them and determine their origin and insertion. The origins and insertions will help you determine if you are looking at the correct muscle. The muscles of the forearm and hand are generally named for their action (*pronate* the hand or *flex* the digits) or for their insertion (*carpi* for inserting on carpals or metacarpals, *digitorum* for fingers, and *pollicis* for thumb). In a few instances, they are named for the shape of the muscle (*teres* for round or *quadratus* for square).

The fingers are controlled by muscles that originate on the humerus or the proximal radius or ulna and extend by long tendons to the digits. The fingers move by a force applied from a distance, analogous to a puppet controlled by puppet strings. If the hand muscles were located on the hand proper, the hands would look like softballs. By having the most of the muscle in the forearm, an efficiency of form occurs that allows for a powerful grip yet precise movements of the fingers. The **brachioradialis** (BRAY-kee-oh-RAY-dee-AL-iss) originates on the distal part of the humerus and is the most lateral muscle of the forearm.

Muscles That Supinate and Pronate the Hand The **supinator** (SOO-pih-NAY-tur) is a muscle that originates on the arm and forearm, wrapping around the radius, and is named for its action of supinating the hand. It is the deepest proximal muscle of the forearm. Examine this muscle in the lab and compare it with figure 11.6.

The **pronator teres** (PRO-nay-tur TERR-eez) is a round muscle named for its action of pronating the hand. It originates on the medial side of the arm and forearm and inserts on the lateral side of the radius. The pronator teres is different from the other superficial forearm muscles in that the pronator teres runs at an oblique angle on the forearm, while the other muscles run parallel along the length of the forearm.

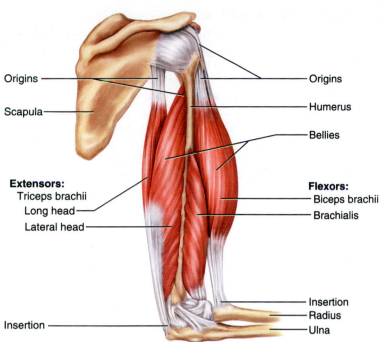

Origins

Scapula

Extensors:
  Triceps brachii
  Long head
  Lateral head

Insertion

(a)

Origins

Humerus

Bellies

Flexors:
  Biceps brachii
  Brachialis

Insertion
Radius
Ulna

FIGURE 11.5

Muscles of the Arm Diagram (a) lateral view; (b) superficial muscles, anterior view; (c) deep muscles, anterior view.

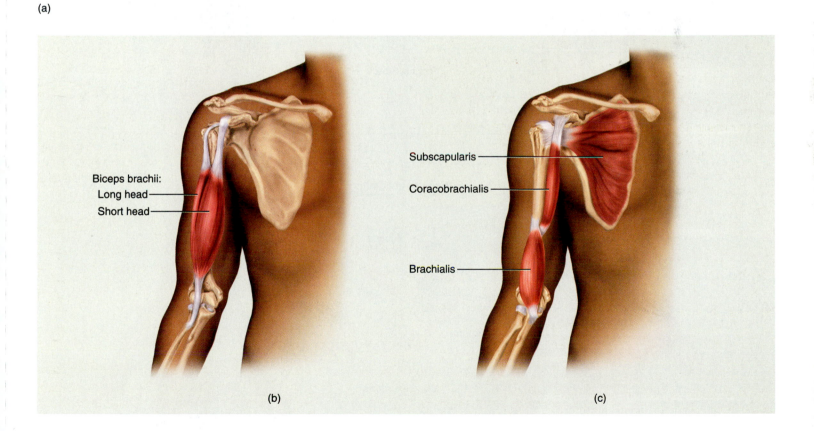

Biceps brachii:
  Long head
  Short head

Subscapularis

Coracobrachialis

Brachialis

(b)

(c)

The **pronator quadratus** is a square muscle that is deep to the other forearm muscles on the distal part of the radius and the ulna. It also pronates the hand. Compare the two pronator muscles in the lab with figures 11.6 and 11.7.

Flexor Muscles   The flexor muscles are grouped by their insertion on the hand. The superficial **palmaris longus** muscle is absent in about 10% of the population. It is centrally located in the middle, anterior forearm and inserts into a broad, flat tendon known as the palmar aponeurosis. This aponeurosis has no bony attachment but attaches to the fascia of the underlying muscles.

The **flexor carpi radialis** muscle inserts on the metacarpals on the radial side of the hand. The pulse of the radial artery is taken at the wrist just lateral to the tendon of the flexor carpi radialis muscle. Most of the flexor muscles of the hand, including the flexor carpi radialis, originate from the medial epicondyle of the humerus. The flexor carpi radialis runs underneath a connective tissue band known as the **flexor retinaculum,** which anchors the tendons to the wrist and prevents the tendons from pulling away from the wrist when the hand is flexed.

The **flexor carpi ulnaris** muscle also has an origin on the medial epicondyle of the humerus and inserts on the carpals and on metacarpal five of the ulnar side of the hand. The flexor carpi ulnaris is a medial muscle of the forearm. Examine these muscles in the lab and in figure 11.7.

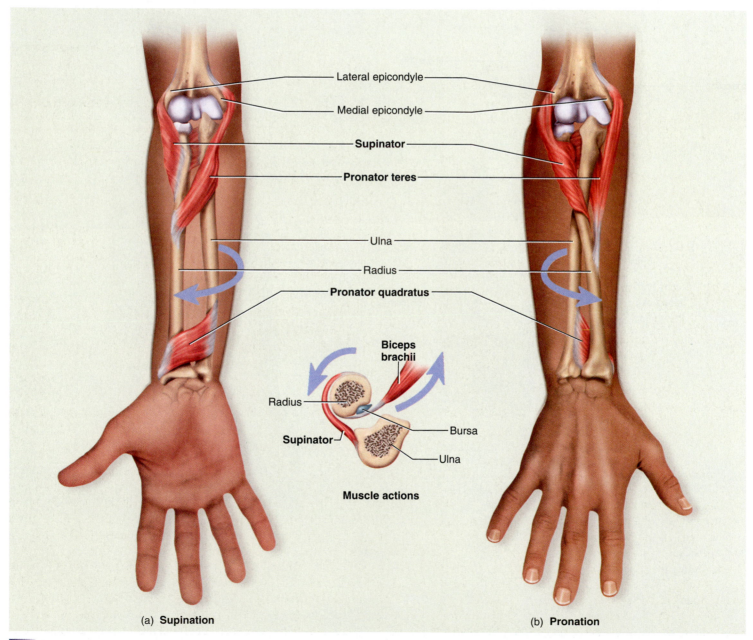

(a) **Supination**

(b) **Pronation**

FIGURE 11.6

Supinator and Pronator Muscles of the Right Forearm, (a) Anterior view, supination (b) pronation.

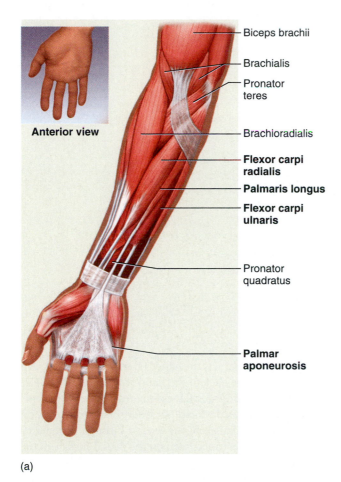

Anterior view

- Biceps brachii
- Brachialis
- Pronator teres
- Brachioradialis
- **Flexor carpi radialis**
- **Palmaris longus**
- **Flexor carpi ulnaris**
- Pronator quadratus
- **Palmar aponeurosis**

(a)

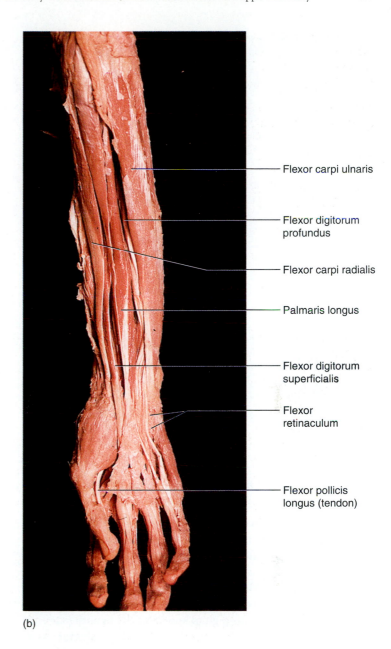

- Flexor carpi ulnaris
- Flexor digitorum profundus
- Flexor carpi radialis
- Palmaris longus
- Flexor digitorum superficialis
- Flexor retinaculum
- Flexor pollicis longus (tendon)

(b)

FIGURE 11.7

Superficial Flexor Muscles of the Right Forearm, Anterior View (a) Diagram; (b) photograph.

The **flexor digitorum superficialis** is a superficial flexor muscle of the digits. It is not the most superficial muscle of the forearm but is actually under the palmaris longus and the flexor carpi radialis and flexor carpi ulnaris. The flexor digitorum superficialis is so named because it is more superficial than the flexor digitorum profundus, discussed next. As with the previous flexor muscles, the flexor digitorum superficialis has an origin on the medial epicondyle of the humerus and the insertion tendons form a V on the middle phalanges of digits 2 through 5.

The **flexor digitorum profundus** is a deep muscle that flexes the digits. It is an exception to the other hand flexors in that it does not originate on the medial epicondyle of the humerus but on the ulna and the membrane between the radius and the ulna (the **interosseous membrane**). It is a deep muscle of the forearm that runs underneath the flexor digitorum superficialis. At the insertion point the tendons of the flexor digitorum profundus run through the split tendons of the flexor digitorum superficialis and insert on the distal phalanges of digits 2 through 5.

The **flexor pollicis** (PAUL-ih-sis) **longus** originates on the radius and the interosseous membrane and runs along the medial side of the radius, inserting on the fingerprint side of the thumb. Flexion of the thumb occurs when you curl your thumb as if you were ready to flip a coin. The flexor pollicis longus is unusual in that it does not originate on the medial epicondyle of the humerus. Examine these muscles and compare them with figure 11.8.

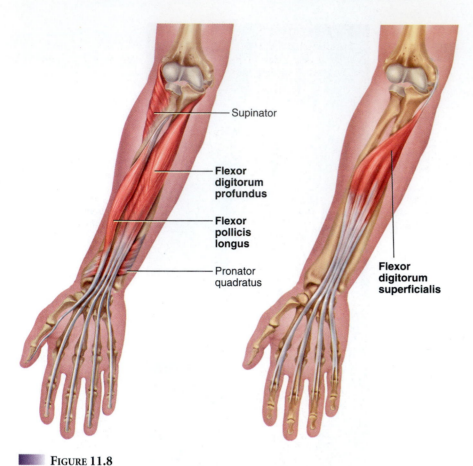

— Supinator

Flexor digitorum profundus

Flexor pollicis longus

— Pronator quadratus

Flexor digitorum superficialis

FIGURE 11.8

Deep Flexor Muscles of the Right Forearm, Anterior View

Intrinsic Muscles of the Hand   There are many muscles that originate on the hand and have an action on the hand. Some of these arise on the pad of muscles at the base of the thumb known as the thenar eminence. Thenar muscles are named for their action on the thumb, including the **flexor pollicis brevis,** the **abductor pollicis brevis,** and the **opponens** (up-POH-nenz) **pollicis.** The pad of tissue at the base of the fifth digit is known as the hypothenar eminence and the muscles that arise there are the **abductor digiti minimi,** the **flexor digiti minimi brevis,** and the **opponens digiti minimi.** Find these muscles on models or cadavers in lab and compare them with figure 11.9.

Extensor Muscles   As a general rule, most of the extensors originate on the lateral epicondyle or lateral supracondylar ridge of the humerus. The extensor muscle tendons are held to the posterior surface of the wrist by a connective tissue band known as the **extensor retinaculum.** The **extensor carpi radialis longus** muscle originates on the humerus and inserts on the second metacarpal (on the radial side) of the hand. The **extensor carpi radialis brevis** is deep to the longus, and it inserts on the dorsum of the third metacarpal. Both of these muscles extend and abduct the hand. Examine these muscles in figure 11.10.

The **extensor carpi ulnaris** originates on the lateral epicondyle and inserts on the fifth metacarpal (on the ulnar side) of

the hand. It extends and adducts the hand. The **extensor digitorum** is a singular muscle on the back of the hand (remember that there are two flexor digitorum muscles). It originates on the lateral epicondyle of the humerus, as do the other extensors, and inserts on the middle and distal phalanges of the second through fifth digits. The tendons of this muscle can be seen on the dorsal surface of the hand and in figure 11.10.

The **abductor pollicis longus** muscle inserts on the metacarpal of the thumb. By pulling your thumb away from your index finger you are abducting the thumb. There are two extensor pollicis muscles, the **extensor pollicis longus** and the **extensor pollicis brevis.** Extension of the thumb is done when you flip a coin. At the end of the flip the thumb is extended. The tendons of the extensor pollicis muscles and the abductor pollicis muscles form a depression in the form of a triangle at the base of the thumb. This depression is known as the anatomical snuff box. These muscles can be seen in figures 11.10 and 11.11.

CAT ANATOMY

If you are using cats, turn to section 2, "Dissection, Overview, and Forelimb Muscles of the Cat," on page 410 of this lab manual.

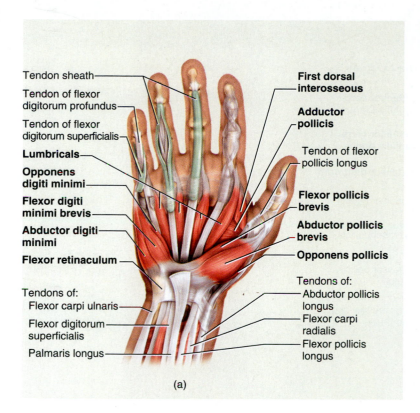

(a)

FIGURE 11.9

Intrinsic Muscles of the Right Hand Diagram (a) palmar aspect, superficial muscles; (b) palmar aspect, deep muscles. Photograph (c) palmar dissection, superficial muscles.

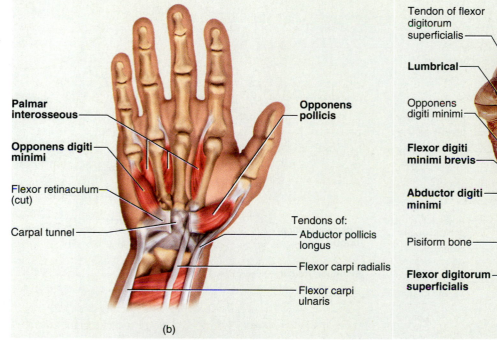

(b)

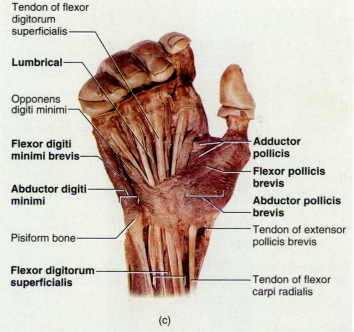

(c)

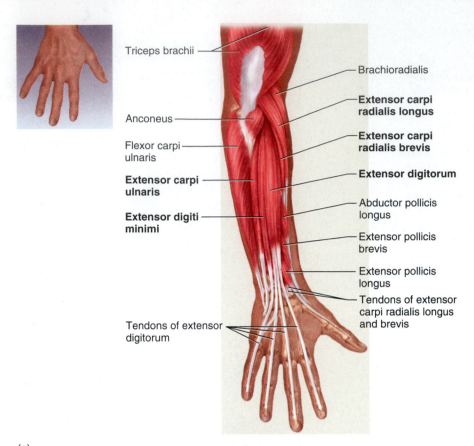

Triceps brachii

Anconeus

Flexor carpi ulnaris

Extensor carpi ulnaris

Extensor digiti minimi

Tendons of extensor digitorum

Brachioradialis

Extensor carpi radialis longus

Extensor carpi radialis brevis

Extensor digitorum

Abductor pollicis longus

Extensor pollicis brevis

Extensor pollicis longus

Tendons of extensor carpi radialis longus and brevis

(a)

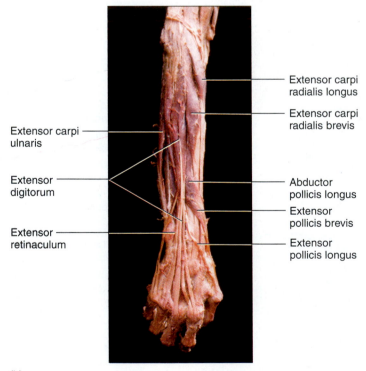

Extensor carpi ulnaris

Extensor digitorum

Extensor retinaculum

Extensor carpi radialis longus

Extensor carpi radialis brevis

Abductor pollicis longus

Extensor pollicis brevis

Extensor pollicis longus

(b)

FIGURE 11.10

Superficial Extensor Muscles of the Right Forearm (a) Diagram; (b) photograph.

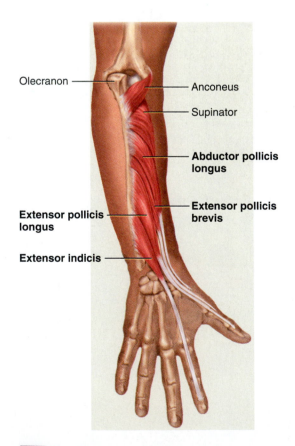

Olecranon

Anconeus

Supinator

Abductor pollicis longus

Extensor pollicis longus

Extensor pollicis brevis

Extensor indicis

FIGURE 11.11

Deep Extensor Muscles of the Right Thumb, Posterior View

Name _____ Date _____

1. What is the insertion of the trapezius muscle?

2. What is the action of the deltoid muscle?

3. What is the insertion of the latissimus dorsi muscle?

4. Name the origin of the supraspinatus muscle.

5. What is the insertion of the pectoralis minor?

6. What is the origin of the pectoralis minor?

7. Does the biceps brachii muscle originate or insert on the humerus?

8. Name the action of the pectoralis major.

9. What is the insertion of the subscapularis?

10. What is the action of the triceps brachii?

11. What is the origin of the brachialis?

12. Name all the muscles that flex the arm.

13. Which muscles are antagonists to the triceps brachii?

14. What is the origin of the flexor carpi ulnaris?

15. What is an antagonist to the supinator muscle?

16. Where does the flexor digitorum superficialis insert?

17. What is the insertion of the extensor carpi ulnaris muscle?

18. What is the action of the flexor carpi radialis muscles?

19. What muscle extends both the hand and the phalanges?

20. What muscles flex the hand?

21. What muscles extend the thumb?

22   Where are the extensor carpi muscles found, on the anterior or posterior side of the forearm?

Muscles of the Hip, Thigh, Leg, and Foot

Muscular System

INTRODUCTION

The muscles of the hip, thigh, leg, and foot are primarily muscles of locomotion. They can be divided into muscles that have an action on the thigh, the leg, or the foot.

The size and shape of the muscles of the hip and thigh in humans are different from those of other mammals in that humans are bipedal. This two-legged walking habit is different from a quadruped's locomotion in terms of balance and forward movement. When a quadruped walks, typically three legs remain on the ground. When a biped walks, the body is placed in an unstable position and the support limb must be stable to maintain the center of gravity. Think about what happens to the mechanics of movement and the center of balance when you move a leg to walk or run.

The muscles of the leg originate typically from the femur, tibia, or fibula and insert on the bones of the foot. These muscles, involving movement of the foot, use some terms of action specific to the foot. The term **dorsiflexion** describes a decrease in the angle between the shin and dorsum (top of the foot). **Plantar flexion** is an increase in the angle between the shin and the dorsum of the foot (seen as you stand on your toes to reach something high). Review other actions of the leg and foot in laboratory exercise 10. Some of the muscles of the feet are similar to those of the hand, such as the extensor digitorum muscles.

LEARNING OBJECTIVES

At the end of this exercise you should be able to
1. locate the muscles of the hip, thigh, leg, and foot on a model, chart, cadaver (if available);
2. list the origin, insertion, and action of each muscle presented;
3. describe what nerve controls each muscle;
4. list what muscles function as synergists or antagonists to the prime mover;
5. name all the muscles that move a joint, such as all the muscles that flex the thigh or those that dorsiflex the foot.

MATERIALS

Human torso model
Human leg models
Human muscle charts
Articulated skeleton
Cadaver (if available)

PROCEDURE

Review the muscle nomenclature and the action of muscles as outlined in laboratory exercises 10 and 11. Examine muscle models or charts in the lab and locate the muscles described in the following section and in table 12.1. Associate the shape of the muscle with the name and begin to visualize the muscles as you study the models or charts. You may want to look at an articulated skeleton as you review the origins and insertions of the muscles, so that you can better see the muscle attachment points. Once you have learned the origin and the insertion, you should be able to understand the action of the muscle. The action of the muscles will be easier to understand when you have identified the origins and the insertions. This is done, in part, by imagining how the bony attachments of the origin and insertion would come together if the muscle pulled them closer to one another. If you have a cadaver, pay particular attention to the origins and the tendons of insertion of the muscles of the leg and foot.

Anterior Muscles of the Hip and Thigh

The **iliopsoas** (IL-ee-oh-SO-az) muscle is a major flexor of the thigh. It originates inside the abdominopelvic cavity and crosses over the pelvic brim to insert on the posterior aspect of the femur. The iliopsoas is actually two muscles, the **iliacus** (ILL-ee-AK-us) and **psoas** (SO-az) **major.** A companion muscle to the iliopsoas is the **psoas minor.** This exercise treats these muscles as one.

The **sartorius** (sar-TORE-ee-us) is a slender muscle that originates on the superior, lateral hip and inserts in an inferior,

TABLE 12.1

Muscles of the Hip, Thigh, Leg, and Foot

Name	Origin	Insertion	Action	Innervation
Iliopsoas	T12, L1–5, sacrum, iliac crest, iliac fossa	Lesser trochanter of femur	Flexes thigh and lumbar vertebrae	Femoral nerve, spinal nerves L1–3
Sartorius	Anterior superior iliac spine	Proximal, medial tibia	Flexes and laterally rotates thigh, flexes leg	Femoral nerve
Gracilis	Symphysis pubis	Proximal, medial tibia	Adducts thigh, flexes leg	Obturator nerve
Pectineus	Pubis	Femur inferior to lesser trochanter	Adducts and flexes thigh	Femoral nerve
Adductor brevis	Pubis	Linea aspera at proximal portion of femur	Adducts, flexes, and laterally rotates thigh	Obturator nerve
Adductor longus	Pubis	Linea aspera at middle portion of femur	Adducts, flexes, and laterally rotates thigh	Obturator nerve
Adductor magnus	Pubis, ischium	Gluteal tuberosity, linea aspera, adductor tubercle of distal femur	Adducts, flexes, extends, and laterally rotates thigh	Obturator nerve
Quadriceps femoris				
Rectus femoris	Anterior inferior iliac spine, margin of acetabulum	Tibial tuberosity by the patellar tendon	Flexes thigh, extends leg	Femoral nerve
Vastus lateralis	Greater trochanter of femur linea aspera of femur	Tibial tuberosity by the patellar tendon	Extends leg	Femoral nerve
Vastus intermedius	Proximal, anterior femur	Tibial tuberosity by the patellar tendon	Extends leg	Femoral nerve
Vastus medialis	Linea aspera, medial side	Tibial tuberosity by the patellar tendon	Extends leg	Femoral nerve
Tensor fasciae latae	Iliac crest, anterior iliac surface	Iliotibial band of fascia lata	Flexes, abducts, and medially rotates thigh	Superior gluteal nerve
Gluteus maximus	Outer iliac blade, iliac crest, sacrum, coccyx	Gluteal tuberosity of femur, iliotibial band of fascia lata	Extends and laterally rotates thigh, braces knee	Inferior gluteal nerve
Gluteus medius	Outer iliac blade	Greater trochanter of femur	Abducts and medially rotates thigh	Superior gluteal nerve
Gluteus minimus	Outer iliac blade	Greater trochanter of femur	Abducts and medially rotates thigh	Superior gluteal nerve
Hamstrings				
Biceps femoris	Ischial tuberosity, linea aspera	Head of fibula, lateral condyle of tibia	Extends thigh, flexes leg	Sciatic nerve
Semitendinosus	Ischial tuberosity	Proximal, medial tibia	Extends thigh, flexes leg	Sciatic nerve
Semimembranosus	Ischial tuberosity	Medial condyle of tibia	Extends thigh, flexes leg	Sciatic nerve
Gastrocnemius	Condyles of femur	Calcaneus by the calcaneal tendon	Flexes leg, plantar flexes foot	Tibial nerve
Soleus	Posterior, proximal tibia and fibula	Calcaneus by the calcaneal tendon	Plantar flexes foot	Tibial nerve
Plantaris	Distal femur	Calcaneus	Flexes leg, plantar flexes foot	Tibial nerve
Popliteus	Lateral condyle of femur	Proximal tibia	Flexes leg, thus "unlocking" knee	Tibial nerve
Tibialis posterior	Proximal tibia and fibula	Second through fourth metatarsals and some tarsals	Plantar flexes and inverts foot	Tibial nerve
Flexor digitorum longus	Posterior tibia	Distal phalanges of second through fifth digits	Flexes toes, plantar flexes and inverts foot	Tibial nerve
Flexor hallucis longus	Midshaft of fibula	Distal phalanx of hallux (big toe)	Flexes hallux, inverts foot	Tibial nerve
Tibialis anterior	Lateral condyle and proximal tibia	First metatarsal, first cuneiform	Dorsiflexes and inverts foot	Deep fibular nerve
Extensor digitorum longus	Lateral condyle of tibia, shaft of fibula	Middle and distal phalanges of second through fifth digits	Extends toes, dorsiflexes foot	Deep fibular nerve
Extensor hallucis longus	Anterior shaft of fibula	Distal phalanx of hallux (big toe)	Extends hallux, dorsiflexes foot	Deep fibular nerve
Fibularis longus	Head and shaft of fibula, lateral condyle of tibia	First metatarsal, first cuneiform	Plantar flexes and everts foot	Superficial fibular nerve
Fibularis brevis	Shaft of fibula	Fifth metatarsal	Plantar flexes and everts foot	Superficial fibular nerve
Fibularis tertius	Anterior, distal fibula	Dorsum of fifth metatarsal	Dorsiflexes and everts foot	Deep fibular nerve

medial location. When hemming a pair of pants a tailor sits cross-legged. This motion mimics the action of the sartorius, which is named for the Latin word, *sartor,* meaning tailor. The sartorius is the most superficial muscle on the anterior thigh. The most superficial muscle of the medial aspect of the thigh is the gracilis. The **gracilis** (GRASS-ih-lis) is a thin muscle that adducts the thigh. Locate these muscles in the lab and in figure 12.1.

The muscles of the anterior aspect of the thigh belong to the **quadriceps femoris** group. All these muscles extend the leg, while only one member of the group flexes the thigh. The quadriceps muscles all have a common ligament of insertion called the patellar ligament. This ligament inserts on the tibial tuberosity. The most superficial muscle of the group is the rectus femoris (figs. 12.1 and 12.2*b*). It is a muscle that originates directly inferior to the

FIGURE 12.1

Muscles of the Right Thigh, Anterior View (a) Diagram (b) photograph.

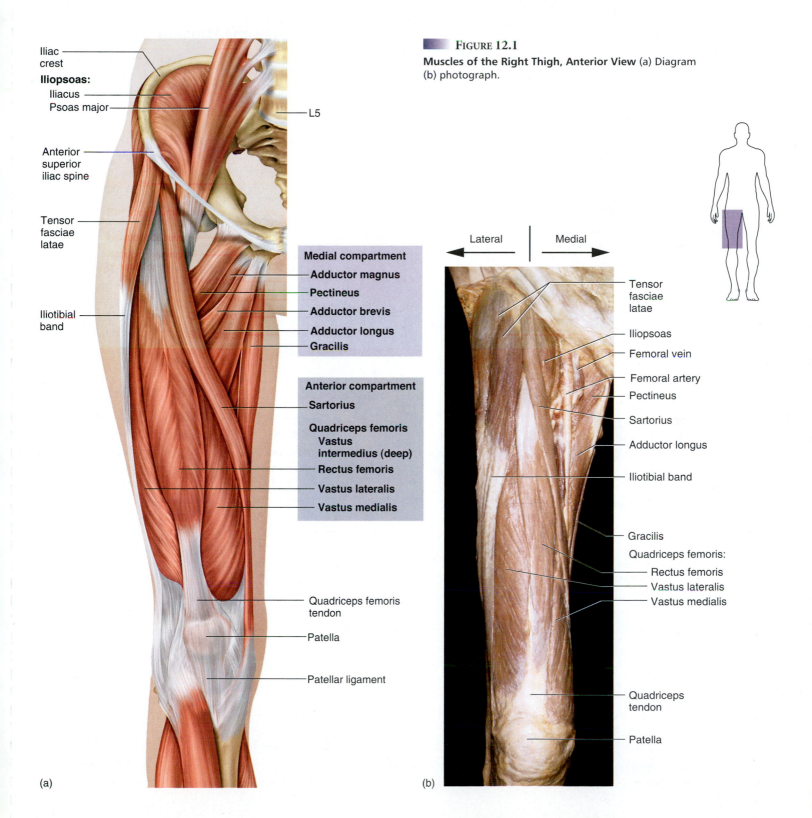

(a)

(b)

The three other quadriceps muscles are the vastus muscles. The word "vastus" means large. The **vastus lateralis** is named due to its lateral position. The **vastus intermedius** is deep to the rectus femoris, and the **vastus medialis** is the most medial of these muscles. These muscles originate on the femur, so they do not have an action on the hip. They insert on the tibia and extend the leg. Locate these muscles in the lab and compare them with figures 12.1 and 12.2.

Medial Muscles of the Hip and Thigh

The muscles in the medial aspect of the thigh are adductor muscles. The gracilis is an example of an adductor muscle, as is the pectineus and the three adductor muscles proper. The **pectineus** (pec-TIN-ee-us) is a small, triangular muscle that is the most proximal of the group. It is superficial to the adductor brevis muscle and proximal to the adductor longus muscle. The gracilis is medial, the adductors are more lateral, and the pectineus is more lateral still. This sequence of the first letters of the muscles spells the word "GAP." Examine the material in the lab and compare the pectineus with figures 12.1 and 12.3.

The **adductor brevis** is the shortest adductor muscle and is found deep in the medial aspect of the thigh. It can be seen if the pectineus is reflected. The adductor brevis originates on the pubis and inserts on the proximal third of the thigh. The **adductor longus** also has a pubic origin but its insertion is in the middle third of the thigh. The largest of the adductor muscles is the **adductor magnus**, and it originates from the pubis to the ischium on the inferior hip bone and inserts along the length of the thigh. Examine these muscles in figure 12.3.

Posterior Muscles of the Hip and Thigh

The **tensor fasciae latae** (TEN-sore FASH-ah LAH-tah) is a lateral muscle of the thigh that attaches to a thick band of connective tissue (the **fascia lata**) on the lateral aspect of the thigh. This muscle abducts the thigh.

The gluteus muscles are different in humans, due to walking upright, than in other mammals. The largest of the gluteal muscles is the **gluteus maximus**. It extends the thigh when standing from a sitting position or when climbing up stairs, and it braces the knee. The **gluteus medius** and the **gluteus minimus** both aid in keeping balance during walking because they maintain the center of gravity. The gluteus medius is superficial to the gluteus minimus but both of these muscles originate on the outer iliac blade. When you

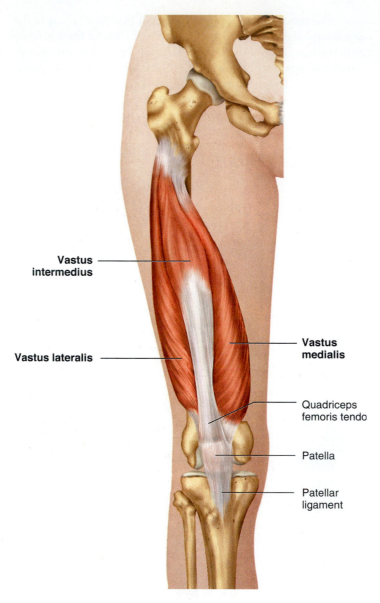

Vastus intermedius

Vastus lateralis

Vastus medialis

Quadriceps femoris tendo

Patella

Patellar ligament

FIGURE 12.2
Quadriceps Muscles of the Right Thigh, Anterior View

sartorius. The **rectus femoris** (FEM-or-us) runs straight down the femur (*rectus* = straight). Because it crosses the hip joint, the rectus femoris is the only quadriceps muscle to flex the thigh.

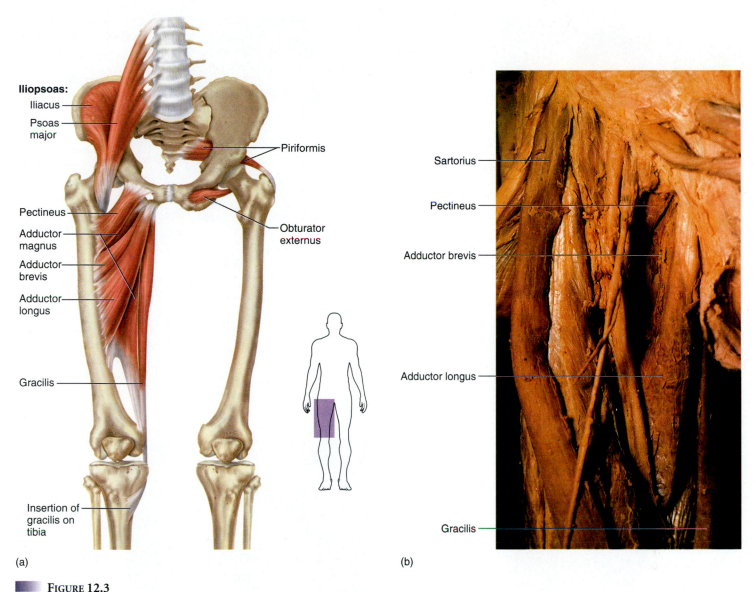

Iliopsoas:
- Iliacus
- Psoas major
- Piriformis
- Pectineus
- Adductor magnus
- Obturator externus
- Adductor brevis
- Adductor longus
- Gracilis
- Insertion of gracilis on tibia

(a)

- Sartorius
- Pectineus
- Adductor brevis
- Adductor longus
- Gracilis

(b)

FIGURE 12.3

Iliopsoas and Adductor Muscles of the Right Thigh, Anterior View (a) Diagram; (b) photograph.

raise one leg to walk, gravity pulls your body in the direction of the lifted leg. The gluteus medius and minimus on the stationary limb contract to maintain upright posture. Compare these muscles with figure 12.4.

Deep to the gluteal muscles are lateral rotators and abductors of the thigh. These are the **piriformis, obturator internus, superior gemellus, inferior gemellus,** and **quadratus femoris.** Find these muscles in lab and compare them with figure 12.4b.

The last group of hip and thigh muscles covered in this exercise are the **hamstring** muscles. The hamstring muscles are named because of an old practice of taking the ham muscles from a pig and tying them together by their tendons (*the hamstrings*) to hang in a smokehouse to cure. Three muscles make up the hamstrings: the biceps femoris, the semitendinosus, and the semimembranosus. All three of these muscles cross the hip joint and are extensor muscles

of the thigh when walking. These three muscles are also walking muscles because they flex the leg.

The **biceps femoris** is literally named the "two-headed muscle of the femur." One head originates, with the other hamstring muscles, on the ischial tuberosity, while the second head originates on the femur. The biceps femoris is a muscle that inserts laterally on the proximal leg. The **semitendinosus** also originates on the ischial tuberosity, and it can be distinguished by the long, pencil-like distal tendon. As opposed to the biceps femoris, the semitendinosus inserts medially. The **semimembranosus** (SEM-ee-MEM-bran-oh-sis) also originates on the ischial tuberosity and has a broad, flat membranous tendon on the proximal part of the muscle. The semimembranosus inserts medially on the leg, along with the semitendinosus. Locate these muscles on the material in lab and compare them with figure 12.5.

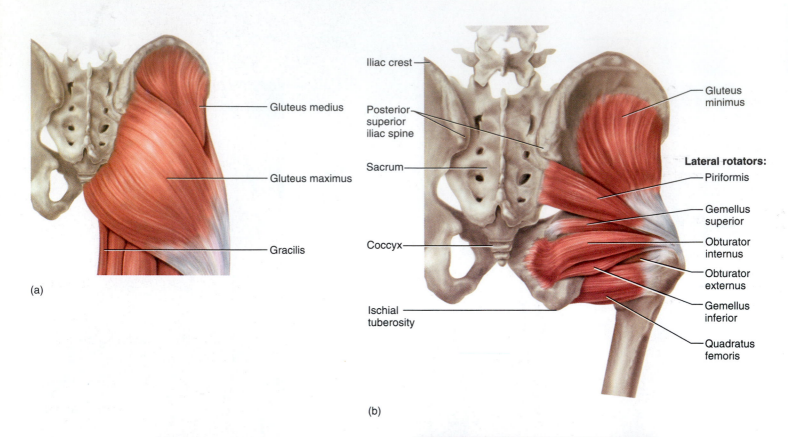

(a)

Gluteus medius

Gluteus maximus

Gracilis

Iliac crest

Posterior superior iliac spine

Sacrum

Coccyx

Ischial tuberosity

Gluteus minimus

Lateral rotators:

Piriformis

Gemellus superior

Obturator internus

Obturator externus

Gemellus inferior

Quadratus femoris

(b)

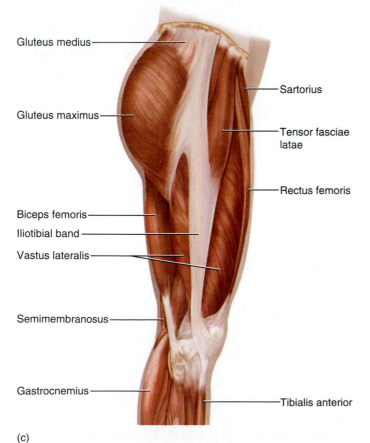

Gluteus medius

Gluteus maximus

Biceps femoris

Iliotibial band

Vastus lateralis

Semimembranosus

Gastrocnemius

Sartorius

Tensor fasciae latae

Rectus femoris

Tibialis anterior

(c)

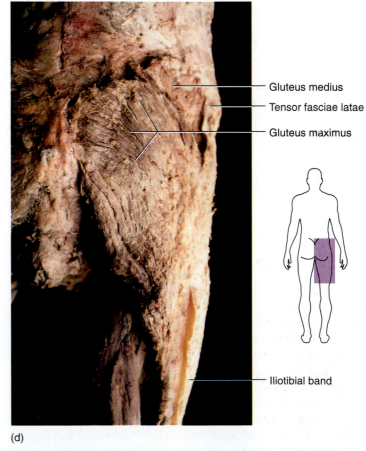

Gluteus medius

Tensor fasciae latae

Gluteus maximus

Iliotibial band

(d)

FIGURE 12.4

Muscles of the Right Hip and Thigh Diagram (a) superficial muscles, posterior view; (b) deep muscles, posterior view; (c) lateral muscles. Photograph (d) superficial muscles of the hip.

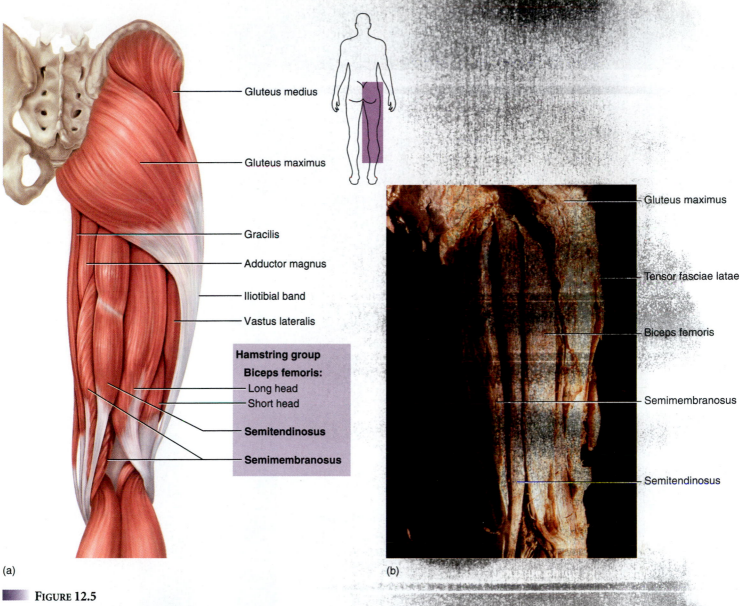

Gluteus medius

Gluteus maximus

Gracilis

Adductor magnus

Iliotibial band

Vastus lateralis

Hamstring group

Biceps femoris:

Long head

Short head

Semitendinosus

Semimembranosus

Gluteus maximus

Tensor fasciae latae

Biceps femoris

Semimembranosus

Semitendinosus

(a)   (b)

FIGURE 12.5

Muscles of the Right Thigh, Posterior View (a) Diagram; (b) photograph.

Muscles of the Leg and Foot

ANTERIOR MUSCLES

The **tibialis anterior** is located just lateral to the crest of the tibia on the anterior side of the leg. The tendon of the tibialis anterior crosses to the medial side of the foot and inserts on the first metatarsal and first cuneiform. As the tibialis anterior contracts, it decreases the angle between the anterior tibial crest and the dorsum of the foot in an action known as dorsiflexion of the foot. Because the insertion of this muscle is medial, it also inverts the foot.

The **extensor digitorum longus** muscle is lateral to the tibialis anterior, and the tendons of this muscle splay out and insert on the middle and distal phalanges of all of the digits of the foot except the hallux. The **extensor hallucis** (HAL-uh-sis) **longus** is the muscle that extends the hallux, as well as acting as a synergist to the extensor digitorum longus in dorsiflexion of the foot. Examine these muscles in the lab and compare them with figure 12.6. The muscles on the dorsum of the foot are held down by the extensor retinaculum.

The **fibularis** (peroneus) muscles are so named for originating on and running the length of the fibula. The **fibularis longus** has a tendon that travels from the lateral side of the foot underneath to the medial side, crossing under the arch of the foot. The **fibularis brevis** parallels the fibularis longus, except that the tendon stops short and inserts on the lateral side of the foot at the fifth metatarsal. Both of these muscles have tendons that hook posterior to the lateral malleolus, and they both plantar flex and evert the foot. The final fibularis muscle, the **fibularis tertius,** does not arch behind the lateral malleolus and it inserts on the dorsum of the fifth metatarsal. When it contracts, it dorsiflexes the foot while everting the foot. Examine these muscles in figures 12.6 and 12.7.

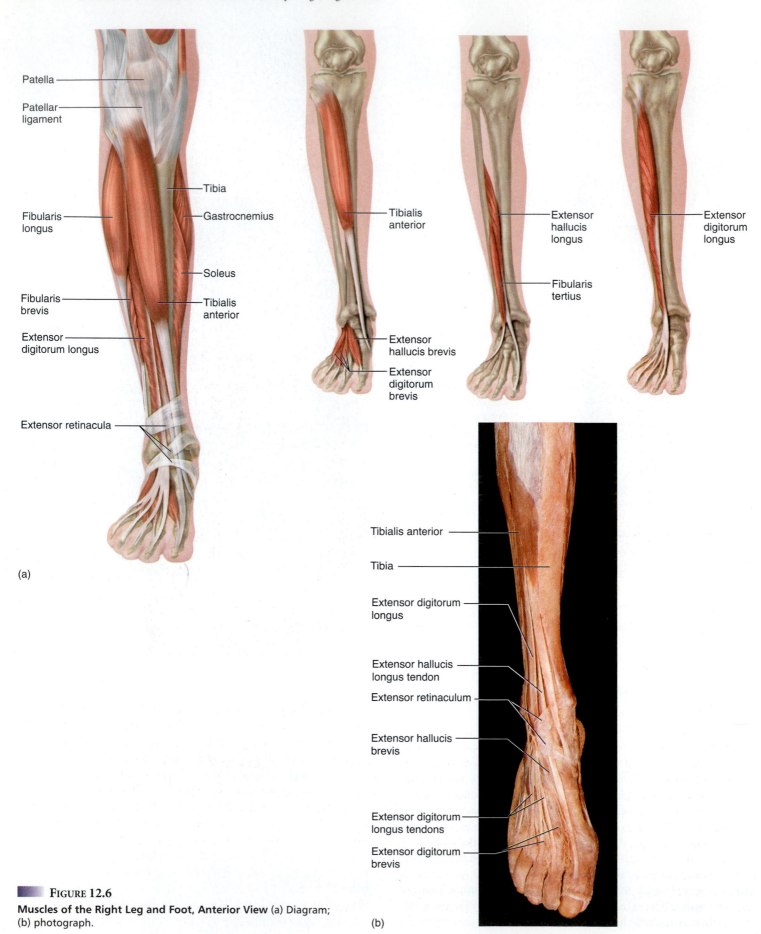

Patella

Patellar ligament

Tibia

Fibularis longus

Gastrocnemius

Soleus

Fibularis brevis

Tibialis anterior

Extensor digitorum longus

Extensor retinacula

(a)

Tibialis anterior

Extensor hallucis brevis

Extensor digitorum brevis

Extensor hallucis longus

Fibularis tertius

Extensor digitorum longus

Tibialis anterior

Tibia

Extensor digitorum longus

Extensor hallucis longus tendon

Extensor retinaculum

Extensor hallucis brevis

Extensor digitorum longus tendons

Extensor digitorum brevis

(b)

FIGURE 12.6

Muscles of the Right Leg and Foot, Anterior View (a) Diagram;
(b) photograph.

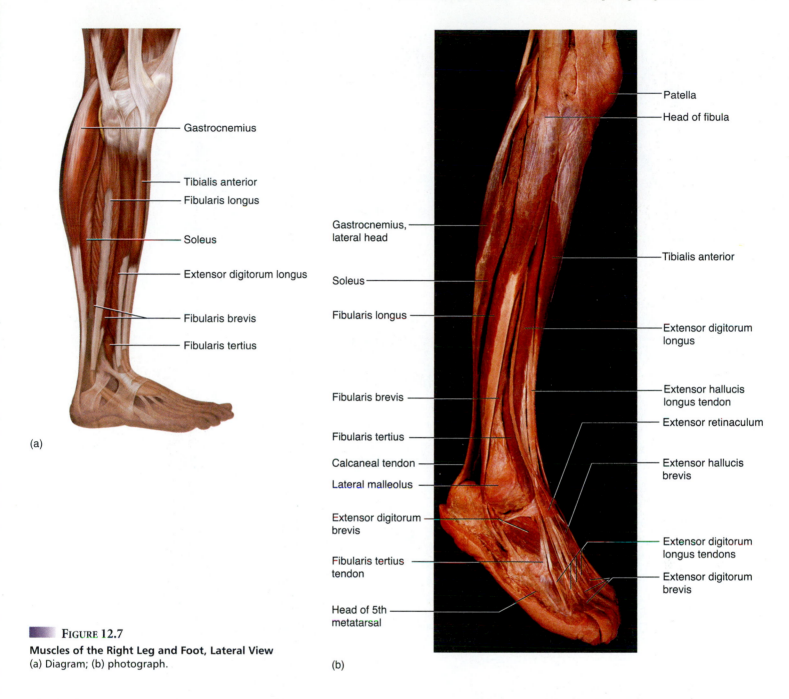

FIGURE 12.7

Muscles of the Right Leg and Foot, Lateral View
(a) Diagram; (b) photograph.

POSTERIOR MUSCLES

The **gastrocnemius** (GAS-trock-NEE-me-us) is a calf muscle with a medial and lateral head; it is the most superficial muscle of the posterior leg group. The gastrocnemius crosses the knee joint and flexes the leg. It inserts on the calcaneus by way of the **calcaneal tendon,** plantar flexing the foot as well. The calcaneal tendon is also known as the Achilles tendon. Deep to the gastrocnemius is the **soleus** (SO-lee-us) muscle. Unlike the gastrocnemius, the soleus does not originate on the femur; therefore, it does not cross the knee and has no action on the leg. It inserts on the calcaneus, sharing the calcaneal tendon with the gastrocnemius, and it plantar flexes the foot. The **plantaris** is a weak plantar flexor of the foot. The **popliteus** (pop-LIT-ee-us) is a small muscle that crosses the knee joint. It unlocks the knee joint. Locate these muscles in figures 12.7 and 12.8.

The **tibialis posterior** is a major muscle deep to the soleus that plantar flexes and inverts the foot. The tendon of the muscle runs along the medial aspect of the ankle. Near the tibialis posterior is the **flexor digitorum longus,** a muscle that inserts on the distal phalanges of all of the digits of the foot except for the hallux (big toe). The flexor digitorum longus flexes (curls) the toes in addition to plantar flexing and inverting the foot. The **flexor hallucis longus** flexes the hallux and aids the flexor digitorum longus in inverting the foot. Locate these muscles in the lab and compare them with figure 12.8.

Cat Anatomy

If you are using cats, turn to section 3 "Hindlimb Muscles of the Cat," on page 418 of this lab manual.

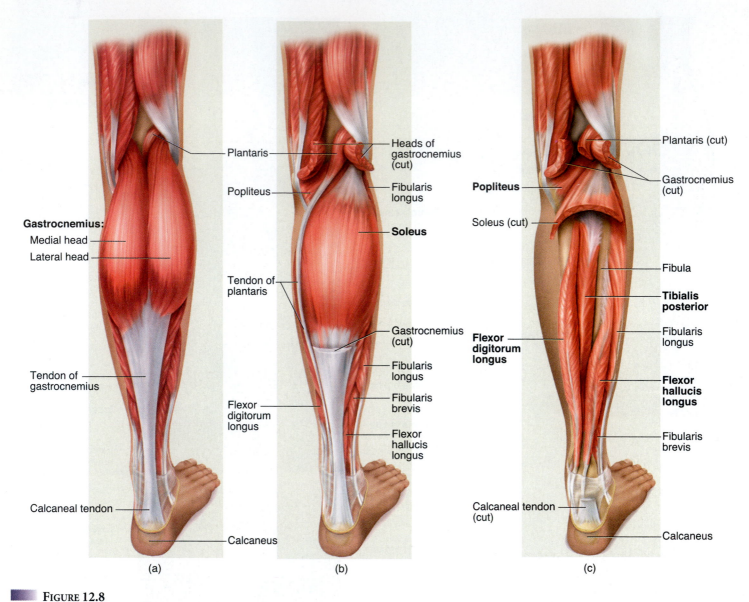

Plantaris

Popliteus

Gastrocnemius:
  Medial head
  Lateral head

Tendon of
gastrocnemius

Calcaneal tendon

Calcaneus

(a)

Heads of
gastrocnemius
(cut)

Fibularis
longus

Soleus

Tendon of
plantaris

Gastrocnemius
(cut)

Fibularis
longus

Fibularis
brevis

Flexor
hallucis
longus

Popliteus

Flexor
digitorum
longus

(b)

Plantaris (cut)

Gastrocnemius
(cut)

Popliteus

Soleus (cut)

Fibula

**Tibialis
posterior**

Fibularis
longus

**Flexor
digitorum
longus**

**Flexor
hallucis
longus**

Fibularis
brevis

Calcaneal tendon
(cut)

Calcaneus

(c)

FIGURE 12.8

Muscles of the Right Leg and Foot, Posterior View (a) Superficial muscles of the leg; (b) middle-level muscles of the leg; (c) deep muscles of the leg.

Name_____ Date _____

1. If you were to ride a horse, what muscles would you use to keep your seat out of the saddle as you ride?

2. How do the gluteus medius and gluteus minimus prevent you from toppling over as you walk?

3. What muscle is an antagonist to the biceps femoris muscle?

4. What are two muscles that are synergists with the biceps femoris muscle?

5. Are all the hamstring muscles identical in action? What is the action of the hamstring muscles?

6. What is the insertion of all the muscles of the quadriceps group?

7. How does the action of the rectus femoris differ from that of the other quadriceps muscles?

8. How many adductor muscles are there?

9. List two muscles in this exercise that are responsible for thigh flexion.

10. Where do the hamstring muscles originate as a group?

11. What is the action of the vastus lateralis?

12. Which muscle group is found on the anterior part of the thigh?

13. Is abduction of the thigh movement toward or away from the midline?

14. What muscle flexes the lumbar vertebrae as part of its action?

15. What is the origin of the gastrocnemius?

16. What is the insertion of the tibialis anterior?

17. How does the action of the fibularis longus differ from that of the fibularis tertius?

18. What is the action of the extensor hallucis longus?

19. The calf is made of what two major muscles?

20. What muscle extends the toes?

21. Name a muscle in this exercise that dorsiflexes the foot.

22. Plantar flexion and eversion of the foot occur by what muscles?

23. What is the insertion of the soleus?

24. What is the action of the tibialis anterior?

25. What is the insertion of the fibularis tertius muscle?

26. What is the insertion of the flexor digitorum longus?

27. Label the muscles in the following illustration using the terms provided.

    extensor digitorum longus   soleus

    fibularis longus   tibialis anterior

    extensor hallucis longus

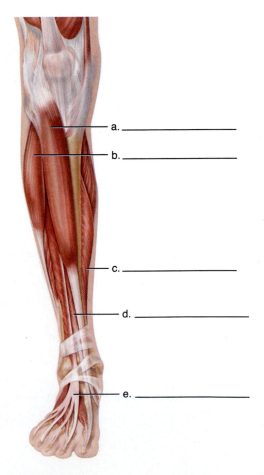

a. _____

b. _____

c. _____

d. _____

e. _____

Muscles of the Head and Neck

Muscular System

INTRODUCTION

In this exercise you continue your study of muscles by studying the muscles of the head and neck. These muscles can be grouped into a few functional classifications. Some of the head muscles provide the strength for chewing food or help manipulate the food in the mouth, so that chewing or swallowing can occur. Other muscles function in facial expression or in the closing of the eyes or mouth. Still other muscles of the neck move the head or neck. Finally, some move the hyoid or the larynx in speech or swallowing.

LEARNING OBJECTIVES

At the end of this exercise you should be able to
1. locate the muscles of the head and neck on a torso model, chart, cadaver (if available);
2. list the origin, insertion, and action of each muscle presented;
3. describe what nerve controls (innervates) each muscle;
4. name all the muscles that have an action on a joint, such as all the muscles that flex the head;
5. reproduce the actions of select muscles on your own head or neck.

MATERIALS

Human torso model, and head and neck model

Human muscle charts

Articulated skeleton

Cadaver (if available)

PROCEDURE

Examine a model or chart of the human musculature and cadaver (if available) as you read the following descriptions. These muscles are listed in table 13.1. Examine a skull or an articulated skeleton and review the bony markings as you study the origins and insertions of the muscles in this exercise. Once you know the origins or insertions the actions should be more comprehensible.

The **levator scapulae** (le-VAY-tur SCAP-u-lay) is named for what it does, elevate the scapula. The levator scapulae originates on the lateral side of the neck and inserts on the upper scapula. If the neck is fixed, the levator scapulae raises the scapulae, as in shrugging. If the scapula is fixed, the muscle rotates or abducts the neck.

The **scalenes** (skah-LEENS) muscles are found on the lateral side of the neck. They are bordered by the sternocleidomastoid in the front and the levator scapulae in the back. They rotate the neck or elevate the ribs. Locate these muscles in lab and examine figures 13.1 and 13.2.

The **sternocleidomastoid** (STER-no-KLY-doh-MAS-toyd) muscle rotates the head in a unique way. Place your hand on the right sternocleidomastoid (figs. 13.1 and 13.2) and turn your head to the right. Notice how the muscle does not contract. Now turn your head to the left, and you can feel the muscle contract. The right sternocleidomastoid turns the head to the left and the left sternocleidomastoid turns the head to the right.

The **sternohyoid, sternothyroid,** and **omohyoid** are all named for their origins and insertions. The sternum anchors the stable part of the muscle (the origin) for sternohyoid and sternothyroid. The thyroid cartilage is depressed when the sternothyroid contracts, and the hyoid bone is depressed when the sternohyoid contracts. The term "omo" means shoulder, and the scapula anchors the stable part of the omohyoid. When the omohyoid contracts, the hyoid is depressed. Examine material in lab and compare it with figure 13.2 for muscles of the anterior neck.

TABLE 13.1
Muscles of the Neck and Head

Name	Origin	Insertion	Action	Innervation
Neck				
Levator scapulae	C1–4	Upper vertebral border of scapula	Elevates scapula, abducts and rotates neck	Spinal nerves C3–5 and dorsal scapular nerve
Scalenes (anterior, middle, posterior)	Transverse process of cervical vertebrae	Ribs 1 and 2	Flexes and rotates neck, elevates ribs 1 and 2	Spinal nerves C4–8
Sternocleidomastoid	Sternum, clavicle	Mastoid process of temporal	Rotates and flexes head	Accessory nerve (XI) Spinal nerves C2–3
Sternohyoid	Manubrium of sternum	Hyoid	Depresses hyoid	Spinal nerves C1–3
Sternothyroid	Manubrium of sternum	Thyroid cartilage of larynx	Depresses thyroid cartilage	Spinal nerves C1–3
Omohyoid	Superior surface of scapula	Hyoid	Depresses hyoid	Spinal nerves C1–3
Platysma	Fascia covering pectoralis major and deltoid	Mandible and skin of lower region of face	Depresses lower lip, opens jaw	Facial nerve (VII)
Digastric	Interior, distal margin of mandible (anterior belly), mastoid notch of temporal (posterior belly)	Hyoid	Elevates, protracts, retracts hyoid; opens mandible	Trigeminal (V) and facial (VII) nerves
Mylohyoid	Inner, inferior margin of mandible	Body of hyoid and median raphe	Elevates hyoid and tongue	Trigeminal nerve (V)
Head				
Frontalis	Galea aponeurotica	Skin superior to orbit	Raises eyebrows, draws scalp anteriorly	Facial nerve (VII)
Occipitalis	Occipital and temporal bone	Galea aponeurotica	Draws scalp posteriorly	Facial nerve (VII)
Temporalis	Temporal fossa	Coronoid process and ramus of mandible	Closes mandible	Trigeminal nerve (V)
Masseter	Zygomatic arch	Angle and ramus of mandible	Closes mandible	Trigeminal nerve (V)
Pterygoids (medial, lateral)	Pterygoid processes of sphenoid bone	Medial ramus of mandible	Medial gliding of mandible (for chewing)	Trigeminal nerve (V)
Orbicularis oculi	Frontal and maxilla on medial margin of orbit	Skin of eyelid	Closes eyelid	Facial nerve (VII)
Orbicularis oris	Fascia of facial muscles near mouth	Skin of lips	Closes lips	Facial nerve (VII)
Corrugator supercilii	Medial portion of frontal bone	Skin of eyebrows	Adducts eyebrows (pulls them medially)	Facial nerve (VII)
Risorius	Fascia of masseter	Skin at angle of mouth	Abducts corner of mouth (draws edge of mouth lateral)	Facial nerve (VII)
Mentalis	Anterior mandible	Skin of chin below lower lip	Protrudes lower lip	Facial nerve (VII)
Buccinator	Maxilla and mandible near molar teeth	Orbicularis oris	Compresses cheek	Facial nerve (VII)
Zygomaticus (major, minor)	Zygomatic bone	Muscle and skin at angle of mouth	Elevates corners of mouth (in smiling and laughing)	Facial nerve (VII)
Depressor labii inferioris	Lateral mandible	Muscles and skin of lower lip	Depresses lower lip	Facial nerve (VII)
Levator labii superioris	Maxilla and zygomatic bones	Muscle and skin of lips	Elevates upper lip, flares nostril	Facial nerve (VII)

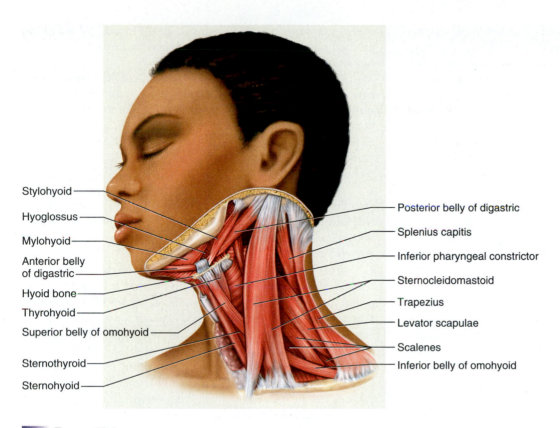

Stylohyoid

Hyoglossus

Mylohyoid

Anterior belly
of digastric

Hyoid bone

Thyrohyoid

Superior belly of omohyoid

Sternothyroid

Sternohyoid

Posterior belly of digastric

Splenius capitis

Inferior pharyngeal constrictor

Sternocleidomastoid

Trapezius

Levator scapulae

Scalenes

Inferior belly of omohyoid

FIGURE 13.1
Muscles of the Neck, Lateral View

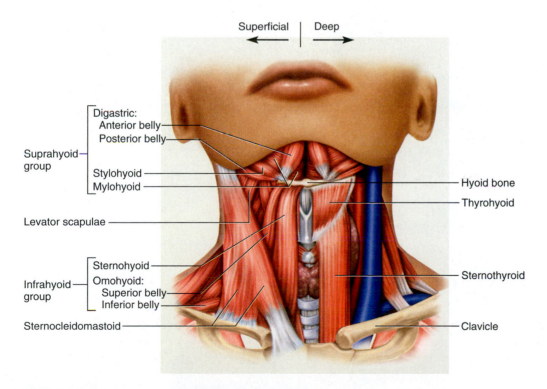

Superficial | Deep

Digastric:
  Anterior belly
  Posterior belly

Suprahyoid
group

Stylohyoid
Mylohyoid

Levator scapulae

Sternohyoid

Omohyoid:
  Superior belly
  Inferior belly

Infrahyoid
group

Sternocleidomastoid

Hyoid bone

Thyrohyoid

Sternothyroid

Clavicle

FIGURE 13.2
Muscles of the Neck, Anterior View

The **platysma** (plah-TISS-mah) is a broad, thin muscle that has a soft origin (on the fascia of the pectoral and deltoid muscles). It inserts on the mandible and skin of the lips and can be seen if you elevate your chin and subsequently pout. The thin wings that stick out on the side of your neck are the edges of the platysma muscle. Examine the platysma in figure 13.3 and in the models or on the cadaver in the lab.

The **frontalis** (fron-TAL-is) muscle originates on a broad, flat tendinous sheet on the superior aspect of the skull known as the **galea aponeurotica** (GALE-ee-uh AP-oh-nu-ROT-ih-KAH). The insertion of the frontalis is on the eyebrow region. The **occipitalis** (AUK-sip-ih-TAL-us) attaches to the posterior part of the galea aponeurotica. Examine these muscles in figures 13.3–13.5. If both muscles contract, the eyebrows are raised. If just the frontalis contracts, the scalp is brought forward, as in frowning. These muscles are sometimes listed as one muscle, the **occipitofrontalis.**

The **orbicularis oculi** is a sphincter muscle that closes the eye and the **orbicularis oris** is one that closes the mouth. Sphincter muscles act as the strings of a drawstring purse. The orbicularis oculi has an origin on the medial portion of the orbit and an insertion on the eyelid. The muscles ring the eye and close the eyelids. The orbicularis oris originates on fascia and facial muscles near the mouth and inserts on the skin of the lips, thus closing the lips. Examine the facial muscles in lab and in figures 13.4 and 13.5. The

corrugator supercilii (SOUP-ur-SIL-ee-ee) muscle has an origin on the frontal bone between the eyebrows and inserts laterally. It furrows the eyebrows.

The **zygomaticus** (ZYE-go-MAH-tih-cus) elevates the corners of the mouth by pulling them superiorly and laterally, as in smiling or laughing. It is named for its origin on the zygomatic bone. The **risorius** (Rih-ZOR-ee-us) is known as the laughing muscle because it pulls the lips laterally. It does not have a bony point of origin but attaches to the fascia of the masseter and platysma.

The **mentalis** (men-TAL-is) originates on the chin (anterior mandible) and is another of the pouting muscles. The **buccinator** (BUX-in-aye-tur) muscle of the cheek runs in a horizontal direction. It puckers the cheeks, as in trumpet playing, and pushes food toward the molars in chewing.

The **depressor labii** (LAYB-ee-eye) **inferioris** pulls the lower corners of the mouth inferiorly when pouting. It is named for its action (depressing the lower lip), while the **levator labii superioris** muscle raises the skin of the lips and expands the nostrils, as in the expression of showing extreme disgust. Look at these muscles in lab and in figures 13.4 and 13.5. The **masseter** (MASS-ih-tur) is a large muscle of the head whose action is to close the jaws. If you place your fingers on the ramus of the mandible and clench your teeth, you can feel the masseter tighten. It is a chewing muscle, or a muscle of *mastication.*

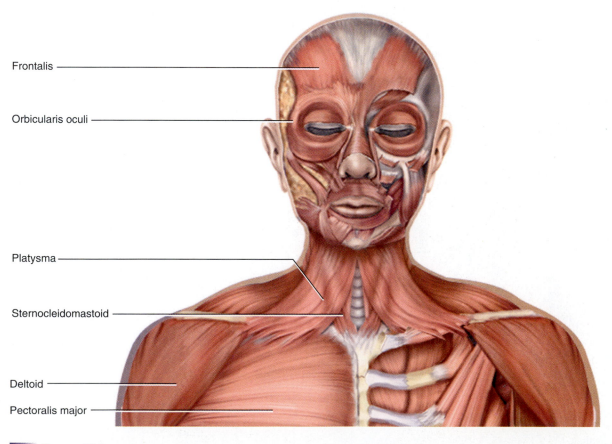

FIGURE 13.3

Platysma, Anterior View

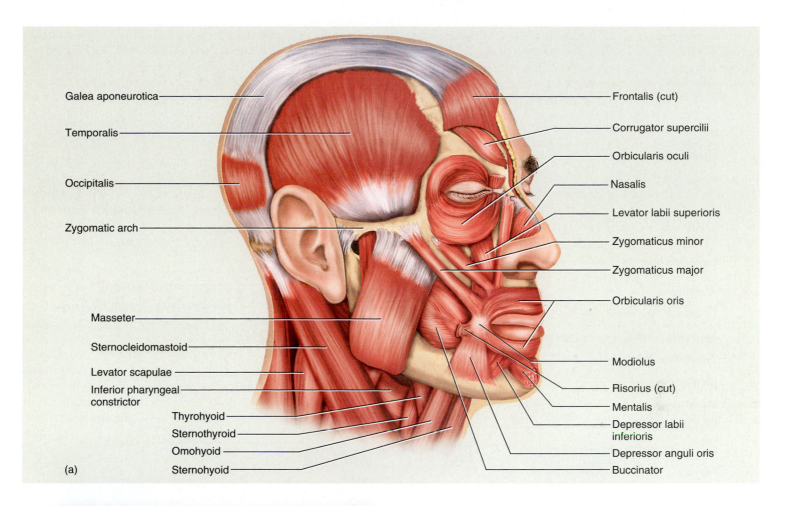

Galea aponeurotica

Temporalis

Occipitalis

Zygomatic arch

Masseter

Sternocleidomastoid

Levator scapulae

Inferior pharyngeal constrictor

Thyrohyoid

Sternothyroid

Omohyoid

Sternohyoid

(a)

Frontalis (cut)

Corrugator supercilii

Orbicularis oculi

Nasalis

Levator labii superioris

Zygomaticus minor

Zygomaticus major

Orbicularis oris

Modiolus

Risorius (cut)

Mentalis

Depressor labii inferioris

Depressor anguli oris

Buccinator

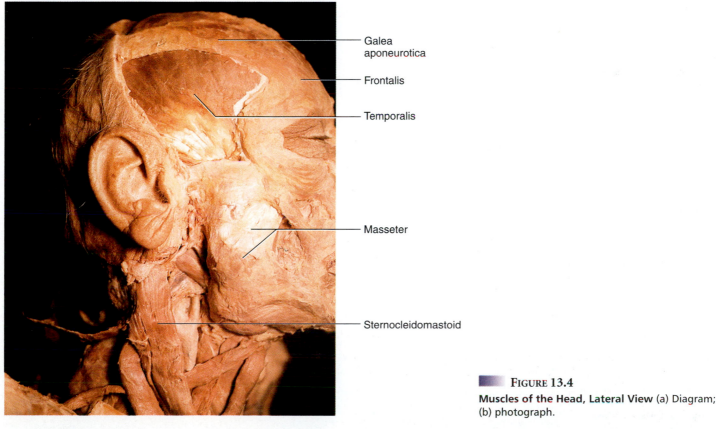

Galea aponeurotica

Frontalis

Temporalis

Masseter

Sternocleidomastoid

(b)

FIGURE 13.4

Muscles of the Head, Lateral View (a) Diagram; (b) photograph.

Superficial | Deep

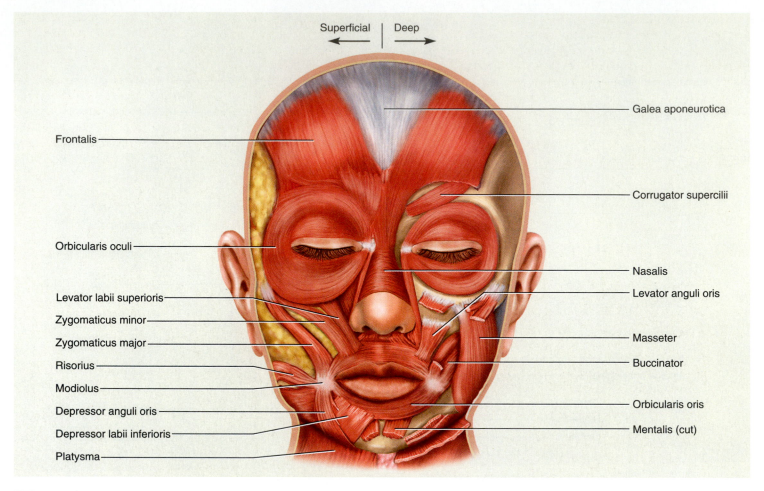

Frontalis

Orbicularis oculi

Levator labii superioris

Zygomaticus minor

Zygomaticus major

Risorius

Modiolus

Depressor anguli oris

Depressor labii inferioris

Platysma

Galea aponeurotica

Corrugator supercilii

Nasalis

Levator anguli oris

Masseter

Buccinator

Orbicularis oris

Mentalis (cut)

FIGURE 13.5
Muscles of the Face, Anterior View

The **temporalis** (TEMP-oh-RAL-us) is a synergist to the masseter and is a powerful muscle that closes the jaw. The temporal fossa is so named because it is a depression superior to the zygomatic arch (though when you examine the skull the temporal fossa appears slightly domed). The temporalis muscle runs underneath the zygomatic arch and inserts on the coronoid process and on the superior and medial ramus of the mandible. These muscles are seen in figure 13.4.

The **digastric** (di-GAS-trik) is a muscle with two bellies. The digastric is one of a few muscles that depresses the mandible. The digastric also moves the hyoid, which is important in tongue move-ment for speech and swallowing. Deep to the digastric is the **mylohyoid,** a broad muscle of the floor of the mouth that aids in pushing the tongue superiorly when swallowing. These muscles can be seen in figure 13.2.

The **pterygoid** (TARE-ih-goyd) muscles are deep muscles that originate on the sphenoid bone and insert laterally on the mandible. They pull the jaw horizontally, which helps in rotatory chewing. The temporalis and masseter close the jaw, while the pterygoids provide the sideways movement characteristic of a person chewing gum. These muscles can be seen in figure 13.6.

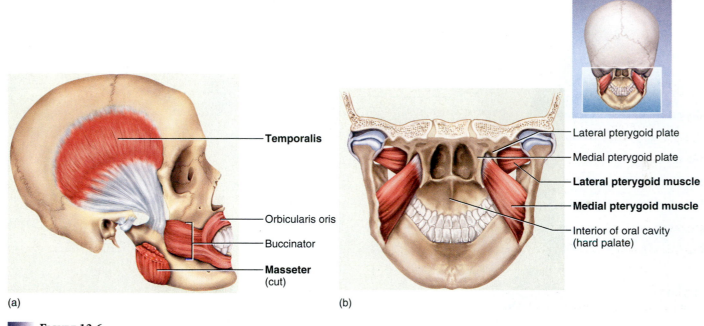

(a)　　　　　　　　　　　　　　　　　　　(b)

FIGURE 13.6

Muscles of the Head (a) Lateral view; (b) posterior view.

ACTIVITIES

1. When you turn your head to the left, which sternocleidomastoid muscle contracts?

2. Tilt your head so your chin is elevated and pout your lower lip. What muscle forms a thin membrane along the anterolateral neck?_____

3. Purse your lips and feel the buccinator muscle as it contracts in the cheek.

4. Clench your teeth and palpate the temporalis muscle and the masseter.

Examine figure 13.7 for surface views of facial muscles. Fill in the responses that represent the muscles in the photographs.

Cat Anatomy

If you are using cats, turn to section 4, "Head and Neck Muscles of the Cat" on page 422 of this lab manual.

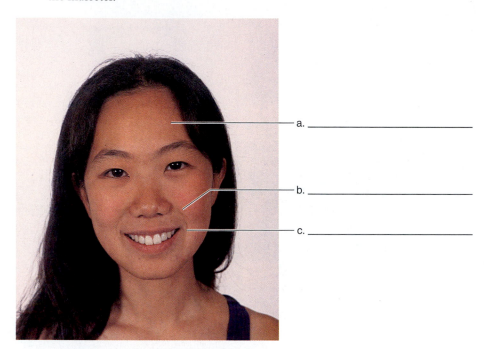

a. _____

b. _____

c. _____

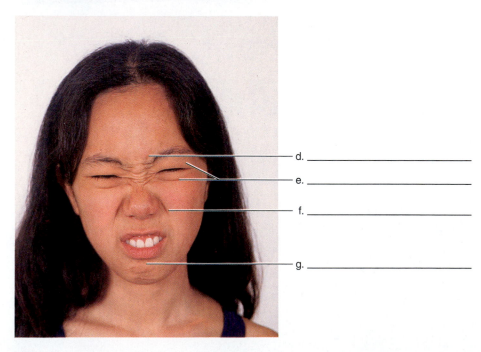

d. _____

e. _____

f. _____

g. _____

FIGURE 13.7

Muscles of Facial Expression

Name _____ Date _____

1. What is the origin of the masseter muscle?

2. What is the action of the levator scapulae muscle?

3. What kind of muscle is the orbicularis oculi or orbicularis oris muscle in terms of action?

4. Fill in the following illustration for the muscles of the head using the terms provided.

masseter   buccinator   temporalis

occipitofrontalis   orbicularis oris   orbicularis oculi

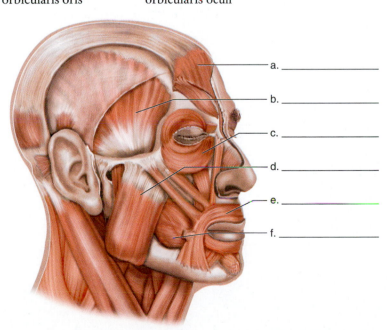

a. _____

b. _____

c. _____

d. _____

e. _____

f. _____

5. What muscle originates on the temporal fossa?

6. Name two muscles that close the jaw.

7. Where does the sternocleidomastoid muscle insert?

8. What muscle closes the lips?

9. Where does the orbicularis oculi insert?

10. What is the insertion of the temporalis?

11. Name a muscle that closes the eye.

12. What is the action of the sternocleidomastoid?

13. What muscle is a synergist with the masseter?

Muscles of the Trunk

Muscular System

INTRODUCTION

In this exercise you continue your study of muscles by studying the muscles of the trunk. These muscles can be grouped into a few functional areas. One group is the muscles that act on the scapula. Another is the abdominal muscles, all of which tighten the abdomen. The respiratory muscles, another group, assist in breathing, and the last set is the postural muscles of the back.

LEARNING OBJECTIVES

At the end of this exercise you should be able to
1. locate the muscles of the trunk on a torso model, chart, cadaver (if available);
2. list the origin, insertion, and action of each muscle;
3. describe what nerve controls (innervates) each muscle;
4. list what muscles function as synergists or antagonists to the prime mover;
5. name all the muscles that have a particular action on the trunk, such as all the muscles that compress the abdomen.

MATERIALS

Human torso model
Human muscle charts
Articulated skeleton
Cadaver (if available)

PROCEDURE

Review the actions as outlined in laboratory exercise 10 and the muscle nomenclature described in laboratory exercise 11. Examine a torso model or chart in the lab and locate the muscles described in this exercise and in table 14.1. Correlate the shape or fiber direction of the muscle with the name and begin to visualize the muscles as you study the models or charts. You may want to look at an articulated skeleton as you review the origins and insertions of the muscles, so that you can better see the muscle attachment points. The descriptions in the text portion of this exercise can help you understand the nature of the muscle, while table 14.1 gives you specific information about the muscle. Once you have learned the origin and the insertion, you should be able to understand the action of the muscle. This is done, in part, by imagining how the bony attachments of the origin and insertion would come together if the muscle pulled them closer to one another.

Anterior Muscles

The muscles of the abdomen compress the viscera, which aids in urination, childbirth, and bowel movements. This is accomplished by the **valsalva maneuver,** which consists of taking a breath and contracting these muscles. The **external abdominal oblique** is a broad, superficial muscle of the abdomen with fibers that run from a superior direction to an inferior, medial direction. The **internal abdominal oblique** is deep to the external abdominal oblique and has fiber directions that run perpendicular to the external abdominal oblique. Locate the abdominal muscles in the lab and compare them with figure 14.1. The deepest of the abdominal muscles is the **transversus abdominis,** which has fibers running in a horizontal direction. The **rectus abdominis** (*rectus* = straight) runs vertically up the abdomen. The rectus abdominis has small connective tissue bands called **tendinous intersections,** which run horizontally across the muscle, dividing it into small segments and the muscle is enclosed by the **rectus sheath.** If the abdominal fat is minimal and the muscles are well developed, the "washboard stomach" (or "six-pack") is apparent due to the muscle fibers increasing in girth while the tendinous intersections remain undeveloped.

The **serratus anterior** is a broad, fan-shaped muscle that has slips of muscle originating on the upper ribs. These slips unite and insert on the medial border and inferior angle of the scapula. As the muscle contracts, it abducts the scapula by pulling it away from the spine. The serratus anterior is shown in figure 14.1.

TABLE 14.1
Muscles of the Trunk

Name	Origin	Insertion	Action	Innervation
Muscles of Thorax, Abdomen, and Pelvis[1]				
External abdominal oblique	Ribs 5–12	Linea alba, iliac crest, pubis	Compresses abdominal wall, laterally rotates trunk	Intercostal nerves from T7–12
Internal abdominal oblique	Inguinal ligament, iliac crest	Linea alba, ribs 10–12	Compresses abdominal wall, laterally rotates trunk	Intercostal nerves from T7–12 and spinal nerve L1
Transversus abdominis	Inguinal ligament, iliac crest, ribs 7–12	Linea alba, crest of pubis	Compresses abdominal wall, laterally rotates trunk	Intercostal nerves from T7–12 and spinal nerve L1
Rectus abdominis	Crest of pubis, symphysis pubis	Cartilages of ribs 5–7, xiphoid process	Flexes vertebral column, compresses abdominal wall	Intercostal nerves from T6–12
Serratus anterior	Ribs 1–8	Vertebral border and inferior angle of scapula	Abducts scapula, rotates scapula to elevate shoulder	Long thoracic nerve
External intercostals	Inferior border of a rib	Superior border of rib below	Elevates ribs (increases volume in thorax)	Intercostal nerves
Internal intercostals	Inferior border of a rib	Superior border of rib below	Depresses ribs (decreases volume in thorax)	Intercostal nerves
Diaphragm	Xiphoid process, lower ribs, upper lumbar vertebrae	Central tendon	Inspiration (contraction), expiration (relaxation)	Phrenic nerve
Deep Muscles of the Back				
Erector spinae:				
Iliocostalis Longissimus Spinalis	Vertebral column, ilium, ribs	Ribs, vertebral column, occipital and temporal bones	Extends and rotates vertebral column and head	Dorsal rami of spinal nerves
Multifidus	Vertebrae C4–L5, sacrum, ilium	Vertebral column	Extends and rotates vertebral column	Dorsal rami of spinal nerves
Quadratus lumborum	Posterior iliac crest	T12 and L1–4, rib 12	Extends and abducts vertebral column	T12, L1–4
Rhomboideus major	Spines of T2–5	Lower one-third of vertebral border of scapula	Adducts scapula (draws scapulae together)	Dorsal scapular nerve
Rhomboideus minor	Ligamentum nuchae and spines C7–T1	Vertebral border of scapula at scapular spine	Adducts scapula (draws scapulae together)	Dorsal scapular nerve
Splenius	Ligamentum nuchae, C7–T6	C2–4, occipital and temporal bone	Extends and rotates head	Middle and lower cervical nerves
Semispinalis	C7–T12	Occipital bone and T1–4	Extends head and vertebral column, rotates vertebral column	Cervical and thoracic spinal nerves

[1]The trapezius and latissimus dorsi are listed in table 11.1

The intercostal muscles and the diaphragm are respiratory muscles. Normally the diaphragm is responsible for about 60% of the resting breath volume, while the external intercostals contribute to the remaining volume. The **diaphragm** is a domed muscle that has a peripheral origin. The insertion of the diaphragm occurs deep on a sheet of fascia called the central tendon at the base of the mediastinum. It does not insert on a bone. If you think of the diaphragm as a trampoline, the outer springs represent the origin, while the cen-ter (where you jump) represents the insertion. Examine the models in the lab and compare them with the diaphragm in figure 14.2.

The intercostal muscles contribute to the breathing volume at rest, but they also contribute to an increase in the movement of the thorax during times of exercise. The **external intercostal** muscles are involved in inhalation, while the **internal intercostal** muscles are involved in exhalation. Locate the intercostal muscles and compare them with figure 14.1b.

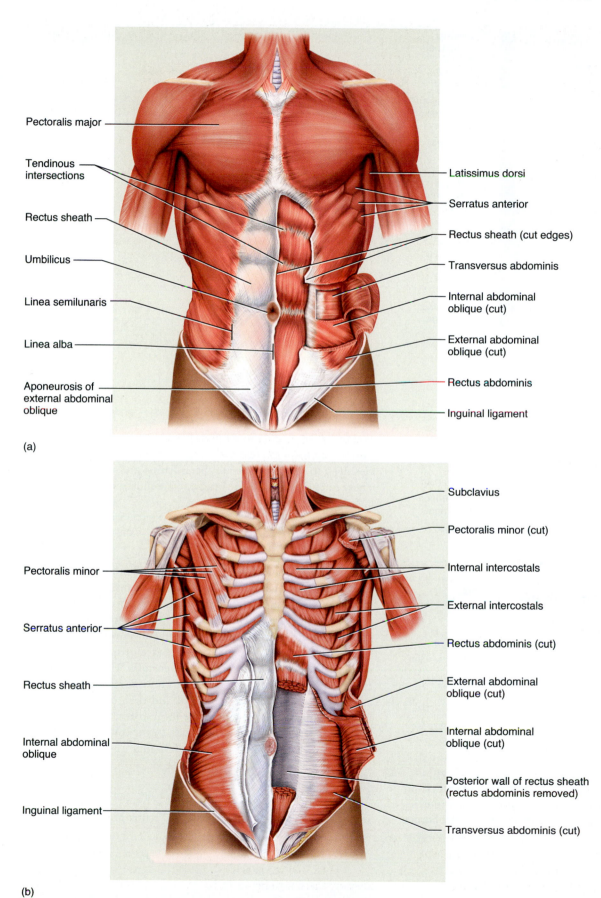

Pectoralis major

Tendinous
intersections

Rectus sheath

Umbilicus

Linea semilunaris

Linea alba

Aponeurosis of
external abdominal
oblique

Latissimus dorsi

Serratus anterior

Rectus sheath (cut edges)

Transversus abdominis

Internal abdominal
oblique (cut)

External abdominal
oblique (cut)

Rectus abdominis

Inguinal ligament

(a)

Pectoralis minor

Serratus anterior

Rectus sheath

Internal abdominal
oblique

Inguinal ligament

Subclavius

Pectoralis minor (cut)

Internal intercostals

External intercostals

Rectus abdominis (cut)

External abdominal
oblique (cut)

Internal abdominal
oblique (cut)

Posterior wall of rectus sheath
(rectus abdominis removed)

Transversus abdominis (cut)

(b)

FIGURE 14.1

Muscles of the Trunk Diagram of anterior view (a) superficial; (b) deep.

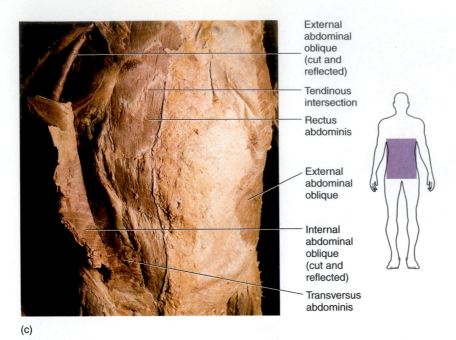

External abdominal oblique (cut and reflected)

Tendinous intersection

Rectus abdominis

External abdominal oblique

Internal abdominal oblique (cut and reflected)

Transversus abdominis

(c)

FIGURE 14.1 *(continued)*
Muscles of the Trunk (c) photograph.

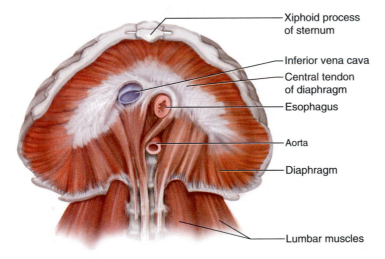

Xiphoid process of sternum

Inferior vena cava

Central tendon of diaphragm

Esophagus

Aorta

Diaphragm

Lumbar muscles

FIGURE 14.2
Diaphragm, Inferior View

Posterior Muscles

The **rhomboideus** muscles are deep to the trapezius and reflection of the trapezius is necessary to see them. The rhomboideus muscles originate on the vertebral column and insert on the medial border of the scapula, and they adduct the scapulae. The **rhomboideus**

major is a larger and more inferior muscle than the **rhomboideus minor.** Compare the material in lab with figure 14.3. Other muscles deep to the trapezius are the **splenius** and the **semispinalis.** These extend and rotate the head and are illustrated in figures 14.3 and 14.4*a*.

Postural Muscles

The postural muscles of the back of the trunk mostly consist of the **erector spinae** muscles. The erector spinae muscles are actually many muscles found between individual vertebrae or between the vertebrae and the ribs. These separate muscles are grouped conveniently into long strap muscles known collectively as the erector spinae. There are three major groups of erector spinae muscles, the **spinalis,** the **longissimus,** and the **iliocostalis.** The **multifidus** is a closely associated muscle (they can be known as the *s.l.i.m.* muscles as you move from medial to lateral and then inferior). Another muscle in the area is the **quadratus lumborum,** which is a square muscle that runs from the iliac crest to the lower vertebrae and twelfth rib. These muscles can be seen in figure 14.4.

Cat Anatomy

If you are using cats, turn to section 5, "Trunk Muscles of the Cat," on page 425 of this lab manual.

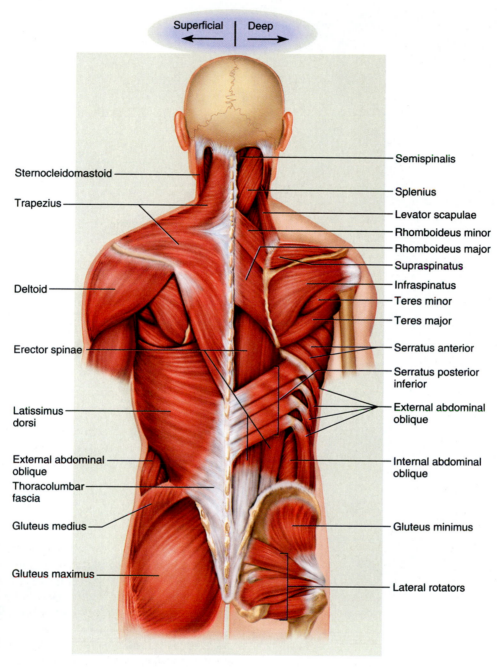

Superficial | Deep

Sternocleidomastoid

Trapezius

Deltoid

Erector spinae

Latissimus dorsi

External abdominal oblique

Thoracolumbar fascia

Gluteus medius

Gluteus maximus

Semispinalis

Splenius

Levator scapulae

Rhomboideus minor

Rhomboideus major

Supraspinatus

Infraspinatus

Teres minor

Teres major

Serratus anterior

Serratus posterior inferior

External abdominal oblique

Internal abdominal oblique

Gluteus minimus

Lateral rotators

FIGURE 14.3

Muscles of the Back, Posterior View

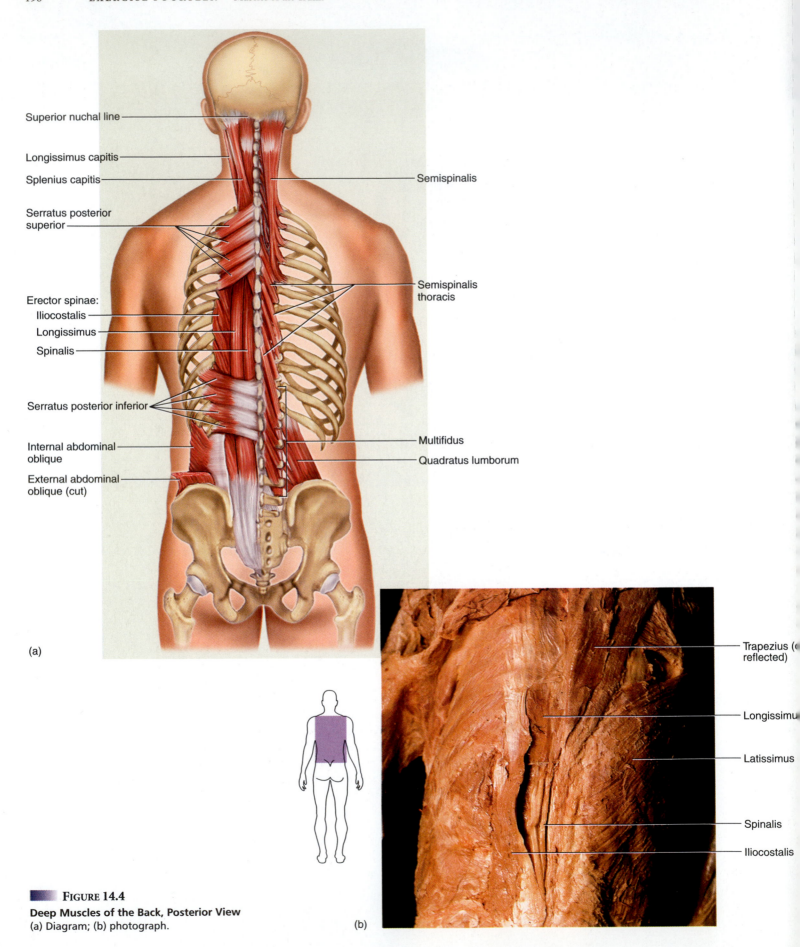

Superior nuchal line

Longissimus capitis

Splenius capitis

Serratus posterior superior

Erector spinae:
  Iliocostalis
  Longissimus
  Spinalis

Serratus posterior inferior

Internal abdominal oblique

External abdominal oblique (cut)

Semispinalis

Semispinalis thoracis

Multifidus

Quadratus lumborum

(a)

(b)

Trapezius (reflected)

Longissimu

Latissimus

Spinalis

Iliocostalis

FIGURE 14.4

Deep Muscles of the Back, Posterior View
(a) Diagram; (b) photograph.

Name_____ Date _____

1. Which is the more superior muscle, the rhomboideus major or the rhomboideus minor?

2. What is the action of the serratus anterior muscle?

3. What is the action of the rhomboideus muscles?

4. How does the serratus anterior function as an antagonist to the rhomboideus muscles?

5. How does the action of the rectus abdominis differ from that of the other abdominal muscles?

6. What is the physical relationship of the intercostal muscles to each other?

7. Compression of the abdominal wall occurs by what four muscles?

8. Extension and rotation of the vertebral column occur by what group of muscles?

9. What muscle is responsible for most of the air inhaled during relaxed breathing?

10. What is the action of the intercostal muscles?

11. What muscle inserts on the central tendon?

12. Abduction of the scapula occurs by what muscle?

13. The tendinous intersections are found in what muscle?

14. Flexion of the vertebral column occurs by what abdominal muscle?

15. Which is the deepest abdominal muscle?

16. What is the origin of the rhomboideus major muscle?

17. Label the muscles in the following illustration using the terms provided.

    rectus abdominis   transversus abdominis   external abdominal oblique

    serratus anterior   internal abdominal oblique   external intercostal

    internal intercostal

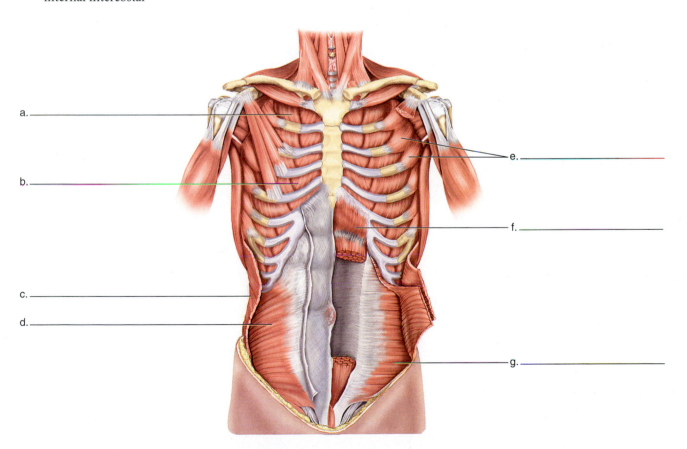

Introduction to the Nervous System

Nervous System

INTRODUCTION

The function of the nervous system, among other things, is **communication** between the various regions of the body, **coordination** of body functions (as in digestion or walking), **orientation** to the environment, and **assimilation** of information. In this exercise you learn about the basic structure of the nervous system and the cells that are part of the nervous system. The functional unit of the nervous system is the neuron. It is the cell that carries out the activity of nervous tissue. Neurons comprise the nerves of the body, the spinal cord, ganglia, and the brain. The processes of neurons form nerves. Neuroglia, or glial cells, make up about 50% of the volume of nervous tissue. They aid the neurons in terms of increasing the speed of neuron transmission, providing nutrients to the neurons, and protecting the neurons.

LEARNING OBJECTIVES

At the end of this exercise you should be able to
1. describe the three parts of the neuron;
2. list the main divisions of the nervous system;
3. group the organs of the nervous system into the main divisions;
4. describe the functions of the various neuroglia.

MATERIALS

Charts or models of the nervous system
Charts or models of neurons
Microscopes
Prepared slides:
    Spinal cord smear
    Longitudinal section of a nerve
    Neuroglia

PROCEDURE

Divisions

The nervous system can be divided into two anatomical subdivisions, the **central nervous system (CNS)** and the **peripheral nervous system (PNS).** The CNS consists of the brain and the spinal cord, and the **peripheral nervous system (PNS)** is composed of all the nervous tissue superficial to the brain and spinal cord. The PNS consists of nerves and ganglia. Ganglia are clusters of nerve cell bodies outside the CNS. Nerves are bundles of nerve fibers in the PNS. The peripheral nervous system has an afferent, or sensory, division and an efferent, or motor, division. The **afferent division** conducts impulses *from* the regions of the body *to* the central nervous system. Sensations, such as touch or sight, travel to the CNS by the afferent division. The **efferent division** conducts impulses *from* the central nervous system *to* the various regions of the body. The movement of muscles occurs when **somatic motor nerves** take impulses from the CNS to the skeletal muscles of the body. Other bodily functions are controlled by **visceral motor nerves**. These are part of the **autonomic nervous system,** and they regulate functions that occur automatically, such as pupil constriction when you move from a shady area to a bright one.

    Visceral motor nerves carry information to the thoracic and abdominal regions. The **visceral motor division** sends impulses to glands, smooth muscle, and cardiac muscle. It can be divided into a **sympathetic division,** which prepares the body for active responses, and a **parasympathetic division,** which slows the body down or allows certain functions, such as kidney function and digestion, to take place. The visceral motor division functions independently and provides automatic controls for activities normally under subconscious direction.

    Somatic nerves receive sensory information from the skin, muscles, joints, and bones and take motor information to skeletal muscles. The term "somatic nerves" generally refers to the nerves in the trunk and extremities. Nerves that are associated with the brain are called **cranial nerves**. They may be sensory, motor, or both.

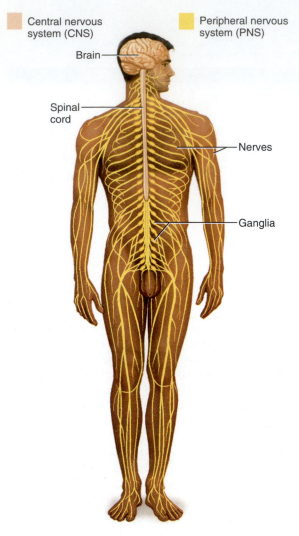

Central nervous
system (CNS)

Peripheral nervous
system (PNS)

Brain

Spinal
cord

Nerves

Ganglia

FIGURE 15.1
Divisions of the Nervous System

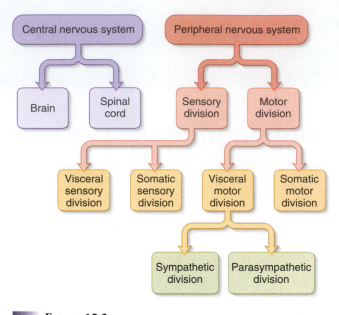

FIGURE 15.2
Schematic Representation of the Nervous System

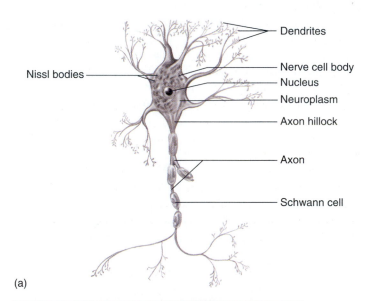

(a)

Dendrites

Nissl bodies

Nerve cell body

Nucleus

Neuroplasm

Axon hillock

Axon

Schwann cell

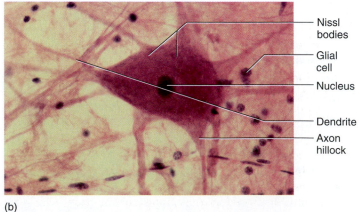

(b)

Nissl bodies

Glial cell

Nucleus

Dendrite

Axon hillock

FIGURE 15.3
Parts of a Multipolar Neuron (a) Diagram; (b) photomicrograph.

Examine the models or charts in the lab for the central and peripheral divisions of the nervous system. Compare the material in the lab with figures 15.1 and 15.2.

Histology of the Neuron

The **neuron,** or **nerve cell,** is a remarkable cell not only for its functional nature but also for the anatomical extremes it exhibits. The nerves in the thigh and leg are composed of many neuron fibers. A neuron in the toe continues as an incredibly long, single cell up the leg and thigh to synapse (join) with other neurons in the lower back. When you look at prepared slides of neurons in the microscope in this exercise, remember that some of these neurons are of great length.

Neurons consist of three main parts: the **dendrite;** the **nerve cell body,** or **soma;** and the **axon.** Neurons communicate with one another at junctions called **synapses.** Examine figure 15.3*a* and models or charts in the lab for the structure of neurons. Dendrites are so named because they have branching structures that resemble a tree (*dendros* = tree). Nerve cell bodies consist of the **neuroplasm** (cytoplasm of the neuron), which contain **Nissl bodies,** or **chromatophilic substances** (rough endoplasmic reticulum of the neuron), and many organelles, including the **nucleus**. The triangular region of the nerve cell body that is devoid of Nissl bodies is the **axon hillock**, and it leads to the axon that exits the nerve cell body. Axons consist of long strands of neurofibrils. In both the PNS and CNS, they may be wrapped in myelin sheaths. These sheaths are discussed later in the exercise.

Examine a prepared slide of a spinal cord smear under the microscope and locate the purple, star-shaped structures under low power. These are the nerve cell bodies of multipolar neurons. Switch to high power and locate the darker-stained Nissl bodies, the nucleus, and the clear, triangular axon hillock. If you find the axon hillock, you should be able to see the axon leading away from this structure. All of the other processes that are attached to the nerve cell body are dendrites. The small nuclei that are scattered throughout the smear belong to glial cells in the spinal cord. Compare your slide with figure 15.3*b*.

Functions of Neurons

There are three types of neurons based on function. **Afferent,** or **sensory, neurons** conduct impulses *to* the central nervous system. They convey information from receptors in the external or internal body environment *to* the spinal cord and/or the brain. **Efferent,** or **motor, neurons** conduct impulses *away from* the central nervous system to organs or glands that carry out an activity. **Association neurons,** or **interneurons,** are located *between* sensory and motor neurons, transmitting information to the brain for processing. The interplay between neurons constitutes **reflex arcs**. When there are only two neurons in an arc (an afferent and an efferent), this is known as a monsynaptic reflex. If there is one or more association neurons in the arc, then it is known as a polysynaptic reflex arc. Examine the different types of neurons in a polysynaptic reflex arc in figure 15.4.

Neuron Shapes

Neurons can be classified according to shape. A **multipolar neuron** consists of several dendritic processes, a single nerve cell body, and a single axon. The majority of the neurons of the body (such as those in the brain and spinal cord) are multipolar neurons. Compare the material in the lab with figure 15.5. **Bipolar neurons** are so named because the nerve cell body has two poles. Dendrites receive information and conduct it to one pole of the nerve cell body. At the other pole, an axon leaves the nerve cell body and transmits the impulse away from the cell body. Bipolar neurons are found in the nose, eye, and inner ear.

Unipolar, or **pseudounipolar, neurons** are cells with just one process attaching to the nerve cell body. Dendrites receive the stimulus and conduct the impulse to the axon, which bypasses the nerve cell body, taking the impulse to the spinal cord. Most of the sensory nerves of the body are composed of unipolar neurons.

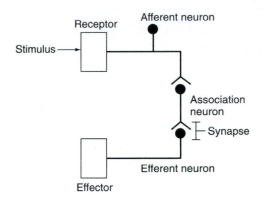

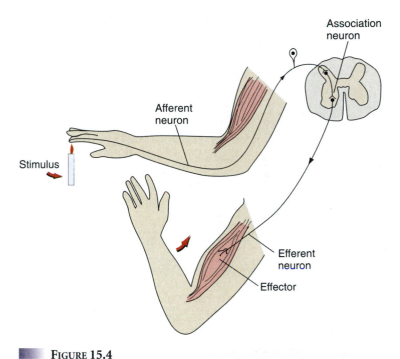

FIGURE 15.4
Afferent, Efferent, and Association Neurons in a Polysynaptic Reflex Arc

Specialized Neurons

The cerebral cortex of the brain contains **pyramidal cells,** which have extensive dendritic branches. Examine a prepared slide of the cerebrum and locate the pyramidal cells. Compare them with figure 15.6.

In the cerebellum, **Purkinje cells** are common. They are large cells in the **gray matter** of the cerebellum. Examine a prepared slide of the cerebellum and compare it with figure 15.6.

Neuroglia

Numerous cells aid the functioning of the neuron. These cells are called **neuroglia** (*neuro* = nerve, *glia* = glue), or **glial cells**. The common glial cell of the peripheral nervous system is the **Schwann cell,** or **neurolemmocyte**. These glial cells wrap around the axon, much as a thin strip of paper can be wrapped around a pencil,

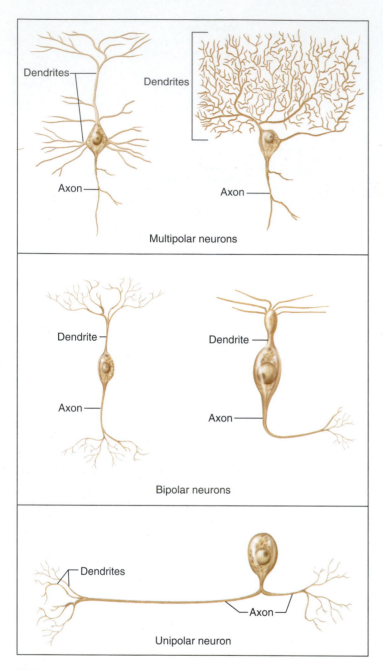

FIGURE 15.5
Neuron Shapes

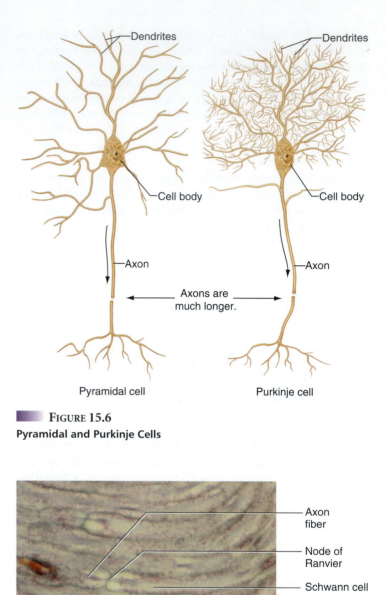

FIGURE 15.6
Pyramidal and Purkinje Cells

FIGURE 15.7
Nerve, Longitudinal Section (400x)

leaving small gaps between successive cells called the **nodes of Ranvier.** The gaps allow for the nerve transmission to jump from node to node, increasing the transmission speed of the neuron. The Schwann cell consists of a significant amount of a lipoprotein material called **myelin**, and the series of Schwann cells produces a **myelin sheath. Myelinated nerve fibers** appear white. Small axons that are not enclosed by myelin sheaths are called unmyelinated fibers. **Nerve cell bodies and unmyelinated fibers** form the portion of the nervous tissue known as gray matter.

Examine a prepared slide of a longitudinal section of nerve under high power. You should be able to see the axon fibers as long,

dark threads in the microscope. What appear to be clear areas on each side of the axon fibers is the myelin sheath. If you scan the slide closely, you should be able to see the junction of two Schwann cells and the node of Ranvier between them. Compare your slide with figure 15.7.

Schwann cells are located in the PNS (fig. 15.8). Other types of neuroglia produce myelin in the CNS. These are known as **oligodendrocytes**. Unlike Schwann cells, oligodendrocytes frequently wrap around several neurons. The white matter of the spinal cord and brain is white due to the lipoprotein sheaths of the oligodendrocytes.

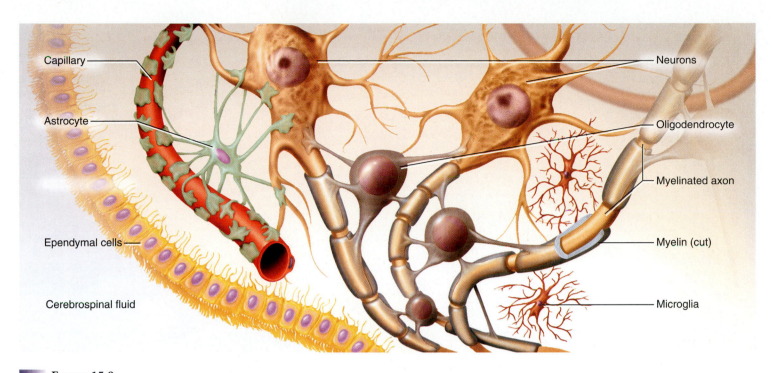

Capillary

Astrocyte

Ependymal cells

Cerebrospinal fluid

Neurons

Oligodendrocyte

Myelinated axon

Myelin (cut)

Microglia

FIGURE 15.8
Neuroglia of the Central Nervous System

Astrocytes are branched glial cells that provide a barrier between the nervous tissue and the blood. Astrocytes, along with capillary endothelial cells, are responsible for the **blood-brain barrier** that protects the nervous tissue from some bloodborne infections. They also inhibit some medications from reaching the brain.

Microglia are small glial cells that are phagocytic—they digest foreign particles that invade the nervous tissue and remove dead or damaged neurons. Examine figure 15.8 for microglial cells.

The last of the glial cells covered in this exercise are the **ependymal cells**. These cells line the spaces of the brain and spinal cord and secrete and circulate cerebrospinal fluid.

Name _____ Date _____

1. Draw a neuron in the space provided and label the axon, dendrite, nerve cell body, Nissl bodies, and axon hillock.

2. Draw a unipolar neuron in the space provided and label the parts.

3. Describe the function of an astrocyte.

4. Describe the function of an ependymal cell.

5. Describe the function of an oligodendrocyte.

6. The brain belongs to what division of the nervous system?

7. A spinal nerve belongs to what division of the nervous system?

8. What adaptive value do you see for long neurons, such as those in the lower extremity, to have myelin sheaths?

9. What does CNS stand for?

10. What kind of cell performs the main function of the nervous system?

11. In what part of the neuron is the nucleus found?

12. What is another name for an efferent neuron?

13. A neuron that has a soma with a dendrite on one side and an axon on the other is what kind of neuron?

14. Two adjacent neurons communicate with one another across a space called a(n)_____.

15. What name is given to cells that support a neuron?

16. Are Schwann cells found in the CNS or the PNS?

17. Myelin is made of what kind of material?

18 Myelin does not conduct impulses along the nerve fiber, but the node of Ranvier does. How does this arrangement increase the speed of transmission?

Brain and Cranial Nerves

INTRODUCTION

Two specific traits distinguish humans from other animals. One is our upright posture, and the other is the extensive development of the brain. In this exercise you examine the anatomy of the human brain and the cranial nerves. The two are studied in lab together because of their close proximity to each other. Sheep brains are commonly used as dissection specimens because of their cost, availability, and similarity in structure to human brains. There are a few anatomic differences in the sheep brain, but these will be described in the exercise.

LEARNING OBJECTIVES

By the end of this exercise you should be able to
1. name the three meninges of the brain and their location relative to one another;
2. locate the three major regions of the brain;
3. name the main structures in each of the three regions of the brain;
4. describe the function of specific areas of the brain, such as the thalamus, Broca area, the occipital lobe, the temporal lobe, the cerebellum, and the medulla oblongata;
5. trace the path of cerebrospinal fluid through the brain;
6. identify each of the 12 pairs of cranial nerves on an illustration, a model, or a real brain;
7. describe whether a cranial nerve is sensory, motor, or both;
8. identify the structure that a particular cranial nerve innervates.

MATERIALS

Models and charts of the human brain

Preserved human brains (if available)

Cast of the ventricles of the brain

Sheep brains

Dissection trays

Scalpels or razor blades

Gloves (household latex gloves work well for repeated use)

Blunt (Mall) probe

First aid kit in lab or prep area

Sharps container

Animal waste disposal container

PROCEDURE

Overview of the Brain

The brain is located in the cranial cavity of the skull and weighs approximately 1.4 kilograms (3 pounds). The brain is derived from three embryonic regions, each of which further develops into more specific areas. The embryonic brain consists of the **forebrain**, or **prosencephalon** (PROZ-en-SEF-uh-lon); the **midbrain**, or **mesencephalon** (MES-en-SEF-uh-lon); and the **hindbrain**, or **rhombencephalon** (ROMB-en-SEF-uh-lon). These regions are illustrated in figure 16.1. Examine table 16.1 for an overview of specific

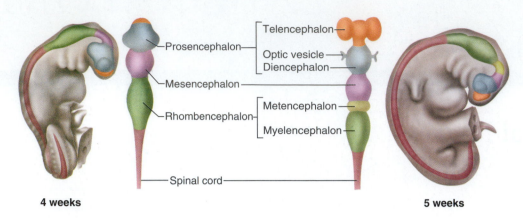

■ **FIGURE 16.1**
Brain Development

TABLE 16.1
Regions of the Brain

Prosencephalon

Telencephalon

   Cerebrum (cerebral hemispheres)

      Cerebral cortex (gray matter)

      Basal nuclei (gray matter)

   Corpus callosum

Diencephalon

   Pineal body

   Thalamus

   Hypothalamus

   Pituitary gland

   Mammillary bodies

Mesencephalon

Peduncles

Tectum

Corpora Quadrigemina

   Superior colliculus

   Inferior colliculus

Rhombencephalon

Metencephalon

   Pons

   Cerebellum

Myelencephalon

   Medulla oblongata

areas of the brain and the regions to which they belong. Figure 16.2 is an illustration of the major areas in an adult brain. Compare this illustration with a model of the brain in lab and locate the **forebrain**. It is the largest region of the brain, consisting primarily of the **cerebrum** (seh-REE-brum) and the **diencephalon** (DY-en-SEF-uh-lon). The **midbrain** is the smallest region of the brain, located between the forebrain and the hindbrain. The **hindbrain** is composed of the **pons**, the **medulla oblongata** (meh-DULL-ah OB-long-GAH-ta), and the **cerebellum** (SER-eh-BEL-um). The hindbrain is the most inferior portion of the brain; it connects to the spinal cord at the foramen magnum.

Meninges

Three membranes surround the brain. These are called the meninges. The outermost membrane is the **dura mater** (DOO-rah MAH-tur), a tough dense connective tissue sheath. The dura resembles a swim cap but is tough and tendinous. The dura mater is divided into an outer, **periosteal layer** and an inner, **meningeal layer**. The dural sinus is a space between these two layers. Deep to the dura mater is the **arachnoid mater**, which is a soft, thin membrane that resembles a spider's web. Between the dura mater and the arachnoid membrane is the **subdural space**. Blood vessels are in the subdural space. Deep to the arachnoid is the **subarachnoid space**, which contains **cerebrospinal fluid (CSF)**. The function of the cerebrospinal fluid is to keep the brain buoyant in the skull, cushion the brain, and remove waste from the central nervous system (CNS). CSF is produced in the subarachnoid space, the ependymal cells, and the choroid plexuses. There are approximately 150 ml of CSF in the central nervous system, and it takes about 6 hours to circulate through the system. The deepest layer is the **pia mater**, which is a membrane that adheres directly to the outer surface of the brain and appears like the surface of the brain. Locate the meninges in preserved brains in the lab (if available) and compare them with figure 16.3.

Ventricles of the Brain

Part of the hollow neural tube that develops in the first trimester of pregnancy becomes the ventricles of the brain in the adult. The two ventricles that occupy the center of each cerebral hemisphere are known as the **lateral ventricles**. These ventricles receive cerebrospinal fluid from tufts of capillaries called **choroid plexuses**. You

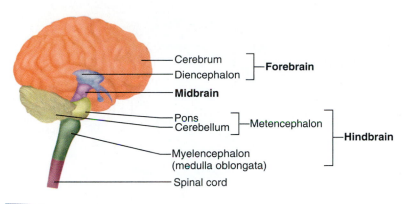

FIGURE 16.2
Major Regions of the Brain

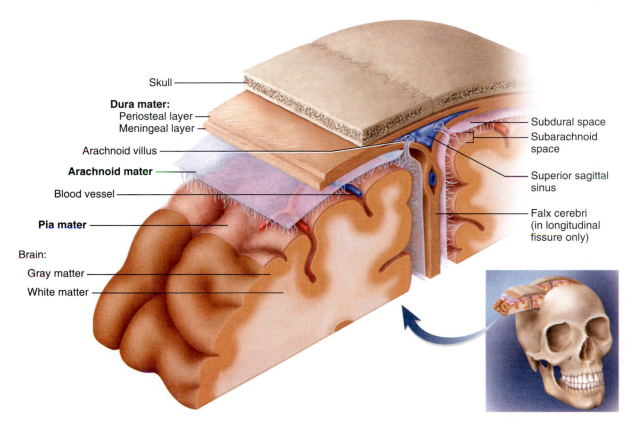

FIGURE 16.3
Meninges of the Brain

can see the choroid plexuses in preserved brains as small, brown areas in the ventricles. Fluid from the lateral ventricles flows through the **interventricular foramina** (**foramina of Monro**) and into the **third ventricle**. The third ventricle lies between the walls of the thalamus and receives CSF from choroid plexuses in that area. The third ventricle drains into the **fourth ventricle** by way of the **cerebral** (**mesencephalic) aqueduct**, which is also known as the **aqueduct of Sylvius**. If this duct becomes occluded, then CSF accumulates in the lateral and third ventricles, causing a condition known as **hydrocephaly**. Locate the ventricles of the brain in figure 16.4.

Normally the cerebral aqueduct is open and passes through the midbrain. Posterior to the cerebral aqueduct is the fourth ventricle, which occupies a space inferior to the cerebellum. The fourth ventricle also has a choroid plexus that secretes CSF. Cerebrospinal fluid flows from the fourth ventricle into the central canal of the spinal cord and the subarachnoid space of the spinal cord and brain. CSF is absorbed by the arachnoid villi under the dura mater of the skull. The villi penetrate the dural sinuses, and the CSF flows to the venous sinuses and returns to the cardiovascular system by the **internal jugular veins** (fig. 16.5).

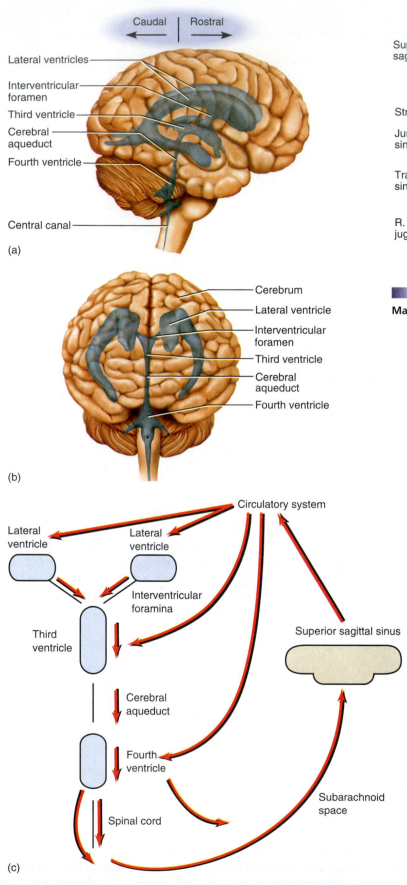

Caudal | Rostral

Lateral ventricles
Interventricular foramen
Third ventricle
Cerebral aqueduct
Fourth ventricle

Central canal

(a)

Cerebrum
Lateral ventricle
Interventricular foramen
Third ventricle
Cerebral aqueduct
Fourth ventricle

(b)

Circulatory system

Lateral ventricle
Lateral ventricle

Interventricular foramina

Third ventricle

Cerebral aqueduct

Superior sagittal sinus

Fourth ventricle

Spinal cord

Subarachnoid space

(c)

FIGURE 16.4

Ventricles of the Brain (a) Lateral view; (b) anterior view; (c) schematic presentation of the flow of cerebrospinal fluid.

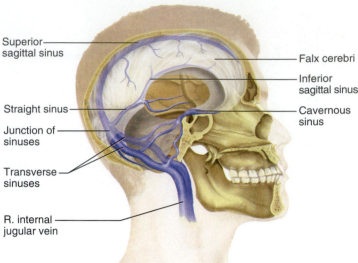

Superior sagittal sinus
Straight sinus
Junction of sinuses
Transverse sinuses
R. internal jugular vein
Falx cerebri
Inferior sagittal sinus
Cavernous sinus

FIGURE 16.5

Major Drainage of the Brain

Blood Supply to the Brain

The blood vessels that supply nutrients and oxygen to the brain, and generate CSF, do not actually penetrate the brain tissue itself but, rather, form a meshwork around the brain. The brain receives blood from four major arteries. These are the **left** and **right vertebral arteries** and the **left** and **right internal carotid arteries**.

The vertebral arteries pass through the foramen magnum and unite to form the **basilar artery** at the base of the brain (fig. 16.6). The two internal carotid arteries pass through the carotid canals and take blood to the base of the brain. The blood from the basilar and the two internal carotid arteries anastomose (join) in a vessel called the **cerebral arterial circle (circle of Willis)**. The cerebral arterial circle forms a loop around the pituitary and is completed by the **anterior** and **posterior communicating arteries.** This union of arteries is called **collateral circulation**, which is important in that congestion of one artery, which might starve the brain of oxygen or nutrients, can be relieved by blood from other collateral arteries. The major arteries that branch off the cerebral arterial circle and take blood directly to the brain are the **anterior**, **middle**, and **posterior cerebral arteries. The cerebellar arteries,** including the **superior artery,** the **anterior inferior artery**, and the **posterior inferior artery** take blood to the cerebellum.

Surface View of the Brain

Examine a model or chart of the brain. Of the major regions of the brain you will be able to easily see the forebrain and the hindbrain. In the forebrain you should examine the large **cerebrum**, which can be seen with folds and ridges known as **convolutions**. The ridges of the convolutions are called **gyri** (singular, **gyrus**) and the depressions are either **sulci** (singu-

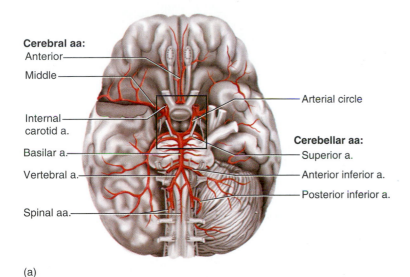

Cerebral aa:
Anterior
Middle
Internal carotid a.
Basilar a.
Vertebral a.
Spinal aa.

Arterial circle

Cerebellar aa:
Superior a.
Anterior inferior a.
Posterior inferior a.

(a)

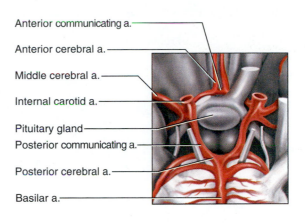

Anterior communicating a.
Anterior cerebral a.
Middle cerebral a.
Internal carotid a.
Pituitary gland
Posterior communicating a.
Posterior cerebral a.
Basilar a.

(b)

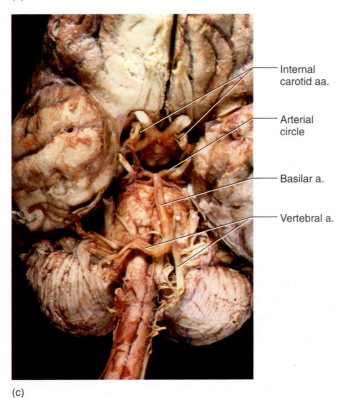

Internal carotid aa.
Arterial circle
Basilar a.
Vertebral a.

(c)

FIGURE 16.6

Brain with Arteries, Inferior View (a) Overview of arterial supply to brain; (b) close-up of arterial circle; (c) photograph. In this illustration a = artery and aa = arteries.

lar, **sulcus**) or **fissures**. Fissures are deeper than sulci. The lateral view of the brain allows you to see the major lobes of each cerebral hemisphere. These lobes are named for the bones of the skull under which they lie. They are the **frontal, parietal, occipital,** and **temporal lobes.** The **lateral fissure** separates the temporal lobe from the frontal and parietal lobes of the brain. Deep to the lateral fissure is a small region of the brain known as the **insula.** These regions can be seen in figure 16.7.

FRONTAL LOBE

The **frontal lobe** is responsible for many of the higher functions associated with being human. Humans have a six-layered neocortex, which is characteristic of primates. The frontal lobe is involved in intellect, abstract reasoning, creativity, social awareness, and language. An important area responsible for controlling the formation of speech is called the **Broca area**, or the **motor speech area.** This area controls the muscles involved in speech. It is located in the lateral aspect of the left frontal lobe. Locate the frontal lobe in figure 16.7. The posterior border of the frontal lobe is defined by the **central sulcus.**

To find the central sulcus, look for two convolutions that run from the superior portion of the cerebrum to the lateral fissure, more or less continuously. The gyrus anterior to the central sulcus is part of the frontal lobe and is known as the **precentral gyrus,** or the **primary motor cortex.** This cortex is important for directing a part of the body to move and has been mapped, as in figure 16.8*a*. An image of the human body constructed on the brain is known as a homunculus. The **motor homunculus** is seen in figure 16.8*a*.

PARIETAL LOBE

The gyrus posterior to the central sulcus is known as the **postcentral gyrus,** or the **primary somatic sensory cortex.** This receives sensory input. The primary somatic sensory cortex pinpoints the part of the body affected, and the association area interprets the sensation (pain, heat, cold, and so on). Locate the association areas of the parietal lobe in figure 16.8*b*. The primary somatic sensory cortex receives initial sensory input, which is integrated in sensory **association areas** just posterior. This area has also been mapped, and you can see the **sensory homunculus** represented in figure 16.8*b*.

OCCIPITAL LOBE

Posterior to the parietal lobe is the **occipital lobe** of the cerebrum. The occipital lobe is considered the **visual area** of the brain, and damage to this lobe can cause blindness. The shape, color, and distance of

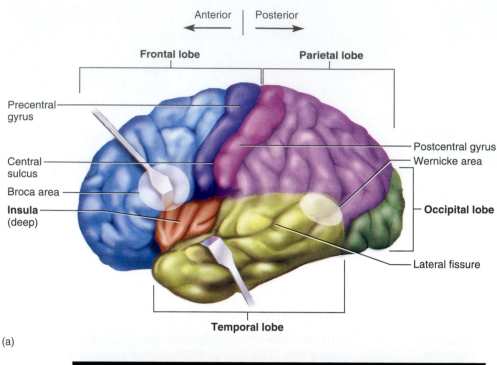

Anterior | Posterior

Frontal lobe   **Parietal lobe**

Precentral gyrus

Central sulcus

Broca area

Insula (deep)

Postcentral gyrus

Wernicke area

Occipital lobe

Lateral fissure

Temporal lobe

(a)

Precentral gyrus

Central sulcus

Frontal lobe

Broca area

Lateral fissure

Temporal lobe

Postcentral gyrus

Parietal lobe

Wernicke area

Occipital lobe

Transverse fissure

(b)

FIGURE 16.7

Brain, Lateral View (a) Diagram; (b) photograph.

objects are perceived here, as is the recollection of past visual images. As you are reading these words, your occipital lobe is receiving the information and transferring it to other regions, which are converting the words to thought. Between the occipital lobe and the cerebellum is a **transverse fissure,** which separates these two regions of the brain. Locate the occipital lobe and the cerebellum in figure 16.7.

TEMPORAL LOBE

The **temporal lobe** is separated from the frontal and parietal lobes by the **lateral fissure,** as seen in figure 16.7. The temporal lobe contains an area, known as the **primary auditory cortex,** that interprets sound impulses sent from the inner ear. The auditory cortex distinguishes the nature of the sound (music, noise, speech), as well

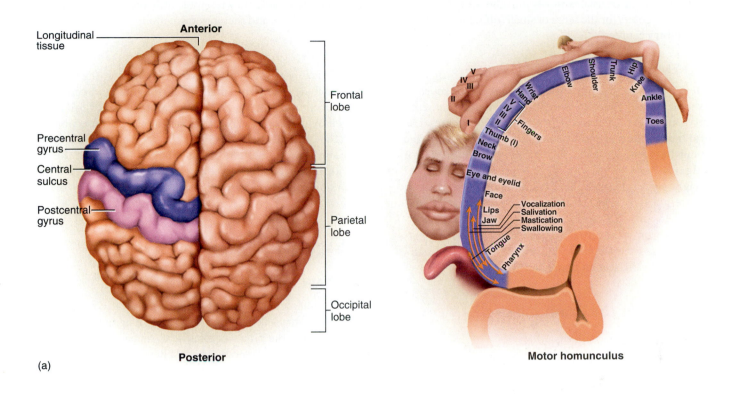

(a)

Labels (top figure a): Anterior; Longitudinal tissue; Precentral gyrus; Central sulcus; Postcentral gyrus; Frontal lobe; Parietal lobe; Occipital lobe; Posterior

Motor homunculus labels: Elbow; Shoulder; Trunk; Hip; Knee; Ankle; Toes; Wrist; Hand; Fingers; Thumb (I); Neck; Brow; Eye and eyelid; Face; Lips; Jaw; Tongue; Pharynx; Vocalization; Salivation; Mastication; Swallowing; II III IV V; I

Motor homunculus

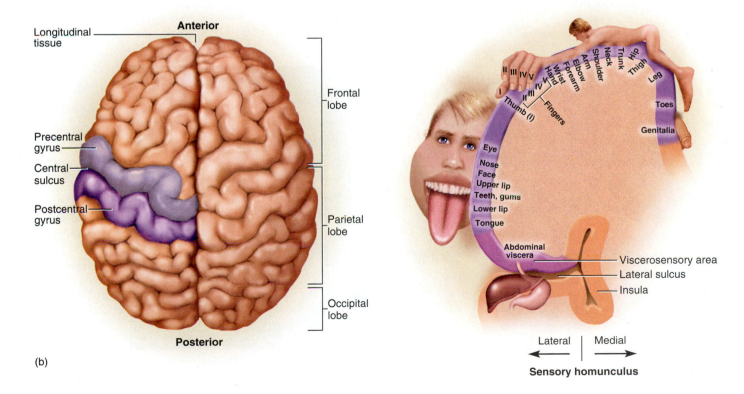

(b)

Labels (bottom figure b): Anterior; Longitudinal tissue; Precentral gyrus; Central sulcus; Postcentral gyrus; Frontal lobe; Parietal lobe; Occipital lobe; Posterior

Sensory homunculus labels: Arm; Elbow; Shoulder; Neck; Trunk; Hip; Thigh; Leg; Toes; Genitalia; Forearm; Wrist; Hand; Fingers; Thumb (I); Eye; Nose; Face; Upper lip; Teeth, gums; Lower lip; Tongue; Abdominal viscera; Viscerosensory area; Lateral sulcus; Insula; II III IV V; I

Lateral ← | → Medial

Sensory homunculus

FIGURE 16.8
Primary Motor and Somatic Sensory Cortex (a) Motor cortex (precentral gyrus); (b) somatic sensory cortex (postcentral gyrus).

as the location, distance, pitch, and rhythm. Wernicke area in the primary auditory cortex translates words into thought. The temporal lobe also has centers for the sense of smell (**olfactory centers**) and taste (**gustatory centers**).

CEREBRAL HEMISPHERES

If you rotate the brain so that you are looking at it from a superior view, you should be able to see the **longitudinal fissure** that separates the cerebrum into the left and right cerebral hemispheres. The **left cerebral hemisphere** in most people is involved in language and reasoning. In most people, the Broca area is on the left side of the brain (figs. 16.7 and 16.9).

The **right cerebral hemisphere** of the brain is involved in space and pattern perceptions, artistic awareness, imagination, and music comprehension. This specialization, in which one hemisphere of the cerebrum is involved in a particular task, is known as cerebral asymmetry, or hemispheric dominance. Examine the surface features of the brain, seen in figure 16.9.

Inferior Aspect of the Brain

FOREBRAIN

If you examine the inferior aspect of the brain (fig. 16.10), you can see the frontal lobes of the cerebrum, as well as the temporal lobes. You may be able to see the **pituitary gland** if it has not been removed. The **optic chiasma** (*chiasma* = cross) is anterior to the pituitary and transmits visual impulses from the optic nerves to the brain. Two small processes posterior to the pituitary are the **mammillary bodies** (so named because they resemble little breasts), which function in olfactory reflexes and memory formation.

HINDBRAIN

The inferior view of the brain provides a look at the hindbrain, which includes the **pons, medulla oblongata,** and **cerebellum.** The cerebellum has fine folds of neural tissue called **folia,** which are seen in an inferior view of the brain. The medulla oblongata is located anterior and inferior to the cerebellum and connects the brain to the spinal cord. The enlarged portion of the brain superior to the medulla is the pons, which serves as a relay center for information. Examine these structures of the brain in figure 16.10.

Midsagittal Section of the Brain

FOREBRAIN

Examine a midsagittal section of a brain, as illustrated in figure 16.11. This section is made by placing a knife in the longitudinal fissure and cutting through the brain. Locate the C-shaped **corpus callosum**, which connects the two cerebral hemispheres. The posterior portion of the corpus callosum is known as the **splenium**, and the anterior portion is the **genu**. Just inferior to the corpus callosum is the **septum pellucidum**, a membrane that separates the lateral ventricles from each other. By removing the septum pellucidum, you will be able to look into the lateral ventricle without obstruction.

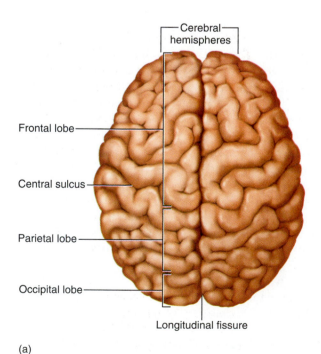

(a)

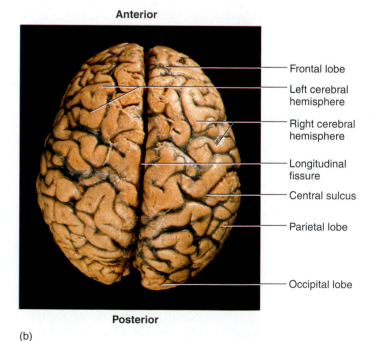

Anterior

Frontal lobe

Left cerebral hemisphere

Right cerebral hemisphere

Longitudinal fissure

Central sulcus

Parietal lobe

Occipital lobe

Posterior

(b)

FIGURE 16.9

Brain, Superior View (a) Diagram; (b) photograph.

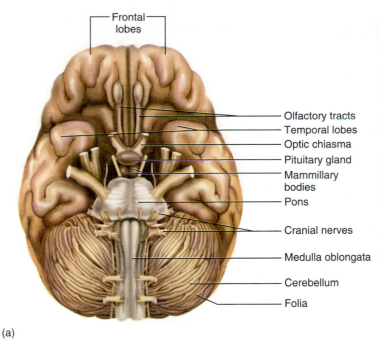

Frontal lobes

Olfactory tracts
Temporal lobes
Optic chiasma
Pituitary gland
Mammillary bodies
Pons
Cranial nerves
Medulla oblongata
Cerebellum
Folia

(a)

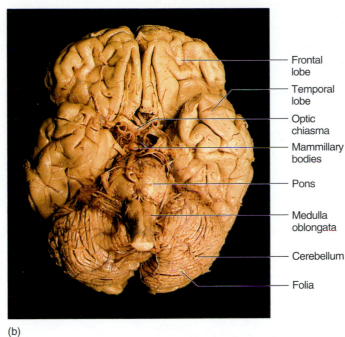

Frontal lobe
Temporal lobe
Optic chiasma
Mammillary bodies
Pons
Medulla oblongata
Cerebellum
Folia

(b)

FIGURE 16.10

Brain, Inferior View (a) Diagram; (b) photograph.

Examine a sectioned brain in the lab and locate the **diencephalon,** which partly consists of the thalamus and the hypothalamus. The **thalamus** forms the lateral wall around the third ventricle; it is a relay center that receives information from various tracts in the CNS and sends them to the cerebral cortex.

Below the thalamus is the **hypothalamus,** which has numerous autonomic centers. The hypothalamus, in part, directs the autonomic nervous system (ANS) and is involved with the **pituitary gland** (in the hypothalamopituitary axis) in many endocrine functions. Centers for thirst, water balance, pleasure, rage, sexual desire, hunger, sleep patterns, temperature, homeostasis, and aggression are located in the hypothalamus.

You should also be able to locate the **mammillary bodies** on the inferior portion of the diencephalon and the **optic chiasma** just anterior to it. The **pineal gland** is located posterior to the thalamus and is an endocrine gland that secretes melatonin. Both the pineal gland and the pituitary gland are covered in more detail in laboratory exercise 19.

MIDBRAIN

The midbrain is a small area inferior and posterior to the diencephalon and is best seen in a midsagittal section. This small area consists of the **cerebral peduncles,** which occupy an area superior to the pons and on the anterior surface of the brain. The **cerebral aqueduct** passes through the midbrain, with the peduncles anterior to, and the **tectum** posterior to, the aqueduct. Locate these features in the material in the lab and in figure 16.11. The tectum consists of four hemispheric processes known as the **corpora quadrigemina,** which consist of the **superior colliculi** (areas of visual reflexes)

and the **inferior colliculi** (areas of auditory reflexes). The midbrain also houses a center known as the **substantia nigra** (not seen in midsagittal sections), which, when not functioning properly, causes Parkinson's disease.

HINDBRAIN

The hindbrain consists of an anterior bulge known as the **pons,** a terminal **medulla oblongata,** and the highly convoluted **cerebellum** (fig. 16.11). The pons is a relay center shunting information from the inferior regions of the body through the thalamus to other areas of the brain. The pons has important **respiratory centers,** which are involved in controlling breathing rate.

The medulla oblongata has centers for respiratory rate, control of blood pressure, and other vital centers. Some information from the right side of the body crosses over to the left brain in the medulla oblongata. Some information from the left side of the body crosses over to the right side of the brain. The area where motor tracts cross over in the medulla oblongata is known as the **decussation of the pyramids.** The medulla oblongata terminates at the foramen magnum where it meets the cervical region of the spinal cord.

The cerebellum is primarily noted for muscle coordination and the maintenance of posture. The cerebellum consists of an outer **cerebellar cortex** consisting of numerous folds called folia and an inner extensively branched pattern of white matter known as the **arbor vitae** (tree of life). The triangular space anterior to the cerebellum is the **fourth ventricle** and can be seen in figure 16.11. Examine the features of the hindbrain as described here and seen in material in the lab.

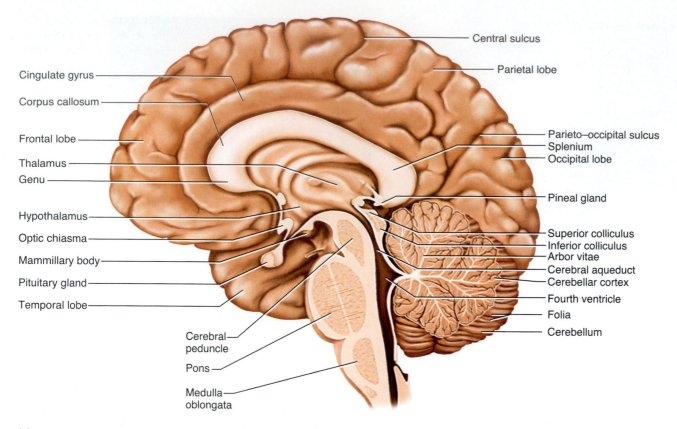

(a)

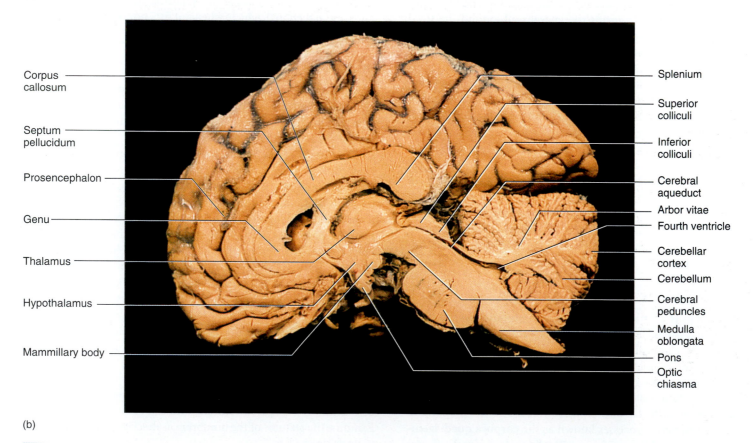

(b)

FIGURE 16.11

Brain, Midsagittal Section (a) Diagram; (b) photograph.

Coronal Section of the Brain

The **gray matter** of the brain consists of nerve cell bodies, is extensive, and forms the superficial **cerebral cortex**. Most of the active, integrative processes of the brain occur in the cerebral cortex (fig. 16.12). The cerebral cortex is approximately 4 mm thick and occupies the superficial regions of the brain. Deep to the cortex is the **white matter** of the brain, consisting mostly of axons that take information from deeper regions of the brain to the cerebral cortex for processing. Sensory information coming from the spinal cord moves through the inferior regions of the brain and through the white matter for integration in the cerebral cortex. White matter also takes information from one region of the cerebral cortex to another for integration or from the cerebral cortex back to the spinal cord and to other parts of the body for action.

Gray matter is not restricted to the cerebral cortex, however. Deep islands of gray matter in the brain compose the **basal nuclei**, also seen in a coronal section. Basal nuclei serve a number of functions in the brain, many of which involve subconscious processes, such as the swinging of arms while walking or the regulation of muscle tone. Basal nuclei are found not only in the cerebrum but in the thalamus and midbrain as well.

LIMBIC SYSTEM

The limbic system is a very complex region of the brain involved in mood and emotion; it has centers for feeding, sexual desire, fear, and satisfaction. The inferior portion of the limbic system has neural fibers that come from the olfactory regions of the brain. These are best seen in a model of the limbic system or in a transected brain. The hippocampus is in the medial temporal lobe of the brain; it is involved in the formation of memories and in spatial understanding. Examine figure 16.13 for the major features of the limbic system.

Brainstem

The brainstem consists of the midbrain, the pons, and the medulla oblongata. Look at a model or section of brain that has had the cerebrum removed. Locate the corpora quadrigemina (fig. 16.14) along with the medulla oblongata and the pons.

Cranial Nerves

There are 12 pairs of cranial nerves. The cranial nerves are part of the PNS, but they are frequently studied along with the brain. The cranial nerves are listed by roman numeral (I–XII), and you should know the nerve by name *and* by number. Examine a model of the brain along with figure 16.15 and note that all the cranial nerves, except nerve XII, are in sequence from anterior to posterior. Nerves may be sensory, motor, or mixed (both sensory and motor). Sensory nerves take information to the CNS, while motor nerves conduct information away from the CNS, resulting in some kind of action (such as skeletal muscle contraction or glandular secretion). When you study nerves you should examine the anatomy of the brain to see where the nerve emerges from the brain. If you know, for example, that the abducens is found between the pons and the medulla oblongata, you can use that nerve as a way to locate other nerves.

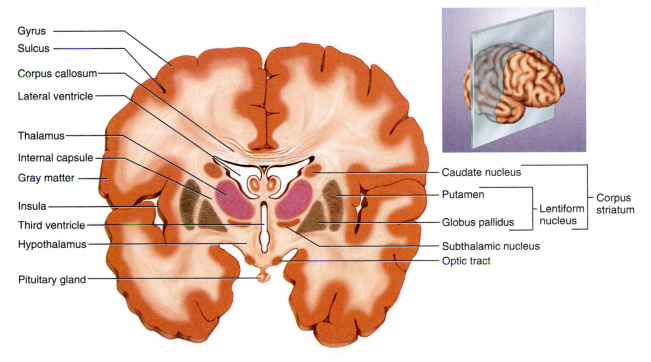

Gyrus
Sulcus
Corpus callosum
Lateral ventricle
Thalamus
Internal capsule
Gray matter
Insula
Third ventricle
Hypothalamus
Pituitary gland

Caudate nucleus
Putamen
Globus pallidus
Subthalamic nucleus
Optic tract
Lentiform nucleus
Corpus striatum

FIGURE 16.12
Brain, Coronal Section

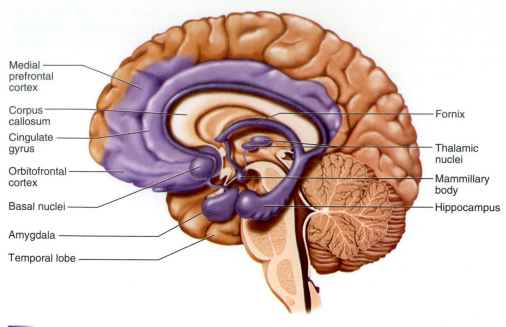

FIGURE 16.13
Limbic System

Medial prefrontal cortex
Corpus callosum
Cingulate gyrus
Orbitofrontal cortex
Basal nuclei
Amygdala
Temporal lobe
Fornix
Thalamic nuclei
Mammillary body
Hippocampus

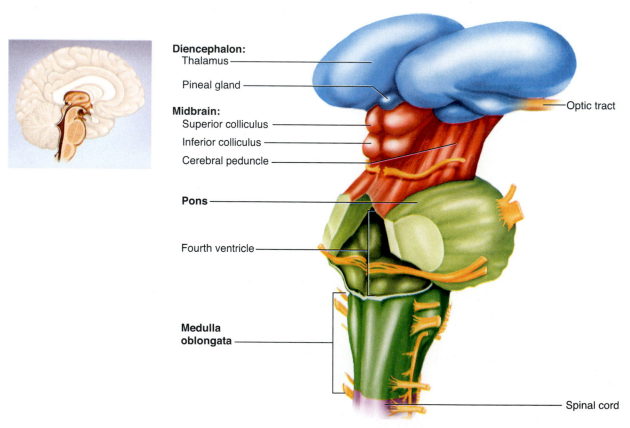

Diencephalon:
Thalamus
Pineal gland

Midbrain:
Superior colliculus
Inferior colliculus
Cerebral peduncle

Pons

Fourth ventricle

Medulla oblongata

Optic tract

Spinal cord

FIGURE 16.14
Brainstem, Posterolateral

FIGURE 16.15
Cranial Nerves, Inferior View (a) Diagram; (b) photograph.

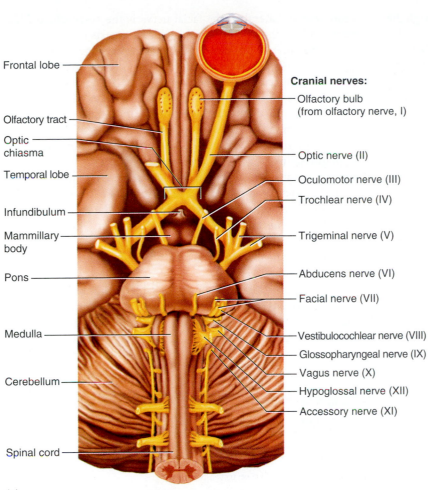

Frontal lobe

Olfactory tract

Optic chiasma

Temporal lobe

Infundibulum

Mammillary body

Pons

Medulla

Cerebellum

Spinal cord

Cranial nerves:

Olfactory bulb (from olfactory nerve, I)

Optic nerve (II)

Oculomotor nerve (III)

Trochlear nerve (IV)

Trigeminal nerve (V)

Abducens nerve (VI)

Facial nerve (VII)

Vestibulocochlear nerve (VIII)

Glossopharyngeal nerve (IX)

Vagus nerve (X)

Hypoglossal nerve (XII)

Accessory nerve (XI)

(a)

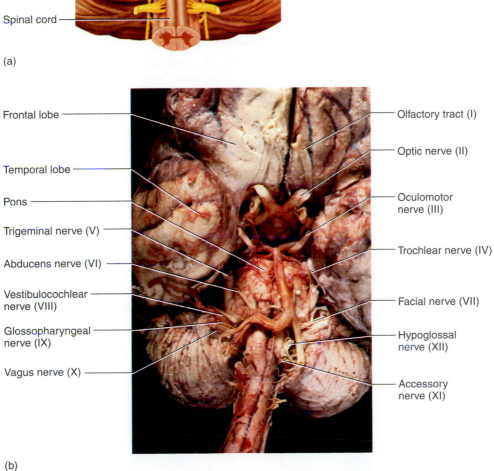

Frontal lobe

Temporal lobe

Pons

Trigeminal nerve (V)

Abducens nerve (VI)

Vestibulocochlear nerve (VIII)

Glossopharyngeal nerve (IX)

Vagus nerve (X)

Olfactory tract (I)

Optic nerve (II)

Oculomotor nerve (III)

Trochlear nerve (IV)

Facial nerve (VII)

Hypoglossal nerve (XII)

Accessory nerve (XI)

(b)

The **olfactory nerves** (I) (*olfaction* = smell) pass through the cribriform plate of the ethmoid, synapsing in the olfactory bulbs. Impulses from the bulbs pass into the olfactory tracts that run along the anterior base of the brain at the inferior aspect of the frontal lobe and transmits the information to the brain. The **optic nerve** (II) (*optic* = sight) is a sensory nerve from the eye that leads to the base of the brain and forms the **optic chiasma**. Some fibers of the optic nerve lead to one side of the brain, while others cross to the other side. The **oculomotor nerve** (III) (*oculo* = eye; *motor* = muscle) controls eye muscles and can be found anterior to the pons, more or less in the midline of the brain, while the **trochlear nerve** (IV) (*trochlea* = hoop that the superior oblique muscle passes through) is found at about a 45° angle from midline on the lateral aspect of the pons. The large **trigeminal nerve** (V) (*trigeminal* = triplets as the nerve divides into three parts) is found at a 90° angle to the pons and is located on the lateral aspect of the pons. It is a mixed nerve (both sensory and motor) that innervates much of the face. The **abducens nerve** (VI) (*abduce* = to pull away; the muscle controlled by this nerve pulls the eye laterally) is found at the midline junction of the pons and medulla oblongata, while the **facial nerve** (VII) (named for its innervation of the face) is more lateral. Lateral to the facial nerve is the **vestibulocochlear nerve** (VIII) (the vestibule and cochlea are parts of the ear), and the nerve inferior to that is the **glossopharyngeal nerve** (IX) (*glosso* = tongue; *pharyngeal* = pharynx). The **vagus nerve** (X) (*vagus* = wandering) is a large nerve or large cluster of fibers on the lateral aspect of the medulla oblongata. The vagus nerve travels to the thorax and abdomen. Inferior to the vagus nerve is the **accessory nerve** (XI), which controls the muscles of the neck and back. Toward the midline of the medulla oblongata is the **hypoglossal nerve** (XII) (*hypo* = below; *glosso* = tongue), which innervates tongue muscles. Locate these nerves in figure 16.15 and note their details in tables 16.2 and 16.3. There have been many mnemonic devices constructed to remember the sequence of cranial nerves. One such mnemonic is "Old Oliver Ogg Traveled To Africa For Very Good Vacations And Holidays." The first letter of each word of the mnemonic represents the first letter of the name of each cranial nerve.

The cranial nerves are listed in table 16.4 as sensory nerves, motor nerves, or both sensory and motor nerves. A mnemonic device is also listed in the table to help you remember the name and function of each nerve.

TABLE 16.2

Cranial Nerves—Function

Number	Name	Function
I	Olfactory	Receives sensory information from the nose, transmitting the sense of smell to the brain
II	Optic	Receives sensory information from the eye, transmitting the sense of vision to the brain
III	Oculomotor	Transmits motor information to move the medial, superior, and inferior rectus muscles and to the inferior oblique muscle of the eye
IV	Trochlear	Transmits motor information to move the superior oblique muscle of the eye
V	Trigeminal	A three-branched nerve; transmits both sensory information from, and motor information to, the head
VI	Abducens	A motor nerve to move the lateral rectus muscle of the eye
VII	Facial	A large nerve that receives sensory information of taste from the anterior tongue and takes motor information to the facial muscles
VIII	Vestibulocochlear	Receives sensory information from the ear; the vestibular part transmits equilibrium information and the cochlear part transmits acoustic information
IX	Glossopharyngeal	A mixed nerve of the tongue and throat, also receives information on taste
X	Vagus	Receives sensory information from the abdomen, thorax, neck, and root of the tongue; transmits motor information to the pharynx, larynx, and controls autonomic functions of heart, digestive organs, spleen, and kidneys
XI	Accessory	A motor nerve to the muscles of the neck and back
XII	Hypoglossal	A motor nerve to the tongue

Dissection of a Sheep Brain

Work in pairs during the dissection of a sheep brain. Take a sheep brain back to your table, along with a dissecting tray and appropriate dissection tools. Work with another group, so that you can dissect two different sheep brains. One of the brains should be sectioned in the midsagittal plane and the other in the coronal plane. If the brains still have the **dura mater**, examine this tough connective tissue coat on the outside of the brain. Cut through this layer to examine the other meninges that are underneath it. Deep to the dura mater is a filmy layer of tissue that contains blood vessels. This is known as the **arachnoid mater**. If you tease some of the membrane away from the brain, you will see that it has a cobweb-like appearance in the **subarachnoid space**. In life, the subarachnoid space contains the **CSF**. Underneath this layer and adhering directly to the brain convolutions is the **pia mater**.

TABLE 16.3

Cranial Nerves—Location

Number	Name	Function
I	Olfactory	Begins in the upper nasal cavity and passes through the cribriform plate of the ethmoid bone. It synapses in the olfactory bulb on either side of the longitudinal fissure of the brain. The fibers take information on the sense of smell and pass via the olfactory tracts to be interpreted in the temporal lobe of the brain.
II	Optic	Takes sensory information from the retina at the back of the eye and transmits the impulses through the optic canal in the sphenoid bone. Some atons cross at the optic chiasma and pass via the optic tracts to the occipital lobe, where vision is interpreted. Other axons do not cross over the optic chiasma.
III	Oculomotor	Emerge from the surface of the brain near the midline and just superior to the pons. It passes through the superior orbital fissure and innervates the inferior oblique muscle and the medial, superior, and inferior rectus muscles and carries parasympathetic fibers to the lens and iris.
IV	Trochlear	Seen at the sides of the pons at about a 45° angle from the midline of the brain. It passes through the superior orbital fissure to the superior oblique muscle.
V	Trigeminal	Seen at a 90° angle from the midline at the lateral sides of the pons. The trigeminal has three branches: (1) the ophthalmic branch passes through the superior orbital fissure; (2) the maxillary branch passes through the foramen rotundum of the sphenoid bone; (3) the mandibular branch passes through the foramen ovale of the sphenoid bone and enters the mandible by the mandibular foramen and exits by the mental foramen.
VI	Abducens	Begins at the midline junction between the pons and the medulla oblongata and passes through the superior orbital fissure to carry motor information to the lateral rectus muscle of the eye.
VII	Facial	Begins as the first of a cluster of nerves on the anterolateral part of the medulla oblongata. It passes through the internal auditory meatus and near the inner ear to the stylomastoid foramen of the temporal bone to innervate facial muscles and glands. It carries sensory information from the anterior tongue. Sensory information of the tongue is interpreted in the parietal and temporal lobe of the brain.
VIII	Vestibulocochlear	Comes from the inner ear and passes through the internal auditory meatus. The conduction passes to the pons, and hearing and balance are interpreted in the temporal lobe.
IX	Glossopharyngeal	Passes through the jugular foramen to innervate muscles of the throat (pharyngeal branches) and sensory receptors of the tongue. Motor portions of the nerve control a muscle of swallowing and a salivary gland, while sensory nerves carry information from the posterior tongue and from baroreceptors and chemoreceptors of the carotid artery.
X	Vagus	Passes through the jugular foramen and along the neck to the larynx, heart, lungs, and abdominal region. The sensory impulses travel in this nerve from the viscera in the abdomen, the thorax, the neck, and the root of the tongue to the brain.
XI	Accessory	Multiple fibers arise from the lateral sides of the superior spinal cord and pass through the jugular foramen to numerous muscles of the neck and back.
XII	Hypoglossal	Begins at the anterior surface of the medulla and passes through the hypoglossal canal to innervate the muscles of the tongue.

TABLE 16.4
Types of Cranial Nerves

Number	Name	Name Mnemonic	Type	Type Mnemonic
I	Olfactory	Old	Sensory	Sally
II	Optic	Oliver	Sensory	Sells
III	Oculomotor	Ogg	Motor*	Many
IV	Trochlear	Traveled	Motor	Mangoes
V	Trigeminal	To	Both	But
VI	Abducens	Africa	Motor	My
VII	Facial	For	Both	Brother
VIII	Vestibulocochlear	Very	Sensory	Sells
IX	Glossopharyngeal	Good	Both	Bigger
X	Vagus	Vacations	Both	Better
XI	Accessory	And	Motor	Mega
XII	Hypoglossal	Holidays	Motor	Mangoes

*Many of the motor nerves have sensory fibers that come from proprioreceptors in the muscles they innervate. Information about the tension of the muscle is sent back to the brain to make adjustments in contractile rate. Since the main function of these nerves is motor, they are listed as motor nerves, even though they have some sensory capabilities.

Find the major lobes of cerebrum, along with the cerebellum, pons, and medulla oblongata (fig. 16.16). Because sheep are quadrupeds, the flexure of the brain does not occur in them as it does in humans. Sheep have a horizontal spinal cord, while humans have a vertical one. Sheep also have a reduced cerebrum. Examine the inferior surface of the sheep brain. Locate the olfactory bulbs, tracts, optic nerve, and optic chiasma. The pituitary gland will probably not be attached, but you should locate the infundibulum, posterior to the optic chiasma. Locate these structures in figure 16.17. For the midsagittal section, divide the brain in the plane of the longitudinal fissure. Your cut should resemble the brain in figure 16.18. Note that the sheep brain has an enlarged corpora quadrigemina, compared with humans. Locate the corpus callosum, lateral ventricles, third ventricle, hypothalamus, pineal gland, superior and inferior colliculi, cerebellum, arbor vitae, pons, medulla oblongata, cerebral aqueduct, and fourth ventricle. After making a coronal section about midway through the cerebrum (fig. 16.19), you should locate the cerebral cortex, cerebral medulla, lateral ventricles, corpus callosum, third ventricle, thalamus, and hypothalamus.

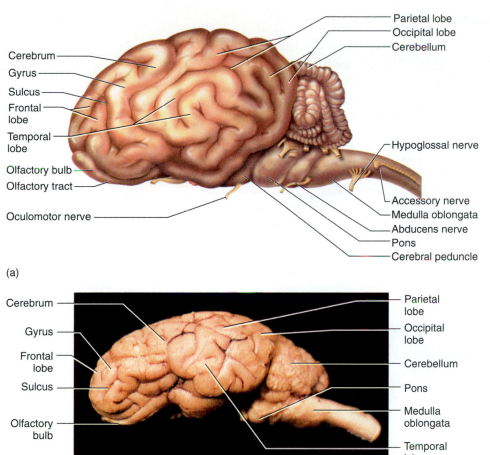

Cerebrum
Gyrus
Sulcus
Frontal lobe
Temporal lobe
Olfactory bulb
Olfactory tract
Oculomotor nerve

Parietal lobe
Occipital lobe
Cerebellum

Hypoglossal nerve

Accessory nerve
Medulla oblongata
Abducens nerve
Pons
Cerebral peduncle

(a)

FIGURE 16.16

Sheep Brain, Lateral View (a) Diagram; (b) photograph.

Cerebrum
Gyrus
Frontal lobe
Sulcus
Olfactory bulb

Parietal lobe
Occipital lobe
Cerebellum
Pons
Medulla oblongata
Temporal lobe

(b)

FIGURE 16.17

Sheep Brain, Inferior View (a) Diagram; (b) photograph.

Olfactory bulb

Optic nerve
Optic chiasma
Optic tract
Oculomotor nerve
Trochlear nerve
Trigeminal nerve
Abducens nerve
Facial nerve
Vestibulocochlear nerve
Glossopharyngeal nerve
Vagus nerve
Accessory nerve
Hypoglossal nerve

Left cerebral hemisphere
Olfactory tract
Infundibulum
Mammillary body
Arterial circle
Pons
Cerebellum
Basilar artery
Medulla oblongata

(a)

Frontal lobe
Optic tract
Infundibulum
Temporal lobe
Cerebellum

Olfactory bulb
Optic nerve
Optic chiasma
Mammillary body
Pons
Medulla oblongata

(b)

227

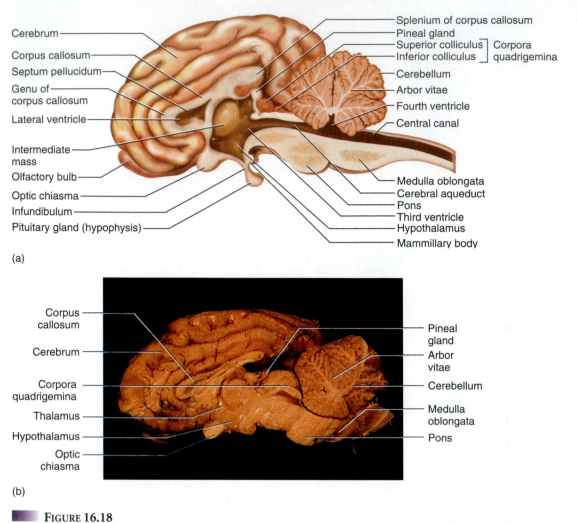

Cerebrum

Corpus callosum

Septum pellucidum

Genu of corpus callosum

Lateral ventricle

Intermediate mass

Olfactory bulb

Optic chiasma

Infundibulum

Pituitary gland (hypophysis)

Splenium of corpus callosum

Pineal gland

Superior colliculus ⎤ Corpora

Inferior colliculus ⎦ quadrigemina

Cerebellum

Arbor vitae

Fourth ventricle

Central canal

Medulla oblongata

Cerebral aqueduct

Pons

Third ventricle

Hypothalamus

Mammillary body

(a)

Corpus callosum

Cerebrum

Corpora quadrigemina

Thalamus

Hypothalamus

Optic chiasma

Pineal gland

Arbor vitae

Cerebellum

Medulla oblongata

Pons

(b)

FIGURE 16.18

Sheep Brain, Longitudinal Section (a) Diagram; (b) photograph.

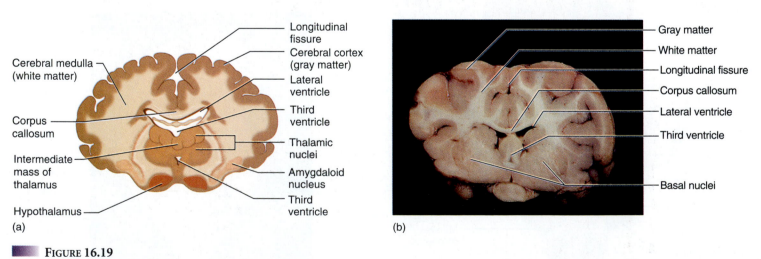

Cerebral medulla (white matter)

Corpus callosum

Intermediate mass of thalamus

Hypothalamus

Longitudinal fissure

Cerebral cortex (gray matter)

Lateral ventricle

Third ventricle

Thalamic nuclei

Amygdaloid nucleus

Third ventricle

(a)

Gray matter

White matter

Longitudinal fissure

Corpus callosum

Lateral ventricle

Third ventricle

Basal nuclei

(b)

FIGURE 16.19

Sheep Brain, Coronal Section (a) Diagram; (b) photograph.

Name _____ Date _____

1. Which of the meninges is located next to the brain?

2. What fluid is found in the ventricles of the brain?

3. Into what space does fluid flow from the cerebral aqueduct?

4. In terms of shape, compare and contrast a sulcus to a gyrus.

5. What are all of the lobes of the cerebrum called?

6. What is the function of the precentral gyrus?

7. What sense does the temporal lobe alone interpret?

8. What physical depression separates the temporal lobe from the parietal lobe?

9. What structure connects the cerebral hemispheres?

10. Name the structures of the midbrain.

11. What function does the cerebellum have?

12. John hit his forehead against the wall. What possible damage might he have done to the function of his brain, particularly the functions associated with the frontal lobe?

13. If a stroke affected all of the sensations interpreted by the brain concerning the face and the hands, what percent of the postcentral gyrus would be affected?

14. One convenient excuse that people often make for their inability to do something is to describe themselves as left-brain or right-brain individuals. Describe what effect the loss of an entire cerebral hemisphere would have on specific functions, such as spatial awareness or the ability to speak.

15. Aphasia is the loss of speech, and different types can occur. If Broca area were affected by a stroke, would the content of the spoken word be affected or would the ability to pronounce the words be affected?

16. Label the following illustration using the terms provided.

pons	thalamus	fourth ventricle	dura mater
corpus callosum	arbor vitae	hypothalamus	corpora quadrigemina
cerebral aqueduct	medulla oblongata	choroid plexus	mammillary body
pineal gland	pituitary gland	optic chiasma	infundibulum

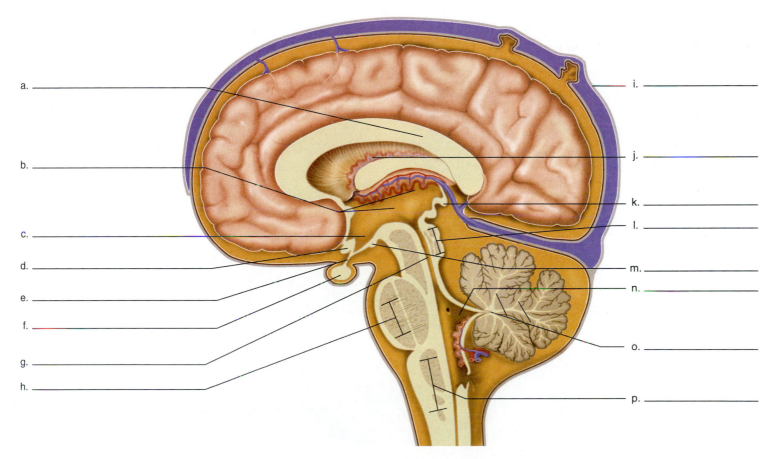

a. _____

b. _____

c. _____

d. _____

e. _____

f. _____

g. _____

h. _____

i. _____

j. _____

k. _____

l. _____

m. _____

n. _____

o. _____

p. _____

17. Where is CSF found in relation to the meninges?

18. Approximately what percent of the precentral gyrus is dedicated to the face?

19. How much of the gyrus is dedicated to the hands?

20. How much of the gyrus is dedicated to the trunk?

21. Label the following illustration using the terms provided.

hypoglossal nerve   medulla oblongata   vagus nerve
pons   trigeminal nerve   optic nerve
olfactory bulb   cerebellum   oculomotor nerve

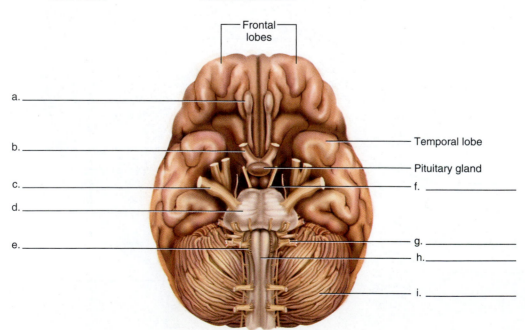

Frontal lobes

a. _____

b. _____

c. _____

d. _____

e. _____

Temporal lobe

Pituitary gland

f. _____

g. _____

h. _____

i. _____

22. Describe these nerves in terms of function (sensory, motor, or both).

optic nerve_____

trochlear nerve_____

glossopharyngeal nerve_____

hypoglossal nerve_____

vagus nerve_____

23. Name the cranial nerve or nerves that would innervate the following areas.

taste buds of the tongue_____

internal ear_____

mandible_____

retina of the eye_____

stomach_____

lateral rectus muscle of the eye_____

Spinal Cord and Somatic Nerves

Nervous System

INTRODUCTION

The spinal cord is part of the central nervous system (CNS); it begins at the foramen magnum of the skull and ends at about vertebra L1 or L2. The spinal cord stops growing early in life, yet the vertebral bodies continue to grow, causing the spinal cord to be shorter than the vertebral canal. The spinal cord receives sensory information from, and transmits motor information to, the spinal nerves, which radiate into the body as peripheral nerves. In this exercise you learn the major features of the spinal cord in longitudinal aspect and in cross section.

The peripheral nervous system consists of the cranial nerves and the somatic nerves. Cranial nerves are associated with the brain and are covered in this lab manual in laboratory exercise 16. Somatic nerves take both sensory and motor information to and from the spinal cord. Nerves such as the radial nerve and the femoral nerve are somatic nerves. These nerves travel throughout the body, receiving sensory information and sending it to the CNS or taking motor information from the CNS to skeletal muscles.

LEARNING OBJECTIVES

At the end of this exercise you should be able to
1. demonstrate the major regions in a cross section of spinal cord;
2. describe the nature of the longitudinal aspect of the spinal cord;
3. list the major nerves that arise from each plexus;
4. name all the major nerves of the upper and lower extremities;
5. list the structures that carry the impulses to and away from the spinal cord.

MATERIALS

Models or charts of the central and peripheral nervous systems
Cadaver (if available)

Prepared slide of a spinal cord in cross section
Model or chart of a spinal cord in cross section and longitudinal section

PROCEDURE

Spinal Cord

LONGITUDINAL ASPECT OF THE SPINAL CORD

Examine a model or chart in the lab of a longitudinal view of the spinal cord and locate the major features illustrated in figure 17.1. The spinal cord terminates inferiorly as the **medullary cone** at approximately vertebra L1 and is attached to the coccyx by a continuation of the pia mater known as the **terminal filum.** The neural continuation of the spinal cord exists as an extension of parallel nerve fibers in the lumbar and sacral regions. These parallel fibers resemble a horse's tail and are called the **cauda equina.** The spinal cord is expanded in two locations. The **cervical enlargement** occurs at vertebrae C3 through T2 and represents a bulge in the spinal cord that has increased neural input and output to the upper extremities. The **lumbar enlargement** occurs at about vertebrae T7 through T11, and this expanse is due to the increased neural input and output to the lower extremity. Compare the material in the lab with figure 17.1.

CROSS SECTION OF SPINAL CORD

Examine a model or chart in the lab of a cross section of the spinal cord. Locate the **gray matter,** which appears as an H pattern or a butterfly pattern in the middle of the spinal cord, with the **white matter** located on the periphery of the cord. Note how the distribution of gray and white matter in the spinal cord is opposite of that in the brain. In a cross section of the spinal cord you should see the gray matter divided in two narrow horns and two rounded horns. The narrow horns are known as the **posterior gray horns,** and these areas receive sensory information from the somatic nerves.

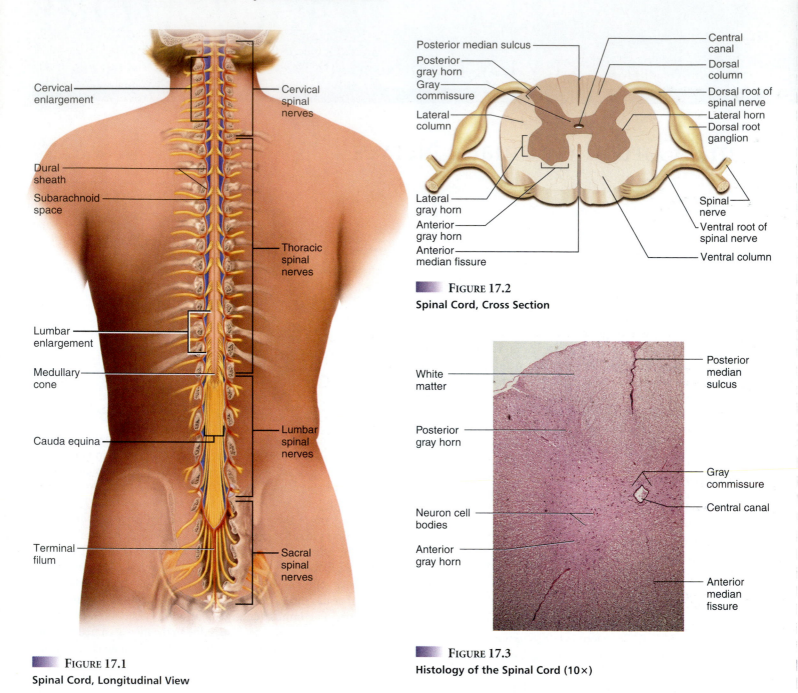

The rounded horns are the **anterior gray horns,** and these send motor signals to the spinal nerves. In some parts of the spinal cord there are additional sections of gray matter known as the **lateral gray horns,** which contain neurons of the sympathetic division.

Each side of the gray matter is connected to the other by a crossbar known as the **gray commissure.** In the middle of the gray commissure is the **central canal,** which runs the length of the spinal cord and contains CSF. The white matter of the cord is divided into **tracts,** or **funiculi,** which take sensory information to the brain or motor information from the brain. **Tracts** are parallel nerve fibers in the CNS, while **nerves** are parallel fibers in the PNS. The tracts that take sensory information to the brain are called **ascending tracts,** and those that receive motor information are called

descending tracts. You should also be able to see a depression in the posterior surface of the spinal cord. This is the **posterior median sulcus,** while the deeper depression on the anterior side is known as the **anterior median fissure.** Examine a model or chart in the lab and compare it with figure 17.2.

HISTOLOGY OF THE SPINAL CORD

Examine a prepared slide of the spinal cord under low power and locate the **anterior** and **posterior horns,** the **gray commissure,** the **central canal,** the **posterior median sulcus,** and the **anterior median fissure.** Examine the anterior horn of the spinal cord and look for the nerve cell bodies of the **multipolar neurons** there. Compare your slide with figure 17.3.

MENINGES

The spinal cord is covered by **meninges,** as is the brain. The outer covering of the cord consists of the **dura mater** (dural sheath). Between the dura mater and the vertebra is the **epidural space,** a site for the injection of anesthetics. The next layer deeper to the dura mater is the **arachnoid mater.** Deep to the arachnoid is the **pia mater,** which is the innermost of the meninges and a thin cover on the spinal cord proper. Between the pia mater and the arachnoid is the **subarachnoid space,** which contains the **cerebrospinal fluid.** Examine figure 17.4 for an illustration of the meninges of the spinal cord.

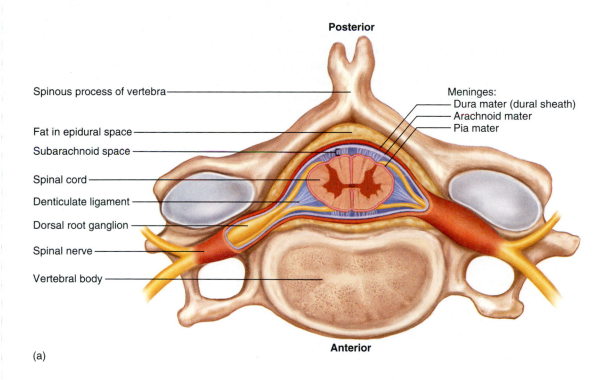

Posterior

Spinous process of vertebra

Fat in epidural space

Subarachnoid space

Spinal cord

Denticulate ligament

Dorsal root ganglion

Spinal nerve

Vertebral body

Meninges:
  Dura mater (dural sheath)
  Arachnoid mater
  Pia mater

Anterior

(a)

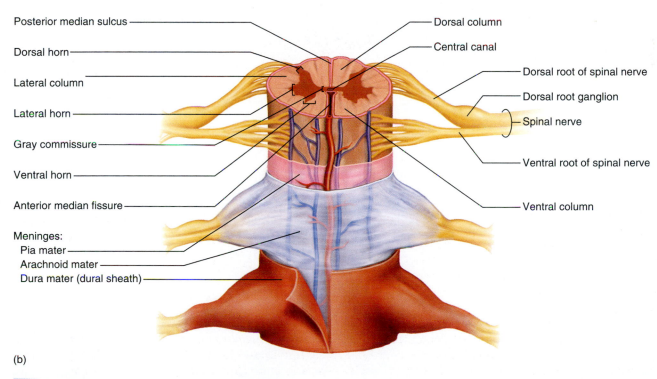

Posterior median sulcus

Dorsal horn

Lateral column

Lateral horn

Gray commissure

Ventral horn

Anterior median fissure

Meninges:
  Pia mater
  Arachnoid mater
  Dura mater (dural sheath)

Dorsal column

Central canal

Dorsal root of spinal nerve

Dorsal root ganglion

Spinal nerve

Ventral root of spinal nerve

Ventral column

(b)

FIGURE 17.4

Spinal Meninges (a) Cross section; (b) longitudinal view.

NERVES ASSOCIATED WITH THE SPINAL CORD

The anatomy of the peripheral nervous system near the spinal cord is complicated. Some nerves, such as those in the middle of the thorax, leave the spinal cord and travel to the ribs. In the neck, arm, thigh, and sacral region, the nerve fibers split, intertwine, and form complex networks called **plexuses** (singular, **plexus**). You can start your study with examining the structure of a single nerve.

Nerve Structure   Nerves have a number of connective tissue wrappings that envelop the individual nerve fibers, clusters of fibers, and the entire nerve. The sheath that wraps around a single nerve fiber, or axon, is the **endoneurium,** and the sheath that wraps around groups of nerve fibers (nerve fascicles) is the **perineurium.** The wrapping that covers the entire nerve is called the **epineurium.** Examine figure 17.5 for these layers.

Spinal Nerves   There are 31 pairs of **spinal nerves,** which pass through the intervertebral foramina, and they are named according to their region of origin. There are **8** pairs of **cervical nerves, 12 pairs of thoracic, 5 pairs of lumbar, 5 sacral pairs,** and **1** pair of **coccygeal nerves.** Examine figure 17.6 and compare it with charts or models in lab to find the spinal nerves. If you look closely at a model of a cross section of spinal cord with nerves, you will see that a spinal nerve is very short and is formed from the union of two roots that attach to the spinal cord. The **dorsal (posterior) root** takes sensory information to the spinal cord. Sensory stimulation in the fingers travels up the nerves in the arm and eventually enters the spinal cord by the dorsal root. The cell bodies of these sensory neurons are in the **dorsal root ganglion** between the spinal nerve and the dorsal root (fig. 17.7).

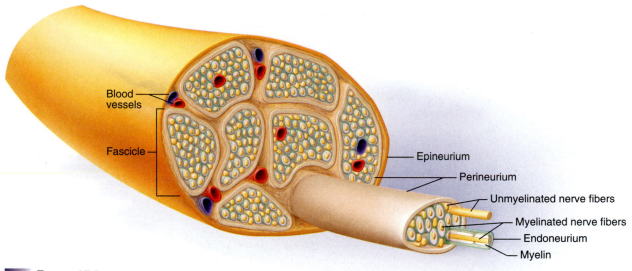

Blood vessels

Fascicle

Epineurium

Perineurium

Unmyelinated nerve fibers

Myelinated nerve fibers

Endoneurium

Myelin

FIGURE 17.5
Coverings of the Nerve

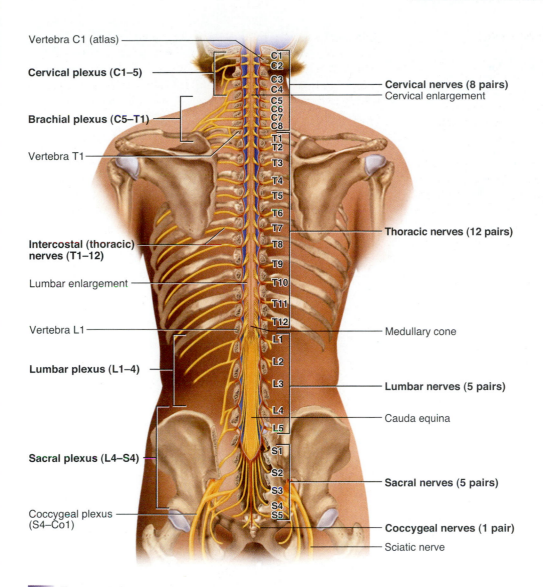

Vertebra C1 (atlas)

Cervical plexus (C1–5)

Brachial plexus (C5–T1)

Vertebra T1

Intercostal (thoracic) nerves (T1–12)

Lumbar enlargement

Vertebra L1

Lumbar plexus (L1–4)

Sacral plexus (L4–S4)

Coccygeal plexus (S4–Co1)

C1
C2
C3
C4
C5
C6
C7
C8
T1
T2
T3
T4
T5
T6
T7
T8
T9
T10
T11
T12
L1
L2
L3
L4
L5
S1
S2
S3
S4
S5

Cervical nerves (8 pairs)
Cervical enlargement

Thoracic nerves (12 pairs)

Medullary cone

Lumbar nerves (5 pairs)

Cauda equina

Sacral nerves (5 pairs)

Coccygeal nerves (1 pair)

Sciatic nerve

FIGURE 17.6
Spinal Nerves and Plexuses

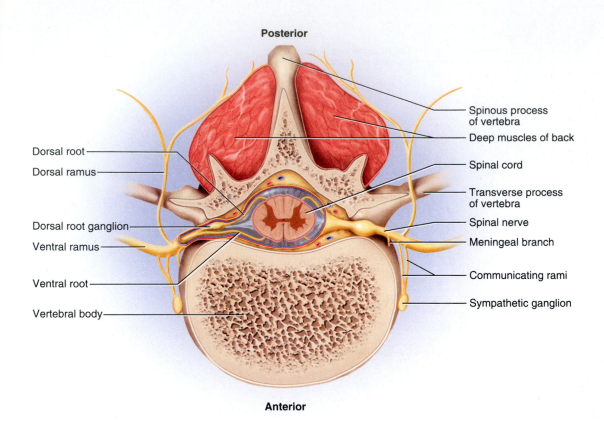

Posterior

Dorsal root

Dorsal ramus

Dorsal root ganglion

Ventral ramus

Ventral root

Vertebral body

Spinous process of vertebra

Deep muscles of back

Spinal cord

Transverse process of vertebra

Spinal nerve

Meningeal branch

Communicating rami

Sympathetic ganglion

Anterior

FIGURE 17.7

Nerves Associated with the Spinal Cord

The **ventral (anterior) root,** adjacent to the dorsal root, takes motor information from the spinal cord to muscles. When you want to move your fingers, impulses travel from the spinal cord through the ventral root and ultimately to the muscles in your fingers.

Superficial to the intervertebral foramen, the spinal nerve divides and forms the **posterior division** (dorsal ramus), which takes sensory and motor fibers to the back, and the **anterior division** (ventral ramus), which takes sensory and motor fibers to the other parts of the body. These divisions form the peripheral nerves that innervate the body.

Where things get complicated with the anatomy is at the level of the plexus. A plexus is made up of a network of nerve fibers. Spinal **roots (spinal nerves)** sometimes divide into **trunks,** which split into **anterior divisions** and **posterior divisions.** These form **cords** in some plexuses, which unite to make **nerves.** The sequence (roots, trunks, anterior divisions, posterior divisions, cords, and nerves) can be remembered by the phrase "Rowdy Tourists Are Playing Cards Now."

There are four generally recognized plexuses. The composition of the plexuses is outlined in table 17.1 and illustrated in figure 17.6.

TABLE **17.1**		
Composition of Plexuses		
Name	**Spinal Nerves Contributing to Plexus**	**Major Nerves of Plexus**
Cervical	C1–5	Phrenic
Brachial	C5–T1	Radial, median, ulnar, musculocutaneous, axillary
Lumbar	L1–4	Femoral, obturator
Sacral	L4–S4	Sciatic (tibial and common fibular)

Cervical and Brachial Plexus Nerves   The cervical plexus exits from the upper spinal nerves of the neck. An important nerve that comes from the cervical plexus is the **phrenic nerve.** This nerve runs to the diaphragm and is responsible for its contraction in breathing. Examine the nerves of the **cervical plexus** in figure 17.8. Many nerves from the cervical plexus innervate the muscles and skin of the neck and the skin of the ear. Since breathing is controlled by the phrenic nerve, the observation that a person is breathing is

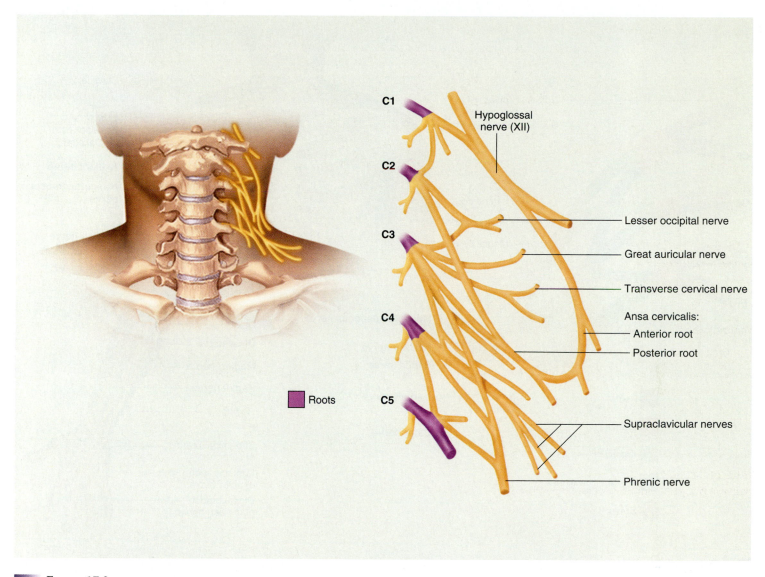

FIGURE 17.8
Nerves of the Cervical Plexus

an obvious test that at least one nerve of the plexus is working. You can also test the function of the nerves by lightly pinching the sides of the neck or the ear for sensory perception.

The **brachial plexus** has branches from C5 to T1 and forms nerves that primarily innervate the upper extremities. The plexus is illustrated in figure 17.9. The muscles innervated from this plexus were covered in laboratory exercise 11. A major nerve from the brachial plexus is the **axillary nerve,** innervating the upper shoulder and seen in figure 17.9. Another nerve, the **radial nerve,** innervates the extensors of the arm, forearm, and hand. The **musculocutaneous nerve** innervates the muscles that flex the arm and forearm, while the **ulnar nerve** crosses posterior to the medial epicondyle of the humerus and is commonly known as the "funny bone." The ulnar nerve innervates some forearm and medial hand flexors. The **median nerve** runs the length of the upper extremity,

innervating lateral hand and forearm flexors. The nerves of this plexus can be tested by pinching the fingers, the medial and lateral aspects of the forearm, and the anterior and posterior aspects of the arm. Sensations from these areas are conducted to the brain via the brachial plexus.

Thoracic Nerves   There are numerous nerves not associated with a plexus. The thoracic nerves are a good example. Many of the **thoracic nerves** exit through the intervertebral foramina of the vertebral column and innervate the ribs, muscles, and other structures of the thoracic wall. Look at models or charts in lab and compare these with the nerves illustrated in figure 17.6.

Lumbar and Sacral Plexus Nerves   The **lumbar** and **sacral plexus** take sensory information from and motor information to the lower limb. Some authors combine the two plexuses into the

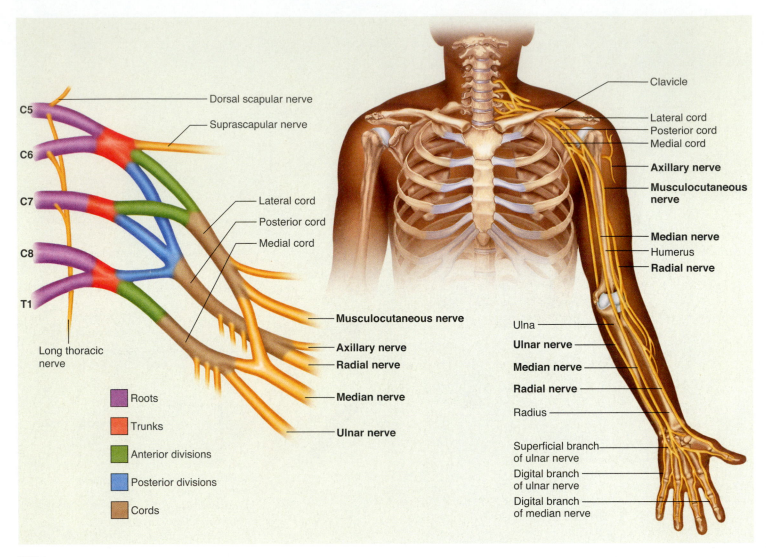

FIGURE 17.9
Nerves of the Brachial Plexus

lumbosacral plexus. The lumbar plexus is composed of spinal nerves L1–4. One of the nerves originating from the lumbar plexus is the **obturator nerve.** It innervates the adductor muscles of the thigh, as seen in figure 17.10. The **femoral nerve** is another nerve arising from the **lumbar plexus.** This large nerve passes posterior to the inguinal ligament and mostly innervates the muscles of the anterior thigh. The femoral nerve is seen in figures 17.10 and 17.11. To test for the nerves of this plexus, you can lightly pinch the anterior thigh for the femoral nerve and the medial thigh for the obturator nerve.

The **sacral plexus** consists of fibers from L4–S4. Many innervate the pelvis and muscles that move the hip, thigh, and leg. Two of the nerves from this plexus, the **tibial** and **common fibular (peroneal) nerve,** unite proximally in one sheath to form the **sciatic nerve** (fig. 17.11). At the distal thigh, the tibial nerve and the common fibular nerve separate. These nerves innervate the posterior thigh, leg, and foot. Test for the sciatic nerve by lightly pinching the posterior aspect of the thigh. Examine the nerves in the lab and compare them with the illustrations.

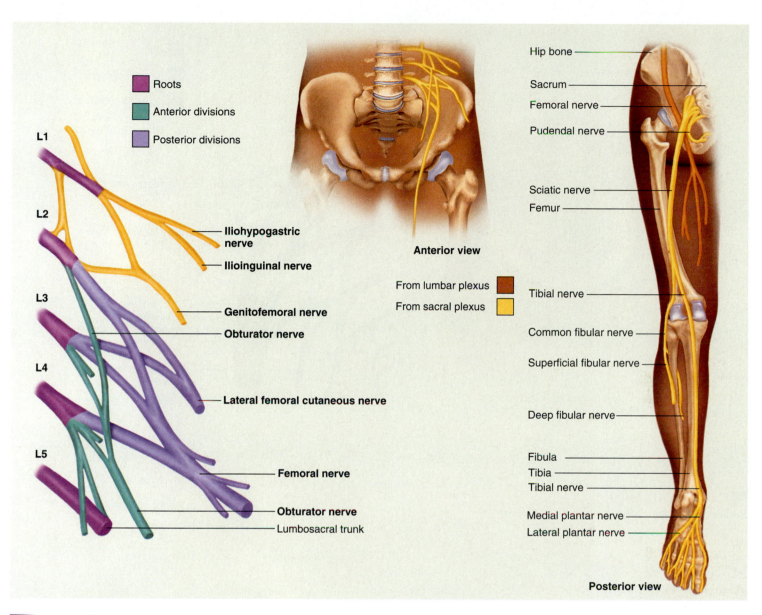

Roots
Anterior divisions
Posterior divisions

L1
L2
L3
L4
L5

Iliohypogastric nerve
Ilioinguinal nerve
Genitofemoral nerve
Obturator nerve
Lateral femoral cutaneous nerve
Femoral nerve
Obturator nerve
Lumbosacral trunk

Anterior view

From lumbar plexus
From sacral plexus

Hip bone
Sacrum
Femoral nerve
Pudendal nerve

Sciatic nerve
Femur

Tibial nerve

Common fibular nerve

Superficial fibular nerve

Deep fibular nerve

Fibula
Tibia
Tibial nerve

Medial plantar nerve
Lateral plantar nerve

Posterior view

▌ **FIGURE 17.10**
Nerves of the Lumbar Plexus

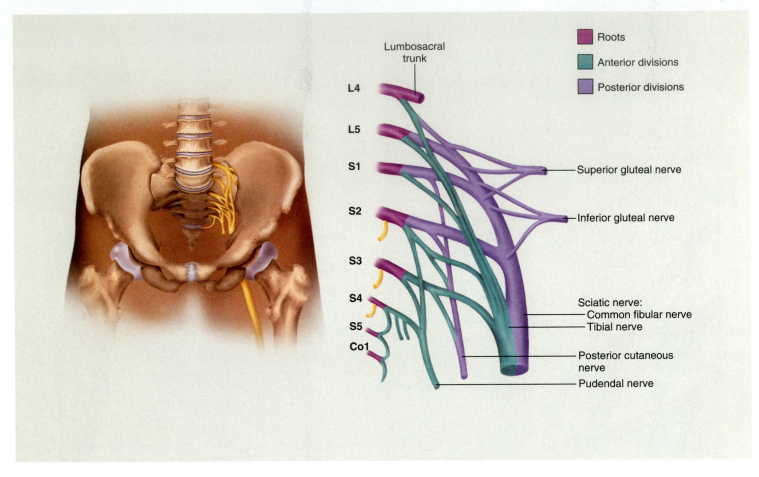

FIGURE 17.11
Nerves of the Sacral Plexus

Name _____ Date _____

1. What anatomic feature is responsible for the cervical enlargement of the spinal cord?

2. Where is the terminal filum found?

3. What is the medullary cone?

4. What is the cauda equina?

5. In the spinal cord, which is deep, the white matter or the gray matter?

6. What is the area of gray matter between the lateral halves of the spinal cord?

7. *In the spinal cord,* what type of impulse (sensory/motor) travels through the

    a. anterior gray horn?

    b. posterior gray horn?

    c. ascending spinal tracts?

    d. descending spinal tracts?

8. How does the dorsal spinal root vary from the ventral spinal root?

9. What is the endoneurium?

10. How do tracts differ from nerves?

11. What is a mixed nerve?

12. What major nerves arise from the following plexuses?

   a. cervical

   b. brachial

   c. lumbar

   d. sacral

13. The diaphragm contractions are regulated by what nerve?

14. The muscles of the arm, such as the biceps brachii, have what innervation?

15. The extensor muscles of the hand are controlled by what nerve?

16. The sciatic nerve is composed of two nerves. What are they?

17 Label the major plexuses in the following figure.

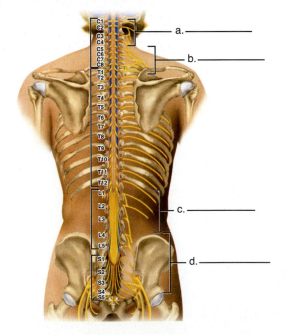

a. ——————

b. ——————

c. ——————

d. ——————

Introduction to Sensory Receptors

Nervous System

INTRODUCTION

The essence of what it means to be human begins with the sense organs. Exquisite sunsets in the western sky, music that reaches the soul, the touch of a loved one's hand stirring the heart, and the experience of a fabulous dinner are all brought to us through the window of the sense organs. Without them we would exist in a cold, dark, soundless world. Our lives would truly be senseless.

Sensory receptors convert stimuli into neural impulses. The stimuli can come from the external environment, such as sound, or they can come from internal areas, such as hunger pangs. These receptors are not uniformly distributed throughout the body but are absent, or few in number, in some areas, while densely clustered in other locations. This pattern of uneven distribution of sense organs is called **punctate distribution.**

For the sensory system to operate, several factors need to be present. There can be no perception without a sensation (an environmental event). The primary types of sensations are called **modalities.** Examples of modalities are light, heat, sound, pressure, and specific chemicals. These environmental modalities stimulate receptors. **Receptors** are receiving entities of the body that respond to specific stimuli. They transform the stimulus to neural signals that are transmitted by sensory nerves and neural tracts to the brain, which interprets the message. If any link in this sensory chain is broken, perception does not occur.

Receptors are sensitive to specific stimuli and can be classified according to the stimuli they receive. The human body has **photoreceptors** in the eyes, which detect light. There are **thermoreceptors,** located in the skin, which detect changes in temperature. **Osmoreceptors** monitor changes in solute concentration. **Proprioreceptors** detect changes in tension, such as those in joints or in muscles. **Pain receptors,** or **nociceptors,** are naked nerve endings in the skin or viscera. **Mechanoreceptors** perceive mechanical stimuli (e.g., touch receptors or receptors that determine hearing or equilibrium in the ear). **Baroreceptors** respond to changes in pressure, such as blood pressure, and **chemoreceptors** respond to changes in the chemical environment (e.g., taste and smell).

The sense of taste, or **gustation,** is received predominantly by taste buds in the tongue, although there are also receptors in the soft palate and pharynx. The sense of smell, or **olfaction,** originates when particles stimulate hair cells in the olfactory epithelium (a specialized neuroepithelium in the upper nasal cavities) and is transmitted to the brain by the olfactory nerves and tracts.

The eye consists of an anterior portion, visible as we look at the face of an individual, and a posterior portion located in the orbit of the skull. Light travels through a number of transparent structures before it strikes the retina, which is the receptive layer of the eye that converts light energy to neural impulses. These impulses travel from the eyes to the optic nerves and then to the brain, where they are interpreted as sight in the occipital lobes. In most people, eyesight accounts for much of their accumulated knowledge.

The ear is a complex sense organ that performs two major functions, hearing and equilibrium. Hearing is a form of **mechanoreception,** because the ear receives mechanical vibrations (sound waves) and translates them into nerve impulses. This process begins with the vibrations reaching the outer ear and ends up being interpreted as sound in the temporal lobe of the brain. Equilibrium (balance) involves receptors in the inner ear, visual cues, and **proprioreception** (a form of mechanoreception involving the perception of gravity or of forces applied to a structure). There are two types of equilibrium sensed by the inner ear. These are static equilibrium and dynamic equilibrium. In **static equilibrium** an individual is able to determine his or her nonmoving position (such as standing upright or lying down). In **dynamic,** or **kinetic, equilibrium,** motion is detected. Sudden acceleration, abrupt turning, and spinning are examples of dynamic equilibrium.

LEARNING OBJECTIVES

At the end of this exercise you should be able to
1. define the terms "modality" and "receptor";
2. list the major receptor types in the body;

3. explain punctate distribution of sensory receptors;
4. list the two major chemoreceptors located in the region of the head;
5. trace the sense of smell from the nose to the integrative areas of the brain;
6. identify the major structures of the mammalian eye;
7. describe the six extrinsic muscles of the eye and their effect on the movement of the eye;
8. explain how mechanical sound vibrations are translated into nerve impulses;
9. list the structures of the outer, middle, and inner ear;
10. describe the structure of the cochlea.

MATERIALS

Microscopes

Prepared slides of thick skin (with Pacinian and Meissner's corpuscles)

Models and charts of the eye

Prepared slides of taste buds

Prepared slides of the eye in sagittal section

Preserved sheep or cow eyes

Dissection trays

Dissection gloves (latex or plastic)

Scalpel or razor blades

Animal waste disposal container

Models and charts of the ear

Prepared slides of the cochlea

Model of ear ossicles

PROCEDURE

Touch Receptors

Examine a slide of thick skin for touch corpuscles. Two of these light touch corpuscles are **Meissner's corpuscles,** in the upper portion of the dermis, and **Merkel discs,** located in the upper dermis and lower epidermis. These two receptors allow for the perception of very slight touch stimuli (such as a fly lightly walking over your cheek). The light touch corpuscles have **receptive fields**. In areas such as the shoulder, the receptive field is large and there are few corpuscles in a given area. In areas with fine touch discrimination, such as the fingers, there are many receptive fields in a given area. In addition, deep touch, or pressure, receptors are found in the dermis, farther away from the epidermis. **Pacinian (lamellated) corpuscles** sense pressure, such as when you lean against a wall or feel a vibration. Other receptors in the skin are warm receptors and cool receptors. When you are at a comfortable temperature, both of these receptors are firing. As the temperature gets warmer, more warm receptors fire and, as you get cooler, more cool receptors fire. If you increase or decrease the skin temperature beyond the perception of these receptors, pain receptors are stimulated. Pain receptors are naked nerve endings in the dermis that respond to numerous environmental stimuli. Locate the Meissner's corpuscles and Pacinian corpuscles in a prepared slide of skin and compare them with figure 18.1.

Examination of Taste Buds

Examine the prepared slide of taste buds and compare them with figure 18.2. Note how they are located on the sides of the papillae on the tongue. Papillae are raised structures on the tongue. The taste

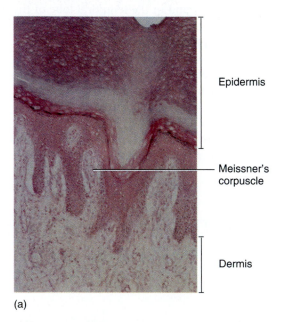

Epidermis

Meissner's corpuscle

Dermis

(a)

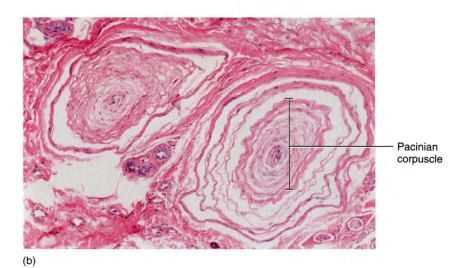

Pacinian corpuscle

(b)

FIGURE 18.1

Skin Receptors (100×) (a) Meissner's corpuscle; (b) Pacinian corpuscle.

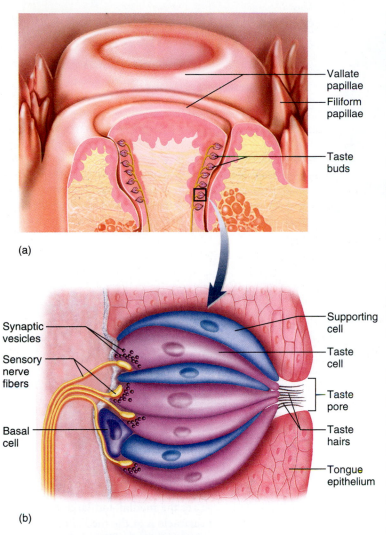

(a)

Vallate papillae

Filiform papillae

Taste buds

Synaptic vesicles

Sensory nerve fibers

Basal cell

Supporting cell

Taste cell

Taste pore

Taste hairs

Tongue epithelium

(b)

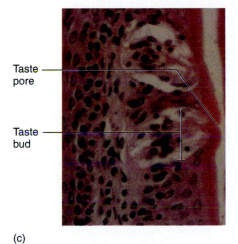

Taste pore

Taste bud

(c)

FIGURE 18.2

Taste Buds (a) Taste buds on sides of a tongue papilla; (b) details of taste buds. Photomicrograph (c) taste buds (100×).

buds appear lighter than the surrounding tissue (like microscopic onions cut in long section). Taste buds are composed of neural tissue and epithelial tissue. The sense of taste is transmitted by the facial, glossopharyngeal and vagus nerves and interpreted in the postcentral gyrus of the parietal lobe, and other parts of the cerebral cortex. There are five primary tastes; sweet, sour, bitter, salty, and umami.

Transmission of the Sense of Olfaction to the Brain

Examine a model, chart, or diagram of a midsagittal section of the head or a model of the brain and locate the **olfactory nerve fibers, cribriform foramina** in the **cribriform plate, olfactory bulb,** and **olfactory tract.** Compare the lab charts or models with figure 18.3.

Odor molecules touch the **olfactory mucosa** and stimulate the hair cells of the olfactory neurons. The transmission of the sense of smell occurs through the cribriform plate of the ethmoid bone to the olfactory bulb at the base of the frontal lobe of the brain. From here the sense of smell is transmitted to two regions—one in the limbic system and another in the temporal lobe of the brain.

External Features of the Eye

Examine a model of the eye and note the external features, as compared with figure 18.4. The eye has a **sclera,** or "white of the eye," which is composed of dense irregular connective tissue. The sclera is a protective portion of the eye that serves as an attachment point for the muscles of the eye and helps maintain the **intraocular pressure** (the pressure inside the eye). This pressure maintains the shape of the eye and keeps the retina adhered to the back wall of the eye. Numerous blood vessels traverse the sclera, and if they become dilated they give the eye the appearance of being "bloodshot." The sclera is continuous with the transparent **cornea** in the front of the eye.

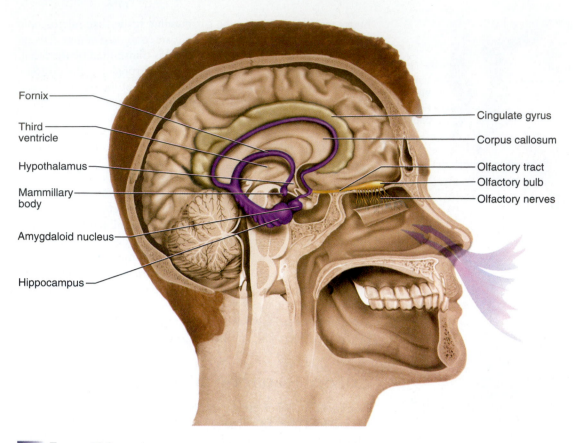

FIGURE 18.3
Olfactory Transmission

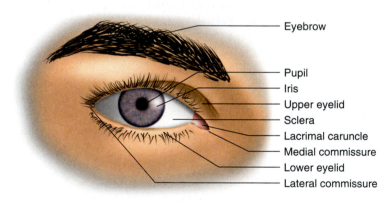

FIGURE 18.4
External Anatomy of the Eye

The corners of the eye are the **medial** and **lateral commissure** (canthi). The **lacrimal caruncle** is at the medial commissure and is the site where tears drain from the eye. **Tarsal glands** secrete an oil, which reduces evaporation from the surface of the eye.

Attached to the sclera are the **extrinsic muscles** of the eye. There are six extrinsic muscles that, in coordination, move the eye in quick and precise ways. Locate these muscles on a model and compare them with figure 18.5. These muscles and their action on the eye are listed in table 18.1.

A structure important in the maintenance of the exterior of the eye is the **lacrimal apparatus**. This consists of the **lacrimal gland** located superior and lateral to the eye (fig. 18.6). Lacrimal secretions bathe and protect the eye and clean dust from its surface. The fluid drains into the lacrimal canal through the **nasolacrimal duct** into the nasal cavity.

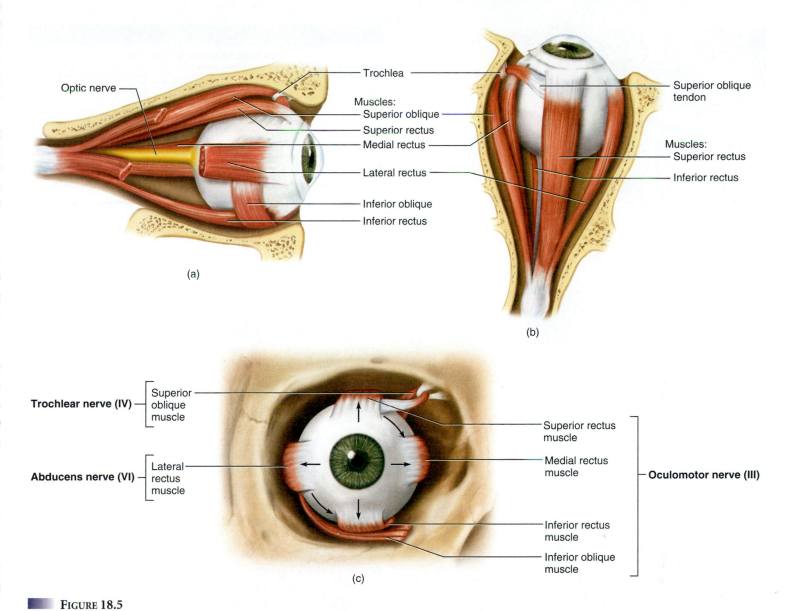

Optic nerve

Trochlea

Muscles:
Superior oblique
Superior rectus
Medial rectus

Lateral rectus

Inferior oblique
Inferior rectus

(a)

Superior oblique
tendon

Muscles:
Superior rectus

Inferior rectus

(b)

Trochlear nerve (IV) — Superior oblique muscle

Superior rectus muscle

Medial rectus muscle

Oculomotor nerve (III)

Abducens nerve (VI) — Lateral rectus muscle

Inferior rectus muscle

Inferior oblique muscle

(c)

FIGURE 18.5

Right Eye, External Features (a) Lateral view; (b) superior view; (c) anterior view.

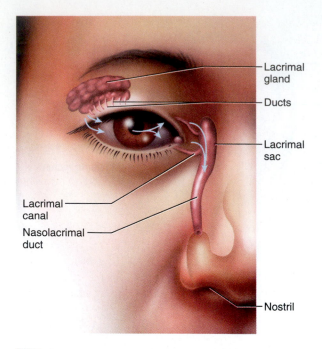

FIGURE 18.6
Lacrimal Apparatus

TABLE 18.1
Extrinsic Muscles of the Eye

Muscle Name	Innervation	Direction Eye Turns
Lateral rectus	VI (abducens)	Laterally
Medial rectus	III (oculomotor)	Medially
Superior rectus	III (oculomotor)	Superiorly
Inferior rectus	III (oculomotor)	Inferiorly
Inferior oblique	III (oculomotor)	Superiorly and laterally
Superior oblique	IV (trochlear)	Inferiorly and laterally

Interior of the Eye

The external tissue covering the sclera is the **conjunctiva.** The conjunctiva is composed of a thin layer of epithelium and is an important indicator of a number of clinical conditions (e.g., conjunctivitis). In the center of the eye is the transparent cornea (fig. 18.7). The cornea is the structure of the eye most responsible for the bending of light rays that strike the eye. It is composed of dense connective tissue and is avascular. Why would the presence of blood vessels in the cornea be a visual liability?

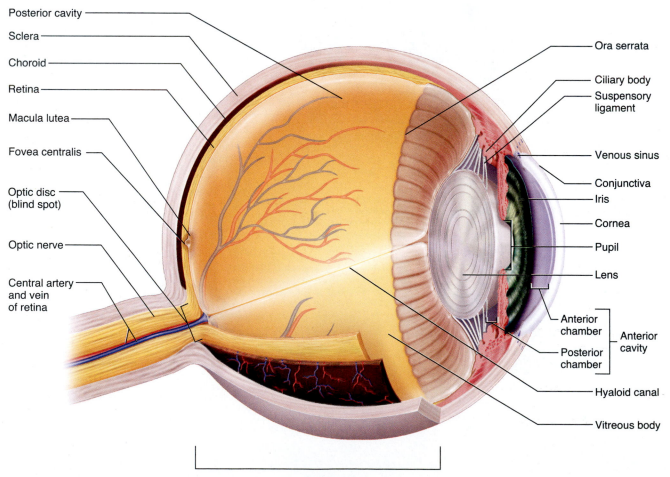

FIGURE 18.7
Sagittal Section of the Eye

Directly posterior to the cornea is the **anterior cavity,** which is subdivided into the **anterior chamber,** between the cornea and the iris, and the **posterior chamber,** between the iris and the lens. The anterior cavity is filled with **aqueous humor,** which is produced by the **ciliary body.** Only a few milliliters of aqueous humor are produced each day, and this amount is absorbed by the **canal of Schlemm (venous sinus),** as illustrated in figure 18.7.

The **iris** is what gives us a particular eye color. People with blue or gray eyes are more sensitive to ultraviolet light than those with brown eyes, due to the protective pigment **melanin** in brown eyes. There are two sets of muscles in the iris. The pupillary constrictor constricts in bright light, reducing the diameter of the **pupil** (the space enclosed by the iris), and the pupillary dilator constricts in dim light, increasing the diameter of the pupil. Posterior to the pupil is the **lens,** which is made of a crystalline protein. The lens is more pliable in youth and becomes less elastic as a person ages, which accounts for the need for reading glasses later in life. **Suspensory ligaments** attach the lens to the ciliary body and flatten the lens by pulling on it, causing the eye to focus on faraway objects.

Deep to the lens is the **posterior cavity,** or **vitreous chamber** (fig. 18.7). This cavity occupies most of the posterior portion, or **fundus,** of the eye. The posterior cavity is filled with **vitreous body,** a clear, jellylike fluid that maintains the shape of the eyeball. Most of the posterior cavity is bounded by three layers. The outermost one, the **sclera,** along with the cornea, make up the **fibrous layer** of the eye. Deep to the sclera is the **choroid,** a pigmented, vascular layer. The blood vessels found in this layer nourish the eye and the pigmentation prevents light from scattering and blurring vision. The choroid, along with the iris and ciliary body, make up the **vascular layer** of the eye. The layer closest to the vitreous humor is the **retina.** This is the inner layer of the eye. These structures are seen in figure 18.7.

Humans have stereoscopic sight, which gives us depth of vision. This is due to the eyes having overlapping visual fields, seen as the green region in figure 18.8. Light strikes the retina at the back of the eye and is transmitted via the optic nerves, optic chiasm, and optic tract to the occipital lobe of the brain, where the impulses are interpreted as sight.

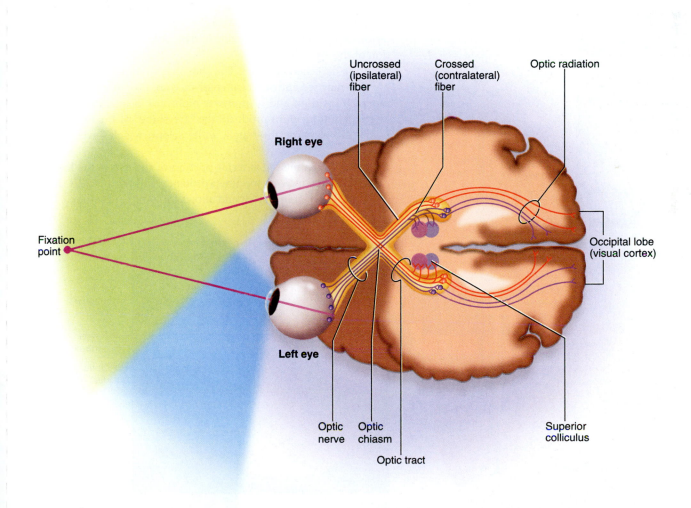

FIGURE 18.8
Visual Pathway to the Brain

The retina is composed of three layers. Examine a slide of the eye in sagittal section (fig. 18.9) and look at the retina under high power. Locate the **ganglionic, bipolar,** and **photoreceptive** layers of the retina. The photoreceptive layer is composed of **rods** and **cones.** Rods are important for determining the shape of objects and for sight in dim light. Cones are involved in color vision and in visual acuity (determining fine detail). Light strikes the photoreceptive cells (rods and cones) at the posterior portion of the retina (fig. 18.9), and these cells transmit visual signals to the **bipolar neurons** located in the bipolar layer. The bipolar neurons are in contact with the **ganglion cells** in the ganglionic layer. Amacrine cells and horizontal cells connect rods, cones, and bipolar cells horizontally. Visual stimulation begins in the posterior portion of the eye and is transmitted anteriorly (toward the vitreous humor) to the ganglion cells. The ganglionic layer transmits the neural impulses via the **optic nerve.**

Examine a model or chart of the eye and locate the **macula lutea** at the posterior region of the eye. "Macula lutea" means yellow spot, and in the center of this structure is the **fovea centralis,** a region where the concentration of cones is greatest. In the fovea, the cone cells are not covered by the neural layers, as they are in the other parts of the retina. When you focus on an object intently, you are directing the image to the fovea. Locate the fovea in models or charts available in the lab and compare them with figures 18.7 and 18.10. You should also be able to see the optic disc. The optic nerve enters the eye at the optic disc. Because there are no photoreceptive cells in this area, it is also known as the blind spot.

Dissection of a Sheep or Cow Eye

Rinse a sheep or cow eye in running water and place it on a dissection tray. Obtain a scalpel, scissors, or a new razor blade and a blunt probe. Be careful with the sharp instruments and cut *away from* the hand holding the eye. Wear latex or plastic gloves while you perform the dissection. Using a scalpel, scissors, or a razor blade, make a coronal section of the eye behind the cornea (fig. 18.11). Do not squeeze the eye with force or thrust the blade sharply, because you may squirt yourself with vitreous humor. Find the **optic nerve.** Cut through the eye entirely and note the jellylike material in the posterior cavity. This is the **vitreous body.** Look at the posterior portion of the eye. Note the beige **retina,** which may have pulled away from the darkened **choroid.** The choroid in humans is very dark, but you may see an iridescent color in your specimen. This is the **tapetum lucidum** and it improves night vision in some animals. The tapetum lucidum produces the "eye shine" of nocturnal animals. Also examine the tough, white **sclera,** which envelops the choroid.

Now examine the anterior portion of the eye. Is the **lens** in place? The lens in your specimen probably will not be clear. Normally the lens is transparent and allows for light penetration.

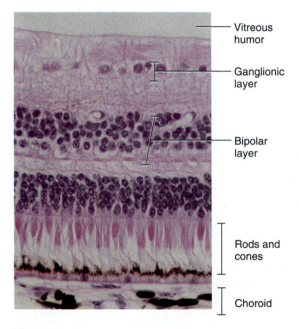

FIGURE 18.9
Retina (400×)

Labels: Vitreous humor, Ganglionic layer, Bipolar layer, Rods and cones, Choroid

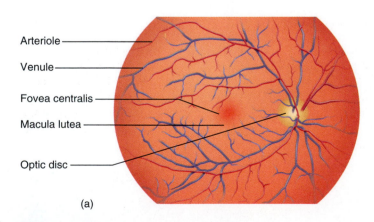

(a)

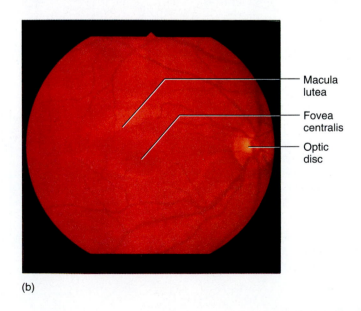

(b)

Labels (a): Arteriole, Venule, Fovea centralis, Macula lutea, Optic disc
Labels (b): Macula lutea, Fovea centralis, Optic disc

FIGURE 18.10
Eye, Posterior View (a) Diagram; (b) photograph.

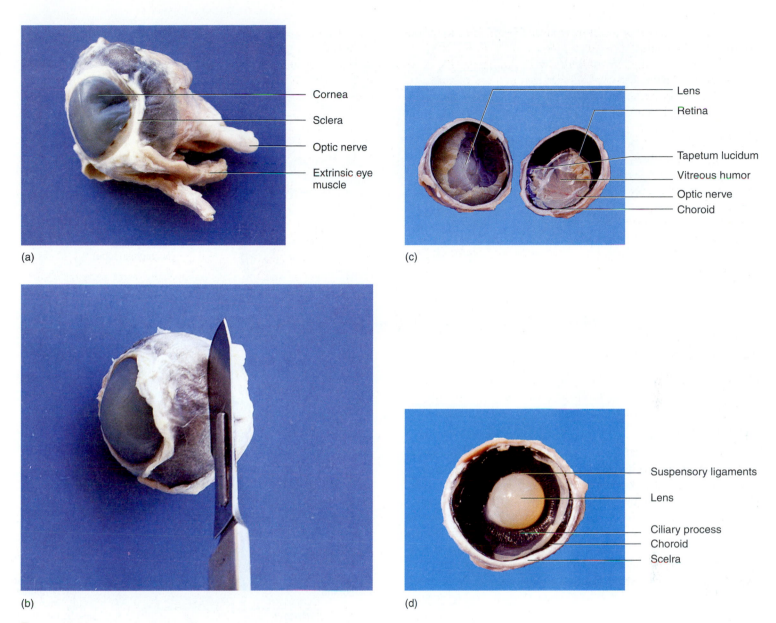

Cornea

Sclera

Optic nerve

Extrinsic eye
muscle

(a)

Lens

Retina

Tapetum lucidum

Vitreous humor

Optic nerve

Choroid

(c)

(b)

Suspensory ligaments

Lens

Ciliary process

Choroid

Scelra

(d)

FIGURE 18.11

Dissection of a Sheep eye. (a) External features; (b) coronal section; (c) eye with vitreous humor; (d) anterior eye without vitreous humor.

Note that the lens is held to the **ciliary body** by the **suspensory ligaments** (figs. 18.7 and 18.11). These ligaments pull on the lens and alter its shape for distant vision. Locate the ciliary body at the edge of the suspensory ligaments.

Now turn the eye over. Is there any **aqueous humor** left in the **anterior cavity?** Make an incision through the **conjunctiva** and **cornea** into the anterior cavity. Can you determine the region of the **anterior chamber** and the **posterior chamber** that make up the anterior cavity? Locate the **iris** and the **pupil.** When you are finished, dispose of the specimen in a designated waste container and rinse your dissection tools.

Anatomy of the Ear

The ear can be divided into three regions—the outer, middle, and inner ear. The **outer ear** consists of auditory structures superficial to the **tympanic membrane** (eardrum). The **middle ear** contains the tympanic membrane, tympanic cavity, ear ossicles, and auditory tube, and the **inner ear,** located in the petrous portion of the temporal bone, consists of the cochlea, vestibule, and semicircular ducts (fig. 18.12).

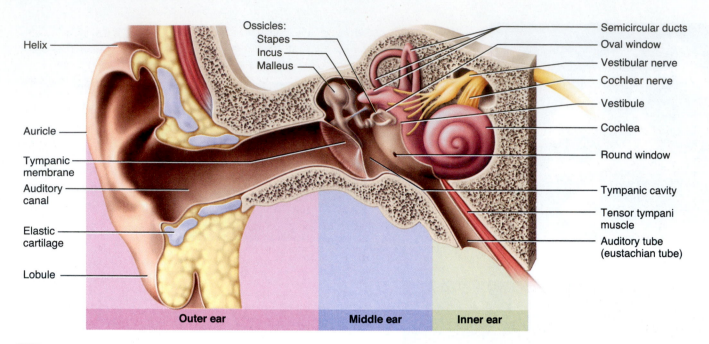

FIGURE 18.12
Anatomy of the Ear

OUTER EAR

The outer ear consists of the **auricle,** or **pinna,** which can further be subdivided into the **helix** and the **lobule (earlobe).** The helix is composed of stratified squamous epithelium overlying elastic cartilage. This cartilage allows the ears to bend significantly. Deep to the pinna the outer ear forms the **auditory canal,** which penetrates into the temporal bone. The **tympanic membrane** is the border between the outer ear and the middle ear. It is composed of connective tissue covered by epithelial tissue. The membrane is sensitive to sound and vibrates as sound is funneled down the auditory canal.

MIDDLE EAR

The middle ear consists of a main cavity known as the **tympanic cavity;** three small bones, or **ossicles;** and the **auditory,** or **eustachian, tube** (fig. 18.12). The ossicle that is attached to the tympanic membrane is the **malleus** (*malleus* = hammer). As the membrane vibrates, the malleus rocks back and forth, amplifying the sound waves. The malleus is attached to the **incus** (*incus* = anvil), which is attached to the **stapes** (*stapes* = stirrup) (fig. 18.12), which is next to the oval window.

The ear ossicles transfer sound to the cochlea. Sound consists of pressure waves. As these waves strike the tympanic membrane, it vibrates. This vibration is conducted by the ossicles to the oval window. The process of moving from a large-diameter structure (tympanic membrane) to a smaller-diameter structure (oval window) magnifies the sound about 20 times. Examine the ossicles in lab and compare them with the model of hearing illustrated in figure 18.13.

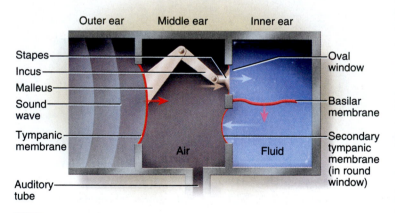

FIGURE 18.13

Model of Hearing Pressure waves of sound vibrate the tympanic membrane, which transfers sound to the oval window and finally to the basilar membrane, where vibration is converted to neural impulses.

The auditory tube connects the middle ear to the nasopharynx (fig. 18.12) and provides for the equalization of pressure between the middle ear and the external environment when changes of pressure occur (such as during changes of elevation). The auditory tube can be a conduit for microorganisms that travel from the nasopharynx to the middle ear and lead to middle ear infections, particularly in young children, in whom the tube is more horizontal than in adults.

INNER EAR

The inner ear is encased in two complex structures and filled with two separate fluids. The outermost structure is the **bony labyrinth** (*labyrinth* = maze). Inside the bony labyrinth is **perilymph,** a clear fluid that is external to the **membranous labyrinth.** The fluid enclosed by the membranous labyrinth is the **endolymph,** which is important in both hearing and equilibrium.

The inner ear is composed of three separate regions—the cochlea, the vestibule, and the semicircular ducts (figs. 18.14 and 18.15).

Cochlea The **cochlea** (figs. 18.14 and 18.15) is involved in hearing. As the sound waves travel down the auditory canal, they cause the tympanic membrane to vibrate. This vibration rocks the ear ossicles, which are connected to the inner ear. As the stapes vibrates, it moves back and forth in the **oval window,** causing fluid to move back and forth in the cochlea. The cochlea also has a **round window,** which allows the vibration from the ossicles to move fluid back and forth. Sound waves are measured by their wavelengths. These wavelengths are measured in cycles per seconds, also known as hertz (Hz).

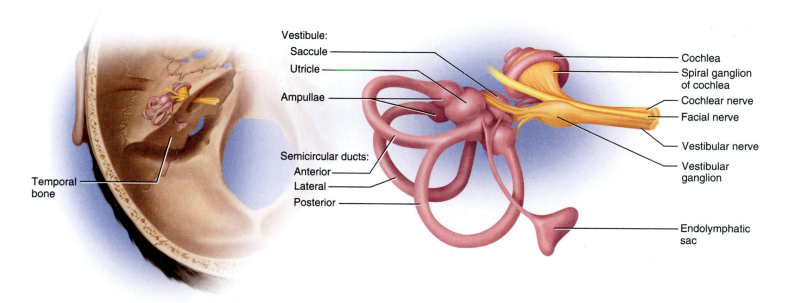

FIGURE 18.14
Anatomy of the Inner Ear

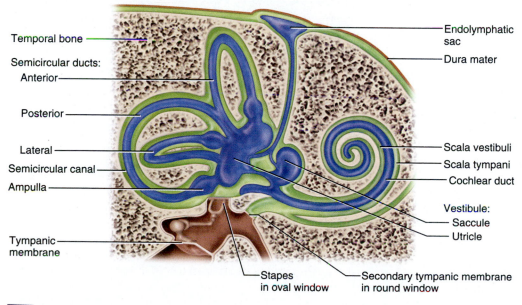

FIGURE 18.15

Inner Ear—Membranous and Bony Labyrinth Perilymph is found in the regions shown in blue and endolymph is found in the regions shown in yellow.

In the human ear, high-pitched sounds with vibrations of short wavelength, up to 20,000 Hz, stimulate the region of the cochlea closest to the middle ear. Low-pitched sounds with longer wavelengths, down to 20 Hz, stimulate the region of the cochlea further from the middle ear. In this way the cochlea can perceive sounds of varying wavelengths at the same time.

Microscopic Section of Cochlea   If you examine a microscopic section of the cochlea in cross section, you will see a number of chambers. The chambers are clustered in threes. Find the **scala vestibuli (vestibular duct), scala media (cochlear duct),** and **scala tympani (tympanic duct)** on the microscope slide. Compare these with figures 18.16 and 18.17.

Note the **spiral organ,** or the **organ of Corti,** which lies on the **basilar membrane.** The spiral organ in the slide is in cross section, but remember that it runs the length of the cochlea. The spiral organ is sensitive to sound waves. As a particular region of the spiral organ is stimulated, the basilar membrane and the gelatinous **tectorial membrane** vibrate independently of one another. This produces a tugging on the cilia that attach the hair cells to the tectorial membrane. The hair cells send impulses to the cochlear branch of the **vestibulocochlear nerve.** These impulses travel to the **auditory cortex** of the **temporal lobe,** where they are interpreted as sound (fig. 18.18).

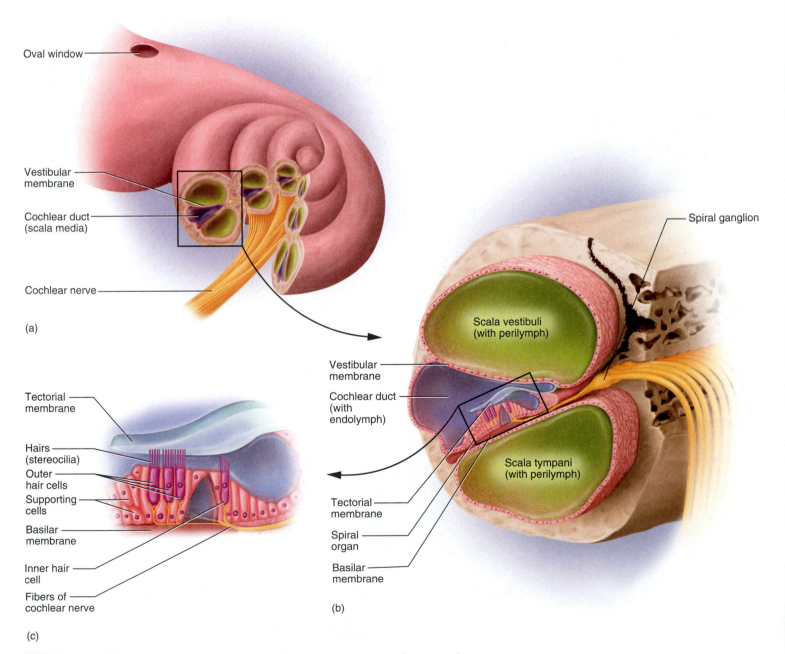

FIGURE 18.16

Anatomy of the Cochlea (a) Overview of cochlea; (b) details of chambers; (c) spiral organ.

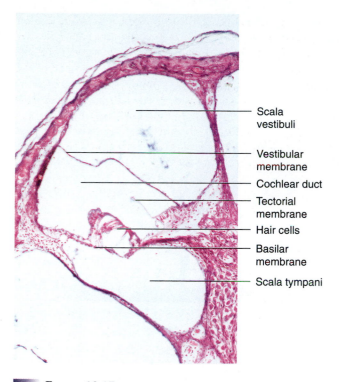

FIGURE 18.17

Photomicrograph of Cross Section of the Cochlea (100×)

Scala vestibuli

Vestibular membrane

Cochlear duct

Tectorial membrane

Hair cells

Basilar membrane

Scala tympani

Vestibule   Another part of the inner ear is the vestibule. The **vestibule** consists of the **utricle** (macula utriculi), and **saccule** (macula sacculi), as seen in figure 18.19. These two chambers are involved in the interpretation of static equilibrium and acceleration. The utricle and saccule have regions known as **maculae,** which consist of **cilia** grouped together with an overlying gelatinous mass and calcium carbonate stones called **otoliths.** These can be seen in figure 18.19. As the head undergoes linear acceleration or is tipped by gravity, the otoliths cause the cilia to bend, indicating that the position of the head has changed. Static equilibrium is perceived not only from the vestibule but from visual cues as well. When the visual cues and the vestibular cues are not synchronized, then a sense of imbalance or nausea can occur.

Semicircular Ducts   The third part of the inner ear consists of the **semicircular ducts** (fig. 18.20), which are involved in determining dynamic equilibrium. There are three semicircular ducts, each at 90° to one another (in the transverse, sagittal, and coronal planes). Each semicircular duct is filled with endolymph and is expanded at the base into an **ampulla.** Inside each ampulla is a **crista ampullaris,** a cluster of **hair cells** (cilia) with an overlying gelatinous mass called the **cupula.** The endolymph that is in the membranous labyrinth has inertia; that is, it tends to remain in the same place. As the head is turned, the cupula bends against the endolymph, the hair cells bend, and **angular,** or **rotational, acceleration** is perceived, as shown in figure 18.20. The three semicircular ducts are at

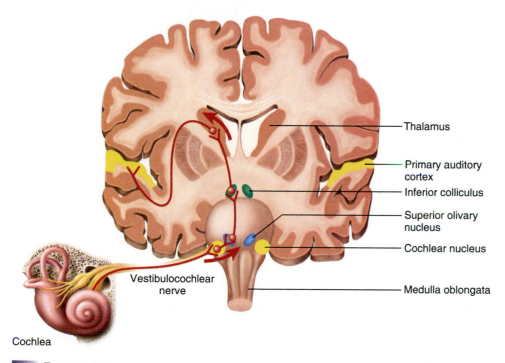

Thalamus

Primary auditory cortex

Inferior colliculus

Superior olivary nucleus

Cochlear nucleus

Medulla oblongata

Vestibulocochlear nerve

Cochlea

FIGURE 18.18

Interpretive Pathway of Hearing, Anterior View Sound impulses from the cochlea travel via nerves and tracts to the primary auditory cortex.

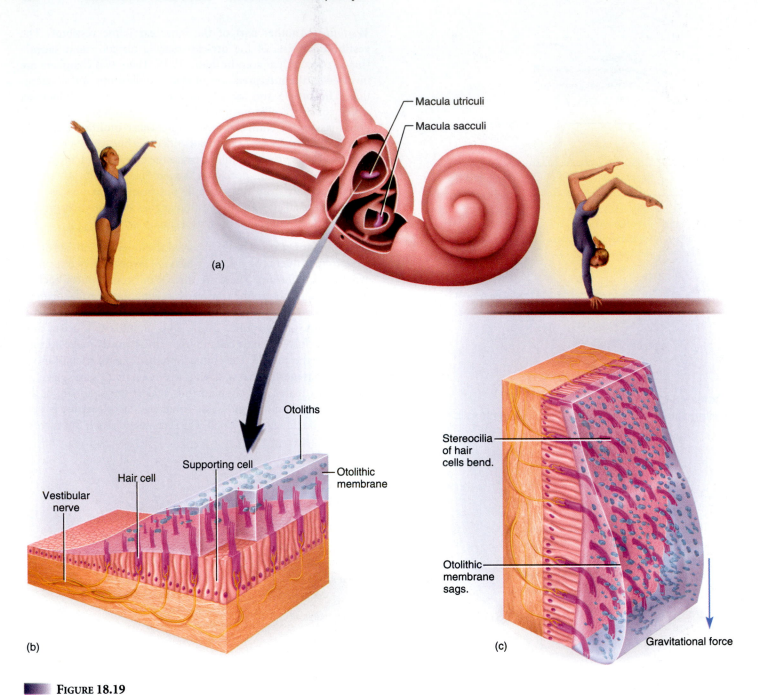

FIGURE 18.19

Vestibule (a) Inner ear; (b) macula when head is upright; (c) macula when head is tilted.

right angles to each other. If motion in the forward plane occurs (such as by doing back flips), the **anterior semicircular ducts** are stimulated. If you turn cartwheels, the **posterior semicircular ducts** are stimulated. If you spin around on your heels, the **lateral semicircular ducts** pick up the information. Motion that occurs in between these areas is picked up by two or more of the ducts and interpreted as movement, due to a combination of impulses from the semicircular ducts.

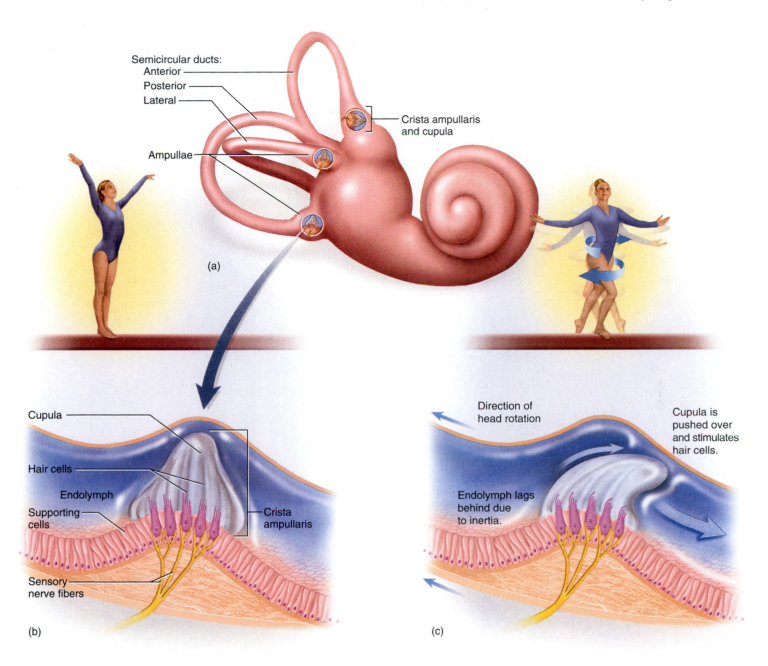

Semicircular ducts:
Anterior
Posterior
Lateral

Ampullae

Crista ampullaris
and cupula

(a)

Cupula

Hair cells

Endolymph

Supporting
cells

Crista
ampullaris

Sensory
nerve fibers

(b)

Direction of
head rotation

Cupula is
pushed over
and stimulates
hair cells.

Endolymph lags
behind due
to inertia.

(c)

FIGURE 18.20

Semicircular Ducts (a) Overview of the ducts with cut away of crista ampullaris; (b) detail of the crista ampullaris at rest; (c) movement of the crista ampullaris in motion.

Name _____ Date _____

1. Distinguish among the functions of Pacinian corpuscles, Meissner's corpuscles, and pain receptors in the skin.

2. What is punctate distribution?

3. What is a modality?

4. What kind of receptors are sensitive to the following modalities?

    a. light

    b. touch

    c. temperature

    d. sound

    e. smell

5. What kind of receptor determines the weight of an object when you pick it up?

6. What nerves transmit the sense of smell to the brain?

7. What nerves transmit the sense of taste to the brain?

8. Where are the taste buds located?

9. Fill in the following illustration using the terms provided.

lens	optic nerve	anterior cavity (anterior chamber)
sclera	choroid	cornea
ciliary body	pupil	suspensory ligaments
retina		

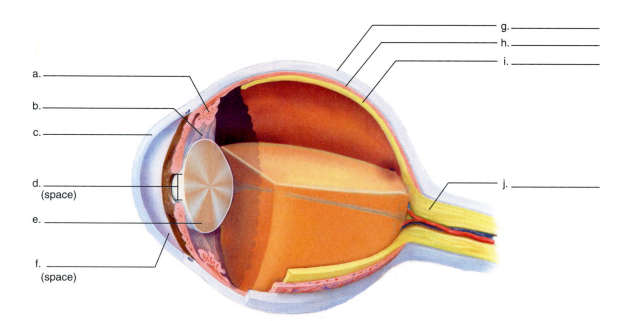

a. _____

b. _____

c. _____

d. _____
   (space)

e. _____

f. _____
   (space)

g. _____

h. _____

i. _____

j. _____

10. Since the lens is made of protein, what effect might the preserving fluid used in lab have on the structure of the lens? How would this affect the clarity?

11. How does the vitreous body differ from the aqueous humor in terms of location and viscosity?

12. What layer of the eye converts visible light into nerve impulses?

13. What nerve takes the impulse of sight to the brain?

14. What is the common name for the sclera?

15. How would you define an extrinsic muscle of the eye?

16. What gland produces tears?

17. What is the name of the transparent layer of the eye in front of the anterior chamber?

18. The iris of the eye has what function?

19. What is the middle tunic of the eye called?

20. Is the lens anterior or posterior to the iris?

21. What is at the area of the eye where the blind spot is found?

22. What are the three general regions of the ear?

23. The ear performs two major sensory functions. What are they?

24. What area is found between the scala vestibuli and the scala tympani?

25. What part of the inner ear is involved in perceiving static equilibrium?

26. What tube is responsible for the equalization of pressure when you change elevation?

27. What is the name of the space that encloses the ear ossicles?

28. Place the ear ossicles in sequence from the tympanic membrane to the oval window.

29. Fill out the illustration using the terms provided.

| earlobe | helix | stapes | auditory canal |
| auditory tube | semicircular ducts | cochlea | tympanic membrane |

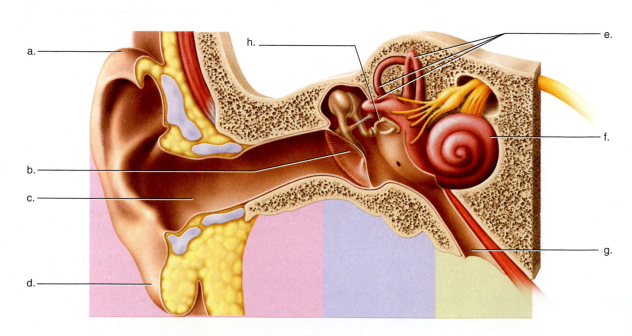

The Endocrine System

Endocrine System

INTRODUCTION

Endocrine glands produce chemical messengers called **hormones** that are picked up by blood capillaries and travel throughout the body. This type of release is called a "ductless secretion," which is characteristic of the endocrine system. Hormones enter the interstitial fluid and then travel by blood vessels, which act as highways to carry them throughout the body. Areas that are receptive to hormones are called **target areas,** and these may be organs or tissues. Many organs of the body, such as the stomach, heart, and kidney, produce hormones and thus have endocrine functions. This exercise focuses on the glands that have a major endocrine component.

Hormones can have many effects. Some of these actions are growth, changes in development (maturation), metabolism, sexual development, regulation of the sexual cycle, and homeostasis.

The glands that secrete material in ducts or tubules (such as sweat and salivary glands) or directly to a surface (ovary) are known as **exocrine glands.** Exocrine glands do not use the cardiovascular system for transport.

LEARNING OBJECTIVES

At the end of this exercise you should be able to
1. discuss how the secretions of the endocrine glands differ from those of the exocrine glands;
2. list the major endocrine organs of the human body;
3. identify endocrine organs in histological slides;
4. name the hormones produced by the endocrine organs.

MATERIALS

Models and charts of endocrine glands

Model or chart of midsagittal section of head

Microscopes

Microscope slides:

    Thyroid

    Pituitary

Adrenal gland

Pancreas

Testis

Ovary

PROCEDURE

Anatomy of the Major Endocrine Organs

Locate the major endocrine glands in charts or models in the lab and compare them with figure 19.1, including the hypothalamus, pineal gland, pituitary gland, thyroid, parathyroids, thymus, pancreas, adrenals, and gonads (testes or ovaries). Once you have noted their location, you can proceed with a more detailed study.

HYPOTHALAMUS

The **hypothalamus** has a major role in controlling endocrine functions. The hypothalamus secretes both releasing hormones and inhibitory hormones that either stimulate hormone secretion from their target areas or prevent the release of hormones from these areas. Hormones such as **gonadotropic-releasing hormone** (GnRH) stimulate the pituitary to release **follicle-stimulating hormone,** and **thyrotropic-releasing hormone** (TRH) stimulates the pituitary to release thyroid-stimulating hormone (TSH). A tropic hormone is one that causes another endocrine gland to release hormones. Locate the hypothalamus in figures 19.1 and 19.2.

PINEAL GLAND

The **pineal gland** develops from the diencephalon of the brain. The gland is so named because it resembles a pine nut. Locate the pineal gland in a model or chart of a midsagittal section of the head and compare it with figure 19.1. The pineal gland secretes the hormone **melatonin,** which is an inhibitory hormone and probably regulates circadian rhythms (daily body rhythms). Melatonin has a psychologically depressing effect in some individuals. It has been named the "hormone of darkness" because it is produced during times of low light levels and its manufacture is inhibited in bright light.

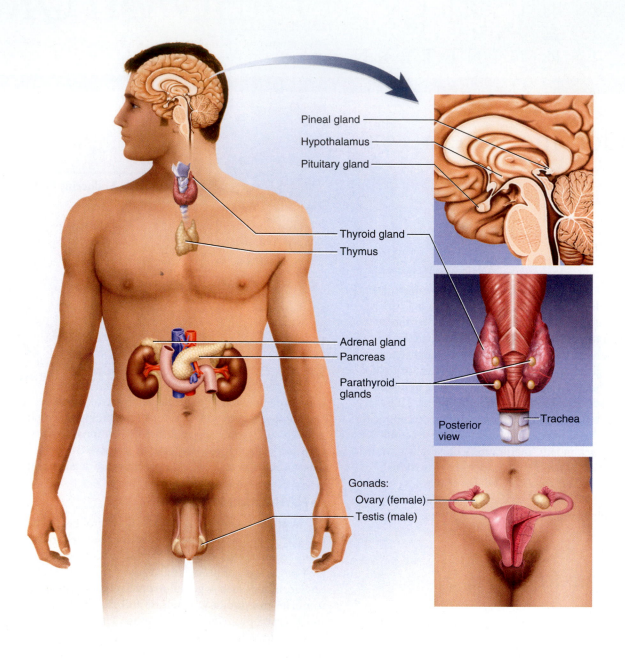

FIGURE 19.1

Major Endocrine Glands

PITUITARY GLAND

Figures 19.1 and 19.2 also illustrate another endocrine gland, called the **pituitary gland,** or **hypophysis.** The pituitary is divided into the **anterior pituitary,** or **adenohypophysis,** and a **posterior pituitary,** or **neurohypophysis.** The pituitary is suspended from the base of the brain by a stalk called the **infundibulum** and is enclosed in the sella turcica of the splenoid bone.

Adenohypophysis   The adenohypophysis originates from the roof of the oral cavity during embryonic development. The cells of the anterior pituitary have the same embryonic origin, yet they have differentiated into several types of specialized cells. The result of this development can be seen if you examine a prepared slide of the pituitary gland. In the histologic section, note that the adenohypophysis is composed of **cuboidal epithelial cells** (fig. 19.3). These

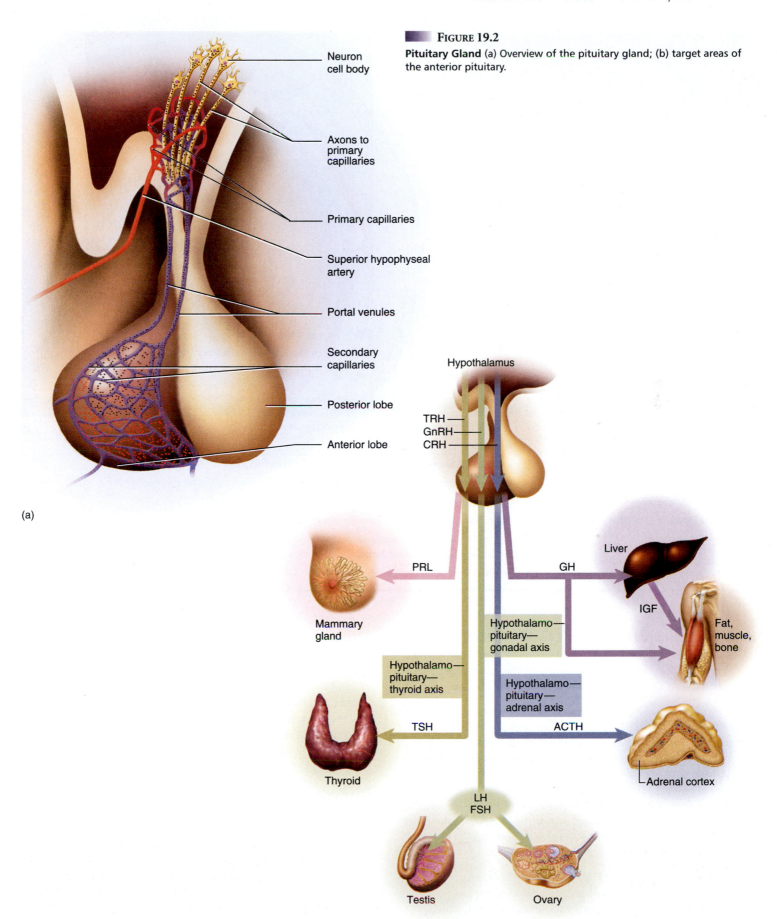

FIGURE 19.2
Pituitary Gland (a) Overview of the pituitary gland; (b) target areas of the anterior pituitary.

(a)

Neuron cell body

Axons to primary capillaries

Primary capillaries

Superior hypophyseal artery

Portal venules

Secondary capillaries

Posterior lobe

Anterior lobe

(b)

Hypothalamus

TRH
GnRH
CRH

PRL

GH

Liver

IGF

Fat, muscle, bone

Mammary gland

Hypothalamo—pituitary—gonadal axis

Hypothalamo—pituitary—thyroid axis

Hypothalamo—pituitary—adrenal axis

TSH

ACTH

Thyroid

Adrenal cortex

LH
FSH

Testis

Ovary

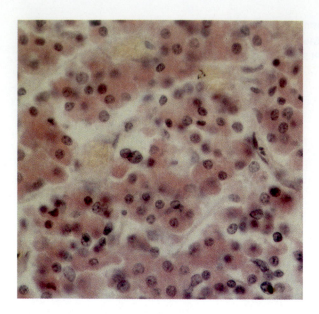

FIGURE 19.3
Histology of the Anterior Pituitary (400×)

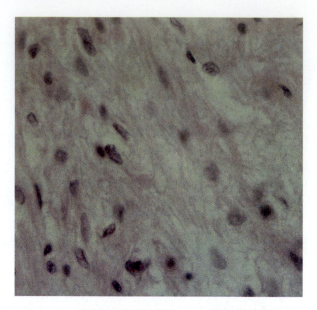

FIGURE 19.4
Histology of the Posterior Pituitary (400×)

TABLE 19.1	
Hormones of the Adenohypophysis	
Name	**Function**
TSH—thyroid-stimulating hormone (thyrotropin)	Increases the secretion of thyroid hormones from the thyroid gland
GH—growth hormone (somatotropin)	Promotes the growth of most cells and tissues in the body
PRL—prolactin (luteotropin—LTH)	Stimulates mammary glands to begin the production of milk
FSH—follicle-stimulating hormone	A gonadotropin causing follicle development in ovaries and spermatozoa production in testes
LH—luteinizing hormone	A gonadotropin causing estrogen and progesterone production in ovaries, follicle maturation, ovulation, and production of testosterone in testes
ACTH—adrenocorticotropin	Regulates hormone production in adrenal cortex

cells, generally grouped into acidophil and basophil cells, produce a number of hormones that have broad effects throughout the body.

Some of the hormones produced in the anterior pituitary are illustrated in figure 19.2 and their functions are described in table 19.1.

Neurohypophysis   The tissue of the neurohypophysis, or posterior pituitary, is very different from that of the adenohypophysis. The neurohypophysis is composed of **nervous tissue** that originated from the base of the brain. Compare the tissue of the neurohypophysis in a prepared slide with figure 19.4.

Hormones released from the posterior pituitary are actually secreted by the **hypothalamus** and flow through axons to be *stored* in the posterior pituitary. These hormones and their actions include the following:

• **Antidiuretic hormone (ADH)** stimulates the reabsorption and retention of water by the kidneys. It is also known as **vasopressin** because it causes arterioles to constrict, which elevates blood pressure.

• **Oxytocin** stimulates the contraction of the cells of the mammary glands, resulting in the release of milk, and causes uterine contractions. External influences on hormone actions can be seen in the release of oxytocin. Look at figure 19.5 and note that the action of an infant suckling the mother's breast sends impulses to the hypothalamus. The hypothalamus stimulates the posterior pituitary to release oxytocin, which causes milk ejection.

THYROID

Figure 19.6 illustrates the **thyroid gland,** which is named for its location near the thyroid cartilage of the larynx. The thyroid gland has two lateral **lobes** and a medial **isthmus** connecting them.

The thyroid has a characteristic histologic structure that can be seen by examining a thin section of the organ. Examine a slide of thyroid gland and compare it with figure 19.7. Locate the **follicle cells** that surround the **colloid,** a storage region for thyroid hormones. Colloid is mostly composed of thyroglobulin, which is a

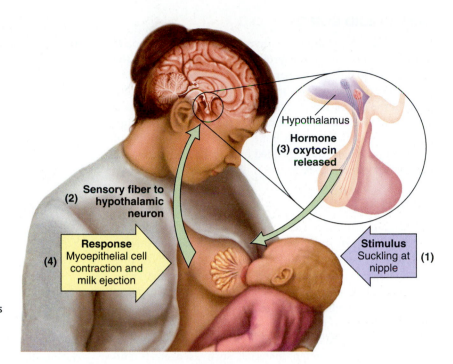

(3) Hypothalamus

Hormone (3) oxytocin released

(2) Sensory fiber to hypothalamic neuron

Response Myoepithelial cell contraction and milk ejection **(4)**

Stimulus Suckling at nipple **(1)**

■ FIGURE 19.5

External Influences on Hormonal Action (1) External stimulus of suckling. (2) Sensory neurons take the impulse to the brain, where (3) the hypothalamus stimulates the posterior pituitary to release oxytocin and (4) milk is released from the breast.

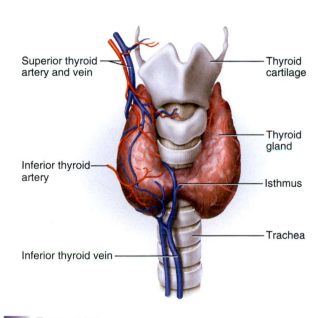

Superior thyroid artery and vein

Thyroid cartilage

Thyroid gland

Inferior thyroid artery

Isthmus

Inferior thyroid vein

Trachea

■ FIGURE 19.6

Anatomy of the Thyroid Gland

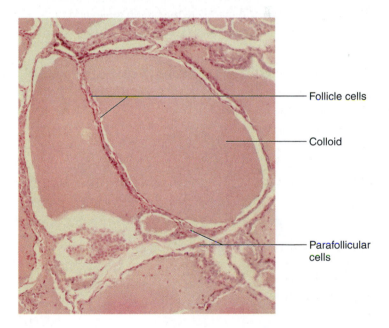

Follicle cells

Colloid

Parafollicular cells

■ FIGURE 19.7

Histology of the Thyroid Gland (100×)

large molecular weight compound. Thyroid hormones are formed inside the larger thyroglobulin molecule. Also locate the **parafollicular cells** in the spaces between the follicles.

The thyroid gland secretes three hormones. Two of these are T_3, or **triiodothyronine** (a molecule that includes three iodine atoms), and T_4, or **thyroxine** (containing four iodine atoms per

molecule). These hormones are involved with **basal metabolic rates** and are stored in the colloid. Another hormone of the thyroid gland is **calcitonin.** Calcitonin causes the deposition of calcium in bone by decreasing osteoclast activity and formation. Calcitonin is produced by the parafollicular, or C, cells. Calcitonin is antagonistic to parathyroid hormone (PTH) (discussed next).

PARATHYROID GLANDS

In the posterior portion of the thyroid are typically two pairs of organs called the **parathyroid glands. Chief cells** in the parathyroid gland secrete **parathyroid hormone (PTH),** or **parathormone,** which is responsible for increasing calcium levels in the blood. This occurs by increasing calcium uptake in the intestines, increasing kidney reabsorption of calcium, and releasing calcium from bone. Examine the location of the parathyroid glands embedded in the posterior surface of the thyroid gland (fig. 19.8).

THYMUS

The **thymus** gland is active in young individuals and plays an important part in immunocompetency. It becomes smaller as we age and develops more fibrous and fatty tissue. The thymus is located anterior and superior to the heart (fig. 19.9). It secretes a hormone, **thymosin,** that causes the maturation of T cells. These cells originate as stem cells in the bone marrow and migrate to the thymus. Under the influence of thymosin, the cells mature to provide "cellular immunity" against **antigens** (foreign material, such as bacteria and viruses, which cause immune reactions). The T cells migrate from the thymus predominantly to the lymph nodes and the spleen to carry out their functions.

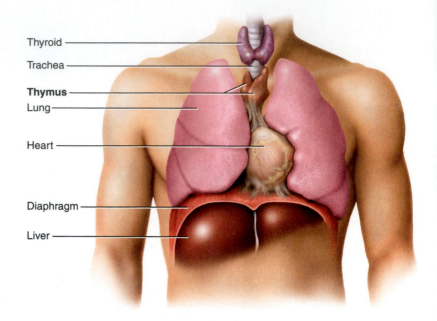

FIGURE 19.9
Thymus Gland

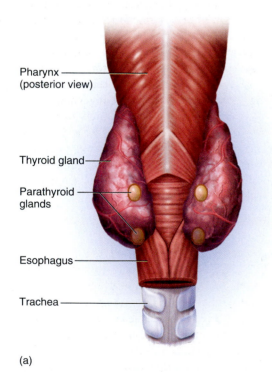

(a)

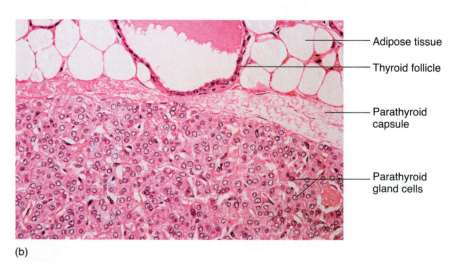

(b)

FIGURE 19.8
Parathyroid Glands (a) Gross anatomy; (b) histology (400×).

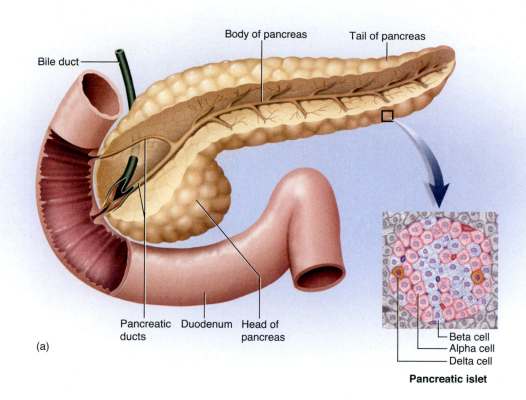

Pancreatic islet

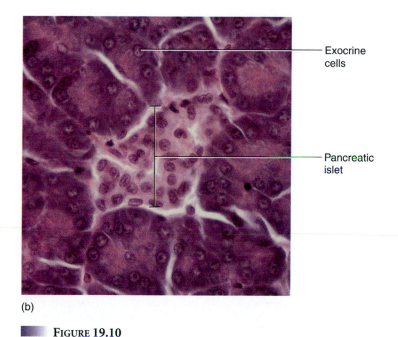

(b)

FIGURE 19.10

Pancreas (a) Gross anatomy; (b) histology (400×).

PANCREAS

The **pancreas** is a mixed gland in that it has an exocrine function and an endocrine function. Locate the pancreas in figures 19.1 and 19.10a. The exocrine function is digestive in nature, because pan-creatic juice secreted through the pancreatic ducts contains both buffers and digestive enzymes. The endocrine function of the pan-creas consists of the secretion of the hormones **insulin** and **glucagon,** which regulate blood glucose levels. When blood glucose levels drop, glucagon converts glycogen (a starch storage product) to glucose. Glucagon is produced in specialized cells (**alpha [α] cells**) in clusters called **pancreatic islets (islets of Langerhans)** (fig. 19.10a). Pancreatic islets also produce insulin, which lowers the blood glucose level. Insulin, which is produced in **beta (β) cells,** stimulates the conversion of glucose to glycogen. The pancreas also contains **delta cells,** which secrete **somatostatin,** which inhibits di-gestive secretions. Locate the pancreatic islets on a microscope slide and compare them with those in figure 19.10b.

ADRENAL GLANDS

The **adrenal glands** (*ad* = next to; *renal* = kidney) are located su-perior to the kidneys. Each adrenal gland is composed of an outer **cortex** and an inner **medulla.** These are illustrated in figure 19.11a.

Examine a microscope slide of an adrenal gland and locate the cortex and the medulla. The hormones secreted from the ad-renal cortex are called **corticosteroid hormones** and are con-trolled by ACTH. They are important in water and ion (Na^+, K^+) balance in the body. They are also important for carbohydrate, protein, and fat metabolism, as well as stress management. The cortex can be divided into three regions, all of which secrete corti-coid hormones. The outermost is the **zona glomerulosa** (fig. 19.11b), which consists of clusters of cells that predominately se-crete **mineralocorticoids** (especially **aldosterone**). Inside this

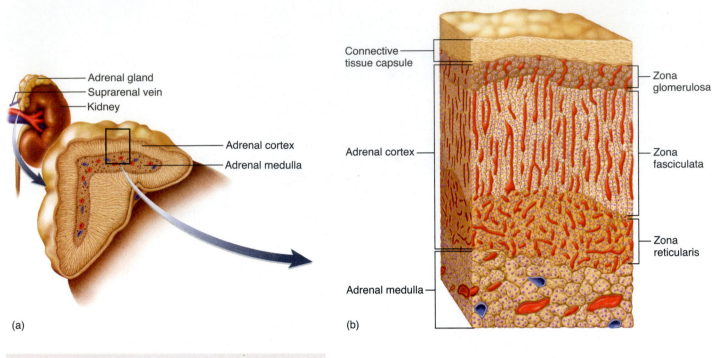

(a)

(b)

Adrenal gland
Suprarenal vein
Kidney

Adrenal cortex
Adrenal medulla

Connective tissue capsule

Adrenal cortex

Adrenal medulla

Zona glomerulosa

Zona fasciculata

Zona reticularis

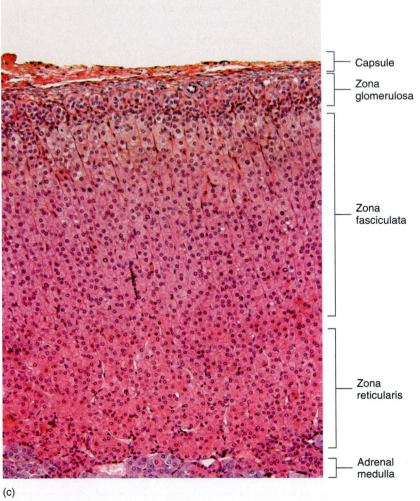

(c)

Capsule

Zona glomerulosa

Zona fasciculata

Zona reticularis

Adrenal medulla

FIGURE 19.11

Adrenal Cortex and Medulla Diagram (a) gross anatomy; (b) histology; (c) photomicrograph (100×).

layer (closer to the medulla) is the **zona fasciculata,** which consists of parallel bundles of cells that primarily secrete **glucocorticoids.** The deepest cortical layer is the **zona reticularis,** which consists of a branched pattern of cells that produce both glucocorticoids and **sex hormones** (**androgens** and **estrogens**). The hormone **androstenedione** is a weak androgen, yet both males and females produce androgens from the adrenal glands in small amounts. Androstenedione is secreted from the adrenal glands but is converted to testosterone in other tissues. Most of the androgens in males are produced by the testes.

The hormones **epinephrine** and **norepinephrine** are produced in the adrenal medulla. Stimulation of the adrenal glands by the sympathetic nervous division causes the release of epinephrine and norepinephrine from the gland.

GONADS

The **gonads** produce sex hormones and are thus considered endocrine glands. The gonads are receptive to follicle-stimulating hormone (FSH) from the anterior pituitary, which causes the production and maturation of the sex cells (**spermatozoa** or **oocytes**). The gonads are also under the influence of luteinizing hormone (LH), which increases the level of hormone production, such as estrogen and testosterone, by the gonads.

Testes   In the male the **testes** produce **testosterone,** which is a hormone responsible for secondary sex characteristics, such as the development of facial and body hair, the expansion of the larynx (which produces a deeper voice), and the increased muscle and bone mass seen in males. The testes are mixed glands that have both an endocrine and an exocrine function. The testes are illustrated in figure 19.1. The endocrine function is testosterone production, and the exocrine function is the production of spermatozoa. The exocrine function of the testis is explored in laboratory exercise 28 on the male reproductive system.

Examine a microscope slide of a testis and find the **seminiferous tubules** and **interstitial cells.** Refer to figure 19.12 for assistance. The interstitial cells are found between the tubules and produce testosterone. Testosterone not only is responsible for secondary sex characteristics but also influences the development of the male genitalia (penis and scrotum) during embryological and fetal development. Testosterone also aids FSH in the production of **spermatozoa,** which occurs in the seminiferous tubules.

Ovaries   In females the **ovaries** produce **estrogen** and **progesterone** (fig. 19.1). Ovaries are also considered mixed glands that produce hormones as an endocrine function and **oocytes** (eggs) as an exocrine function. The ovaries are illustrated in figure 19.13. Female hormones are also responsible for secondary sex characteristics in women, such as enlarged breasts, the development of an additional subcutaneous adipose layer, and a higher voice. Estrogen and progesterone also influence the development of the endometrium, cause maturation of the oocytes, and regulate the menstrual cycle. "Estrogen" is a generic term for several hormones produced by the female, including **estradiol.** During pregnancy the placenta has a major role as an endocrine gland in the secretion of estrogen and progesterone.

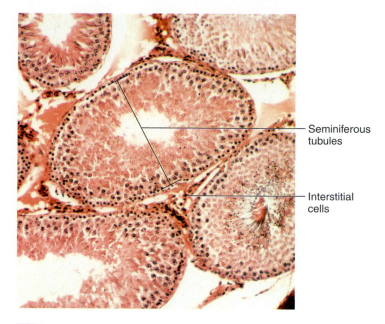

FIGURE 19.12
Histology of the Testis (100×)

Seminiferous tubules

Interstitial cells

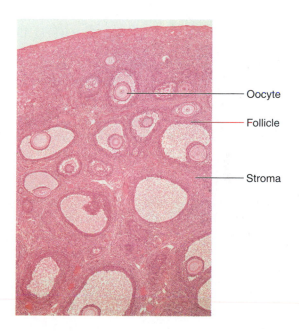

FIGURE 19.13
Histology of the Ovary (40×)

Oocyte

Follicle

Stroma

Name _____ Date _____

1. What name is given to regions that are receptive to hormones?

2. Melatonin is secreted by what gland?

3. In what specific part of what gland is ADH stored?

4. What is the effect of TSH and where is it produced?

5. What does glucagon do as a hormone and where is it produced?

6. Which hormones in the adrenal gland control water and electrolyte balance?

7. What is the primary gland that secretes epinephrine?

8. Where is growth hormone produced?

9. What is another name for T_3?

10. What connects the two lobes of the thyroid gland?

11. What impact does parathormone have on calcium levels in the blood?

12. Interstitial cells produce which hormone?

13. What hormone causes a *decrease* in calcium levels in the blood?

14. Label the endocrine glands indicated in the following illustration using the terms provided.

parathyroid glands

adrenal glands

pancreas

testes

thyroid gland

ovaries

pituitary gland

thymus

pineal gland

hypothalamus

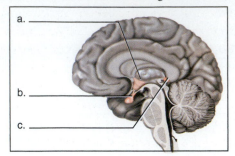

a. _____

b. _____

c. _____

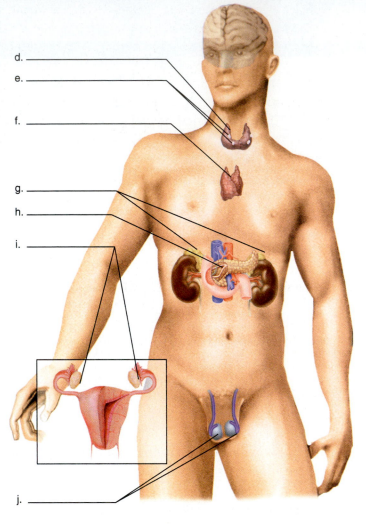

d. _____

e. _____

f. _____

g. _____

h. _____

i. _____

j. _____

15. Identify the three layers of the adrenal cortex as illustrated and list a hormone produced by each layer.

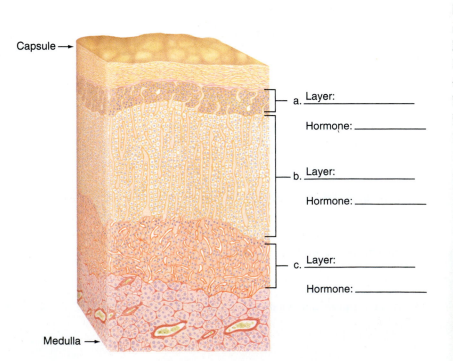

Capsule →

Medulla →

a. Layer: _____

Hormone: _____

b. Layer: _____

Hormone: _____

c. Layer: _____

Hormone: _____

16. Target cells must have the correct receptor site in order for hormones to influence those cells. Cut out the following 12 outlines, which represent either hormones or an effect caused by a hormone. In the structures, hormones are circled above the endocrine glands that produce them. The effects caused each hormone are not circled and the organ where the effect occurs is listed below the effects. Assemble the hormones and effects according to the sequence in which they fit. There may be hormones or cells that do not fit the sequence. Once the sequences are assembled, color each one with a different color. Two of the sequences represent an axis, which is a relationship between the hypothalamus, pituitary, and a more distant gland. There are three axes in the body: the hypothalamo-pituitary-gonadal axis (HPG), the hypothalamo-pituitary-thyroid axis (HPT), and the hypothalamo-pituitary-adrenal axis (HPA). Write HPG, HPT, or HPA on the corresponding sequence. Notice that, in some sequences, a hormone can have more than one effect (though not all the effects of the hormones are listed on this page). Select a hormone and target gland from your lab text and make a sequence of your own.

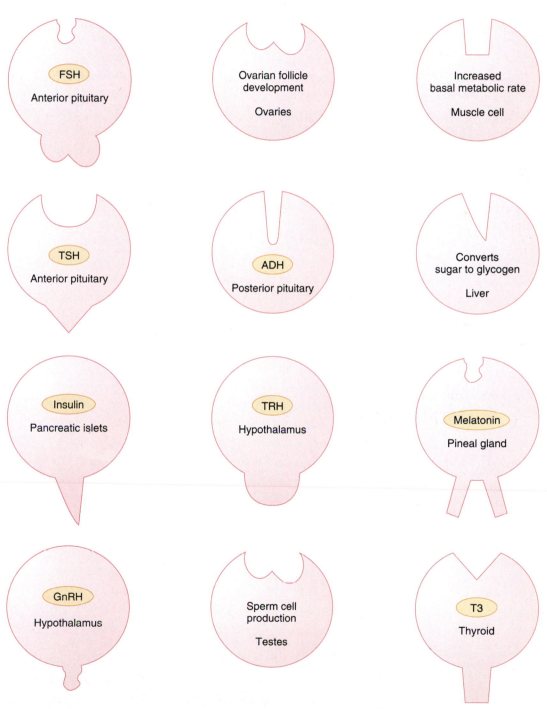

Blood Cells

Cardiovascular System

INTRODUCTION

Blood is made of two major components—formed elements and plasma. Formed elements make up about 45% of the blood volume and can further be subdivided into red blood cells (erythrocytes), white blood cells (leukocytes), and platelets (thrombocytes). Plasma constitutes approximately 55% of the blood volume and contains water, lipids, dissolved substances, colloidal proteins, and clotting factors.

The study of blood is important because it is the fluid medium of the cardiovascular system and because it has significant clinical implications. Changes in the numbers and types of blood cells may be used as indicators of disease. In this exercise you examine the nature of blood cells.

Caution   The risk of bloodborne diseases has been significantly minimized by the use of sterilized human blood and nonhuman mammal blood in this exercise. However, use the same precautions as if you were handling fresh and potentially contaminated human blood. Your instructor will determine whether to use animal blood (from nonhuman mammals), sterilized blood, or synthetic blood. In any of these cases, follow strict procedures for handling potentially pathogenic material.

1. Wear protective gloves during the procedures.
2. Do not eat or drink in lab.
3. If you have an open wound do not participate in this exercise, or make sure the wound is *securely* covered.
4. Your instructor may elect to have you use your own blood. Due to the potential for disease transmission, such as AIDS or hepatitis, in fresh blood samples, **make sure you keep away from other students' blood and keep them away from your blood!**
5. Place all disposable material in the biohazard bag.
6. Place all used lancets in the sharps container and place all glassware that is to be reused in a 10% bleach solution.
7. After you have finished the exercise, clean and disinfect the countertops with a 10% bleach solution.

LEARNING OBJECTIVES

At the end of this exercise you should be able to
1. discuss the composition of blood plasma;
2. distinguish among the various formed elements of blood;
3. describe hematopoiesis;
4. determine the percent of each type of leukocyte in a differential white cell count;
5. describe what an elevated level of a particular white blood cell may indicate as far as a disease state or an allergic reaction.

MATERIALS

Prepared slides of human blood with Wright's or Giemsa stain
Compound microscopes
Lab charts or illustrations showing the various blood cell types
Latex gloves
Roll of paper towels
Vial of mammal blood or
      Sterile cotton balls
      Alcohol swabs
      Sterile, disposable lancets
      Adhesive bandages
Dropper bottle of Wright's stain
Squeeze bottle of distilled water or phosphate buffer solution
Large finger bowl or staining tray
Toothpicks
Clean microscope slides
Coverslips
Pasteur pipette and bulbs
Hand counter
Biohazard bag or container
10% bleach container
Sharps container

PROCEDURE

Plasma

Plasma is the fluid portion of blood and is about 92% water. The remainder mostly consists of proteins, such as albumins, globulins, and fibrinogen. **Albumins** are produced by the liver and make up the majority of the plasma proteins. Some **globulins** are made by the plasma cells and make up the next largest amount of proteins. **Fibrinogen** is a clotting protein and this and other clotting factors are produced by the liver. Plasma also contains ions (Na^+, K^+, and Cl^-), nutrients, hormones, and wastes.

Examination of Blood Cells

In this part of the exercise you will need to distinguish among erythrocytes, leukocytes, and platelets. If you are using a prepared slide of blood, you can move to the section "Microscopic Examination of Blood Cells." If you are to make a smear, be sure to read the following directions.

PREPARING A FRESH BLOOD SMEAR

Your lab instructor will direct you to use either fresh, nonhuman mammal blood or your own blood. If you are using provided mammal blood, wear latex gloves and withdraw a small amount of blood from the vial with a clean Pasteur pipette. Place a drop of the blood on a clean microscope slide. If you are using your own blood, be sure you follow the safety precautions as discussed at the beginning of the exercise.

WITHDRAWING YOUR OWN BLOOD

Obtain the following materials: a clean, sterile lancet; an alcohol swab; an adhesive bandage; latex gloves; sterile cotton balls; two clean microscope slides; and a paper towel.

1. Arrange the materials on the paper towel in front of you on the countertop.
2. Clean the end of the donor finger with the alcohol swab and let the hand from which you will withdraw blood hang by your side for a few moments to collect blood in the fingertips. You will be puncturing the pad of your fingertip (where the fingerprints are located).
3. Peel back the covering of the lancet and hold on to the blunt end as you withdraw it from the package. Do not touch the sharp end or lay the lancet on the table before puncturing your finger.
4. Jab your finger quickly and wipe away the first drop of blood that forms with a sterile cotton ball. Throw the lancet in the sharps container. Never reuse the lancet or set it down on the table.
5. Place the second drop of blood that forms on the microscope slide about 2 cm away from one of the ends of the slide (fig. 20.1) and place another cotton ball on your finger.

6. You can now put an adhesive bandage on your finger.
7. Whether you are using prepared blood or your own blood, use another clean slide to spread the blood by touching the drop with the edge of the slide and *push the blood* across the slide (fig. 20.1). This should produce a smooth, thin smear of blood. Let the blood smear dry completely. Place the slide used to spread the blood in the 10% bleach solution.
8. After the slide is dry, place it in a large finger bowl elevated on toothpicks or on a staining tray.
9. Cover the blood smear with several drops of Wright's stain from a dropper bottle. Let the stain remain on the slide for 1 to 2 minutes.
10. After this time add water or a prepared phosphate buffer solution to the slide. You can rock the slide gently with gloved hands or blow on it to stir the stain and water. A metallic green material should come to the surface of the slide. Let the slide remain covered with stain and water for 3 to 8 minutes.
11. Wash the slide gently with distilled water until the material is light pink and then stand it on edge to dry. You may also stain blood using an alternate stain (such as Giemsa stain), following your lab instructor's procedure.

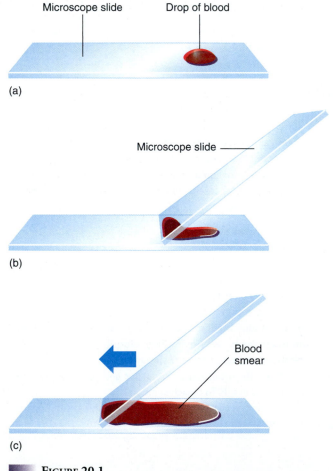

(a)

(b)

(c)

FIGURE 20.1

Making a Blood Smear (a) Placing blood on slides; (b) touching glass slide to front of blood drop; (c) spreading blood across slide.

12. Place all blood-contaminated disposable material in the biohazard container.
13. Once the slide is completely dry, it can be examined under the high-power or oil immersion lens of your microscope.

Microscopic Examination of Blood Cells

OVERVIEW

As you examine the blood slide you should refer to figures 20.2 to 20.4. You will have to look at many blood cells in order to see all the cells that are included in this exercise.

ERYTHROCYTES

Examine a slide of blood stained with either Wright's or Giemsa stain. **Erythrocytes** are the most common cells you will find on the slide. There are about 5 million erythrocytes per cubic millimeter. They do not have a nucleus but appear as pink, biconcave discs (like donuts with the holes filled in). This shape increases surface area and provides for greater oxygen-carrying capabilities. Blood formation is called **hematopoiesis** or **hemopoiesis.** The specific production of erythrocytes is called **erythropoiesis.** Erythrocytes are about 7.5 μm in diameter, on average, and have a life span of about 120 days, after which time they are broken down by the spleen or liver. Compare what you see under the microscope with figure 20.2.

PLATELETS

There are about 130,000 to 360,000 platelets per cubic millimeter of blood. Platelets, or thrombocytes, are involved in clotting and consist of small fragments of megakaryocytes. Examine the slide for small, purple fragments that may be single or clustered and compare them with the platelets in figure 20.2.

LEUKOCYTES

There are far fewer **leukocytes,** or **white blood cells,** in the blood than erythrocytes. The number of leukocytes in a healthy adult is about 7,000 cells per cubic millimeter, with a range of 5,000 to 10,000 cells per cubic milliliter of blood. Leukocytes are formed in bone marrow tissue. The life span of a white blood cell varies from a few hours to several months. Many are capable of ameboid movement as they squeeze between cells (a process termed *diapedesis*), engulfing foreign particles or cellular debris.

Leukocytes can be divided into two groups based on the presence or absence of granules in their cytoplasm. These two groups are the **granular** and **agranular leukocytes.** Examine the slide under high power or oil immersion and identify the different leukocytes as discussed in the following sections.

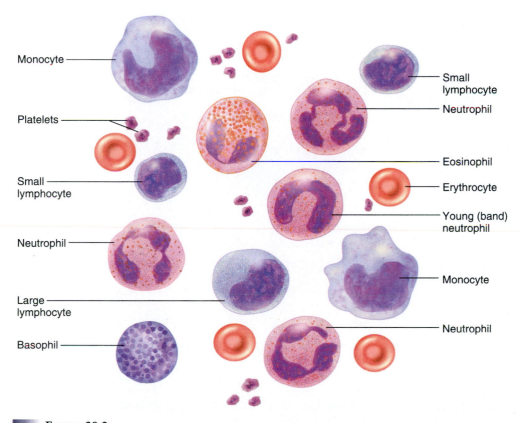

Monocyte

Platelets

Small lymphocyte

Neutrophil

Large lymphocyte

Basophil

Small lymphocyte

Neutrophil

Eosinophil

Erythrocyte

Young (band) neutrophil

Monocyte

Neutrophil

FIGURE 20.2
Formed Elements

Granular Leukocytes   Granular leukocytes (**granulocytes**) are so named because they have granules in their cytoplasm. They are also known as **polymorphonuclear (PMN) leukocytes** due to the variable shape of their nuclei (which are lobed, not round). The three types of granular leukocytes are **neutrophils, eosinophils,** and **basophils.**

Neutrophils are the most common of all leukocytes. Neutrophils typically live for about 10 to 12 hours and have ameboid capabilities. They move by diapedesis between blood capillary cells and in the interstitial areas between cells of the body. Neutrophils move toward infection sites and destroy foreign material. They are the major phagocytic white blood cell. The granules of neutrophils absorb very little stain, but neutrophils can be distinguished from other leukocytes by their three- to five-lobed nucleus. They are about one and a half times the size of erythrocytes. Examine your slide for neutrophils and compare them with figure 20.3*a*.

Eosinophils typically have a two-lobed nucleus with pink-orange granules in the cytoplasm. The term "eosinophil" actually means eosin-loving (eosin is a pink-orange stain and is picked up by the granules). They are about twice the size of erythrocytes. Compare figure 20.3*b* with your slide as you locate the eosinophils.

Basophils are rare. The granules stain very dark (blue-purple), and sometimes the nucleus is obscured because of the dark-staining granules. The nucleus is **S**-shaped and the cell is about twice the size of erythrocytes. Their granules contain histamines (vasodilators) and heparin (an anticoagulant) that allow for movement of other leukocytes out of the capillaries to infection sites. You may have to look at 200 to 300 leukocytes before finding a basophil. Compare the basophil in your slide with figure 20.3*c*.

Agranular Leukocytes   Agranular leukocytes (**agranulocytes**) are so named because they lack cytoplasmic granules. The nuclei are not lobed but may be dented or kidney bean–shaped. There are two types of agranular leukocytes—**lymphocytes** and **monocytes**.

Lymphocytes have a large, unlobed nucleus that usually has a flattened or dented area. The cytoplasm is clear and may appear as a blue halo around the purple nucleus. In terms of function, lymphocytes do not need prior exposure to recognize antigens, and they are remarkable in that separate cells have specific antibodies for specific antigens. There are two groups of lymphocytes, the B cells and T cells. These cells cannot normally be distinguished from one another in standard histologic preparations (for example, Wright's stain) and are considered simply as lymphocytes in this exercise. Both B cells and T cells arise from fetal bone marrow. **B cells** probably mature in the fetal liver and spleen and **T cells** mature in the thymus gland. B cell lymphocytes mature into **plasma cells**, which make **antibodies.** Plasma cells provide **humoral im-**

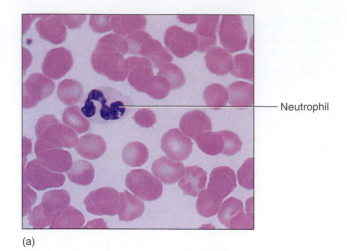

(a)

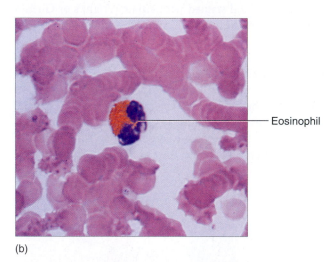

(b)

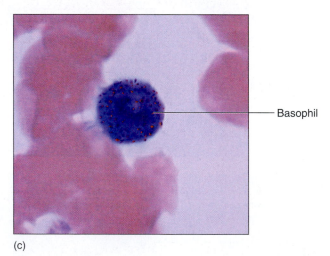

(c)

FIGURE 20.3

Granular Leukocytes (a) Neutrophil (1,000×); (b) eosinophil (1,000×); (c) basophil (1,500×).

munity (i.e., the plasma cells secrete antibodies that travel in the "humor," or fluid portion, of the blood).

T cells provide **cellular immunity.** In cellular immunity the cells themselves (not antibodies in the blood plasma) move close to

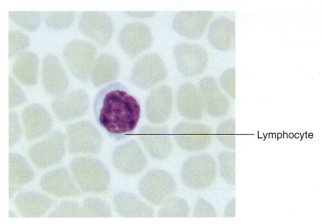

(a)

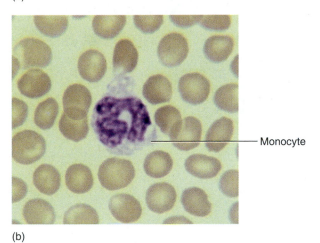

(b)

FIGURE 20.4
Agranular Leukocytes (1,000X) (a) Lymphocyte; (b) monocyte.

and destroy some types of bacteria or virus-infected cells. T cells also attack tumors and transplanted tissues. Most of the T cells are found in the lymph nodes, thymus, and spleen. They enter the bloodstream via the lymphatic tissue. Compare your slide with figure 20.4 for lymphocytes.

Monocytes are very large and have a kidney bean– or horseshoe-shaped nucleus. They are about three times the size of erythrocytes and are activated by T cells. Locate the large cells on your slide and compare them with figure 20.4.

Differential White Blood Cell Count

Once you have identified the various leukocytes on the blood slide, conduct a differential white blood cell count. This procedure is frequently used in clinical settings because it is easy to perform and relatively inexpensive. Changes in the relative percentages of leukocytes may indicate the presence of a particular disease.

1. Use a hand counter and count 100 leukocytes. One lab partner should look into the microscope and methodically call out the names of the different types of leukocytes seen.

2. The other lab partner should record how many of each type are found and keep track of the overall number with the use of a hand counter until 100 cells are counted.

3. Scan the slide in a systematic way, so that you don't count any cell twice. One method is illustrated in figure 20.5. Tally your results.

FIGURE 20.5
Counting Leukocytes

Neutrophils: _____

Eosinophils: _____

Basophils: _____

Lymphocytes: _____

Monocytes: _____

Neutrophils represent about 60% to 70% of all the leukocytes. They increase in number in appendicitis or acute bacterial infections.

Eosinophils represent about 2% to 4% of all leukocytes. Eosinophils increase in number during allergic reactions and parasitic infections (e.g., trichinosis).

Basophils represent about 0.5% to 1% of all leukocytes. They increase in number during allergies and radiation.

Lymphocytes make up 25% to 33% of the leukocytes. These cells increase in times of viral infection, antibody-antigen reactions, and infectious mononucleosis.

Monocytes make up about 3% to 8% of the leukocytes. They increase in times of chronic infections, such as tuberculosis.

Review the general properties of blood in Table 20.1 and the characteristics of formed elements in Table 20.2.

TABLE 20.1	
General Properties of Blood	
Volume	Female: 4–5 L
	Male: 5–6 L
pH	7.35–7.45
Mean Salinity	0.9%
Hematocrit	Female: 37% to 48%
	Male: 45% to 52%
Hemoglobin	Female: 12–16 g/dL
	Male: 13–18 g/dL
Platelet Count	130,000–360,000/uL
Total WBC* Count	5,000–10,000/uL

*White blood cell.

TABLE 20.2

Characteristics of Formed Elements

Formed Element	Granules (If Present)	Shape of Nucleus (If Present)	Cause of Increase
Erythrocyte	No granules	No nucleus	Exercise, increase in elevation
Neutrophil	Poor staining	Three- to five-lobed	Appendicitis, acute bacterial infections
Eosinophil	Orange-staining	Bilobed	Allergic reactions, parasitic infections
Basophil	Purple-staining	S-shaped	Allergic reactions, radiation
Lymphocyte	Not obvious	Indented	Viral infection, antibody reaction, mononucleosis
Monocyte	Not obvious	Kidney bean–shaped	Chronic infection

Clean Up   Make sure the lab is clean after you finish. Place any slide with fresh blood on it or any material contaminated with bodily fluid in the bleach solution. Place all gloves or contaminated paper towels in the biohazard container. All sharps material (broken slides or coverslips, lancets, etc.) should be placed in the sharps container. Clean the counters with a towel and a 10% bleach solution. If you used immersion oil on the microscope, make sure the objective lenses are wiped clean (use clean lens paper only).

Name _____ Date _____

1. Formed elements consist of three main components. What are they?

2. What is the most common plasma protein?

3. What is another name for a platelet?

4. Which is the most common blood cell?

5. What is another name for a leukocyte?

6. What leukocyte is most numerous in a normal blood smear?

7. How many erythrocytes are normally found per cubic millimeter of blood?

8. What is an average number of leukocytes found per cubic millimeter of blood?

9. B cells and T cells belong to what class of agranular leukocytes?

10. What value is there to a change in the percentage of leukocytes to diagnostic medicine?

11. In counting 100 leukocytes, you are accurately able to distinguish 15 basophils. Is this a normal number for the white blood cell count, and what possible health implications can you draw from this?

12. What is the function of the thrombocytes of the blood?

13. Formed elements constitute what percent of the total blood volume?

14. Label the formed elements in the following illustration using the terms provided.

platelet   eosinophil   lymphocyte

erythrocyte   neutrophil

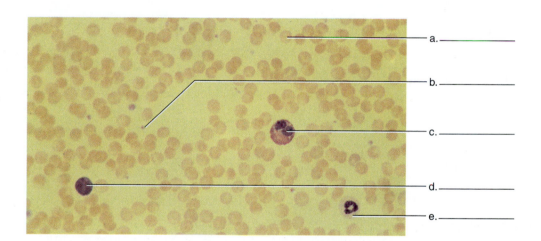

a. _____

b. _____

c. _____

d. _____

e. _____

The Heart

Cardiovascular System

INTRODUCTION

The heart is a muscular pump, which provides the force for circulating blood throughout the body. Blood has two major circulation routes. One is the circulation from the heart to the pulmonary trunk, to the lungs, and back to the heart. This is the **pulmonary circulation.** The other is from the heart to the aorta and throughout the rest of the body (head, upper extremities, trunk, lower extremities) and back to the heart. This is the **systemic circulation.** Blood travels from the heart in vessels called **arteries** and blood returns to the heart through **veins.** The blood vessels and the heart together constitute the **cardiovascular system.**

In this exercise you examine models of the heart and preserved sheep and human hearts, if available. As you look at the preserved material try to see how the structure of the heart relates to its function.

LEARNING OBJECTIVES

At the end of this exercise you should be able to
1. list the three layers of the heart wall;
2. describe the position of the heart in the thoracic cavity;
3. describe the significant surface features of the heart;
4. describe the internal anatomy of the heart;
5. find and name the anatomic features on models of the heart and in the sheep heart;
6. describe the blood flow through the heart and the function of the internal parts of the heart;
7. discuss the functioning of the heart valves and their role in circulating blood through the heart.

MATERIALS

Models and charts of the heart

Preserved sheep hearts

Preserved human hearts (if available)

Blunt probes (Mall probes)

Dissection pans

Razor blades or scalpels

Sharps container

Disposable gloves

Waste container

Microscopes

Prepared slides of cardiac muscle

PROCEDURE

Heart Location and Membranes

The heart is located deep in the thorax between the lungs in a region known as the mediastinum. The **mediastinum** contains the heart, the coverings of the heart (the pericardia), and other structures, such as the esophagus and descending aorta. The mediastinum is located between the sternum, lungs, and thoracic vertebrae and is illustrated in figure 21.1.

If you were to open the chest cavity, the first structure you would see is the **parietal pericardium.** This tough, outer connective tissue sheath encloses the heart. Deep to the parietal pericardium is the **pericardial cavity,** which contains a small amount of **serous pericardial fluid.** This fluid reduces the friction between the outer surface of the heart and the parietal pericardium. The heart wall itself has an outer layer known as the **epicardium,** or **visceral pericardium.** Locate these pericardial layers in figure 21.1.

Heart Wall

The heart wall is composed of three major layers. The outermost layer is the epicardium, or visceral pericardium, which is composed of epithelial and connective tissue. The middle layer, the **myocardium,** is the thickest of the three layers. It is mostly made of cardiac muscle. You may wish to review the slides of involuntary

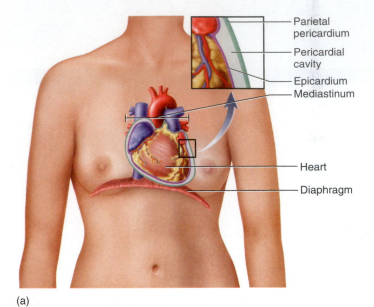

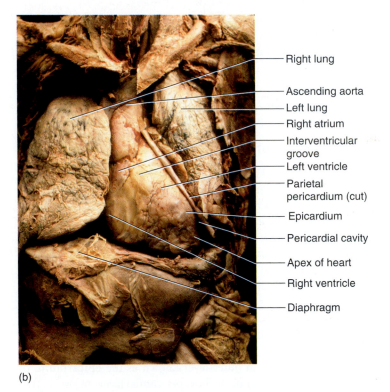

(a)

(b)

FIGURE 21.1
Heart in Thoracic Cavity and Heart Coverings (a) Coronal section;
(b) photograph of cadaver.

cardiac muscle and note the intercalated discs, branching fibers, and fine striations of cardiac tissue (as described in laboratory exercise 4). The cardiac muscle is arranged spirally around the heart, and this arrangement provides a more efficient wringing motion to the heart. The inner layer of the heart wall is known as the **endocardium,** which is a serous membrane consisting of **endothelium** (simple squamous epithelium) and connective tissue. These layers are seen in figure 21.2.

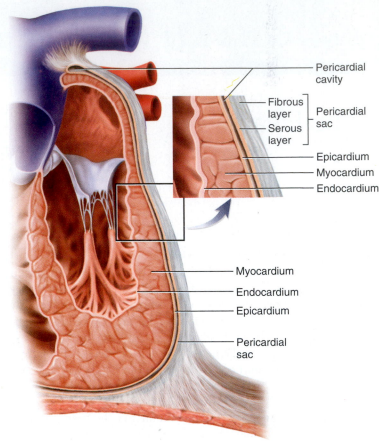

FIGURE 21.2
Pericardial Layers and Heart Wall

It is best to examine heart models before dissecting a sheep heart unless your instructor directs you to do otherwise. Heart models are color coordinated and labeled to make the structures easier to locate.

Examination of the Heart Model

OVERVIEW

The heart has four chambers, two upper atria and two lower ventricles. Blood with low oxygen levels enters the heart in the right atrium (fig. 21.3), and is pumped into the right ventricle. Once in the right ventricle, the contraction of the ventricular wall sends blood to the lungs. The blood is oxygenated in the lungs and returns to the heart by entering the left atrium. Blood moves from the left atrium to the left ventricle and is then pumped from there to the rest of the body.

INTERIOR OF THE HEART

Examine a model of the interior of the heart and locate the right and left ventricles. Note that the wall of the **right ventricle** is thinner than the wall of the left ventricle. The ventricles are separated by the **interventricular septum,** which forms a wall between the two ventricular chambers. Compare the model with figure 21.3.

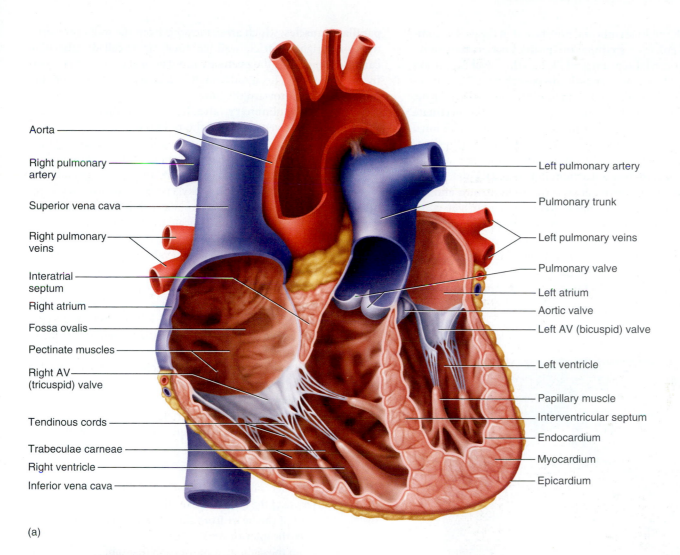

Aorta

Right pulmonary
artery

Superior vena cava

Right pulmonary
veins

Interatrial
septum

Right atrium

Fossa ovalis

Pectinate muscles

Right AV
(tricuspid) valve

Tendinous cords

Trabeculae carneae

Right ventricle

Inferior vena cava

Left pulmonary artery

Pulmonary trunk

Left pulmonary veins

Pulmonary valve

Left atrium

Aortic valve

Left AV (bicuspid) valve

Left ventricle

Papillary muscle

Interventricular septum

Endocardium

Myocardium

Epicardium

(a)

Right atrium

Right
ventricle wall

Tricuspid
valve

Tendinous
cords

Papillary
muscle

Left
atrium

Interatrial
septum

Bicuspid
valve

Tendinous
cords

Papillary
muscle

Interventricular
septum

Trabeculae
carneae

Apex of heart

(b)

FIGURE 21.3

Heart, Frontal Section (a) Diagram; (b) photograph.

Examine the **right atrium** and note how thin the wall is, compared with the ventricles. Examine the medial wall of the atrium, known as the **interatrial septum,** and locate a thin, oval depression in the atrial wall. This depression is the **fossa ovalis** (fig. 21.4). In fetal hearts this is the site of the **foramen ovale,** but closure of the foramen usually occurs just after birth. Note the extensive **pectinate muscles** on the wall of the atrium. These provide additional strength to the atrial wall. Blood in the superior vena cava, the inferior vena cava, and the coronary sinus returns to the right atrium. Examine the features of the right atrium in figures 21.3 and 21.4.

Now examine the valve between the right atrium and the right ventricle. This is the **right atrioventricular valve,** or **tricuspid valve,** and it prevents backflow of blood from the right ventricle into the right atrium during ventricular contraction. Examine the valve for three flat sheets of tissue. These are the three **cusps** of the tricuspid valve (figs. 21.3 and 21.4). The tricuspid valve has thin, threadlike attachments called **tendinous cords** (chordae tendineae). These tough cords are attached to larger

papillary muscles, which are extensions from the wall of the ventricle. The right ventricle wall has small struts called **trabeculae carneae,** which, like the pectinate muscles of the atria, strengthen the ventricle wall. The blood from the right ventricle flows into the pulmonary trunk toward the lungs.

Locate the **pulmonary valve.** It appears as three small cusps between the right ventricle and the pulmonary trunk and keeps blood from flowing backward from the pulmonary trunk into the right ventricle during ventricular relaxation. Examine the details of the right ventricle in models in the lab and in figures 21.3 and 21.4.

The **pulmonary veins** carry oxygenated blood from the lungs into the **left atrium.** These vessels are located in the superior, posterior portion of the left atrium. Locate the two large cusps of the **bicuspid valve** between the left atrium and left ventricle. The bicuspid valve is also known as the **mitral valve** (named for the shape of a bishop's miters hat) or **left atrioventricular valve.** It also has attached tendinous cords and papillary muscles, which you should locate in the models in the lab. The left ventricle has trabeculae carneae as well. Note the thickness of the left ventricle wall, compared with the wall of the right ventricle. Compare the left side of the heart with figure 21.3.

The **aortic valve** is located at the junction of the left ventricle and the ascending aorta. It has the same basic structure and general function as the pulmonary valve in that it prevents the backflow of blood from the aorta into the left ventricle. Blood from the left ventricle moves into the aorta and subsequently to the rest of the body.

BLOOD FLOW THROUGH THE HEART

On the heart model follow the pathway that blood takes through the heart as you read the remainder of this paragraph. Blood enters the right atrium of the heart from three sources. These are the superior vena cava, the inferior vena cava, and the coronary sinus. The blood flows from the right atrium through the right atrioventricular valve into the right ventricle. When the right ventricle contracts, blood flows past the pulmonary valve and into the pulmonary trunk. From there it moves to the pulmonary arteries and then to the lungs, where the blood in oxygenated. The return of the blood from the lungs to the heart is by the pulmonary veins, which carry blood to the left atrium. Blood then flows through the left atrioventricular valve into the left ventricle. When the left ventricle contracts, the blood flows past the aortic valve and into the aorta. This is illustrated in figure 21.5.

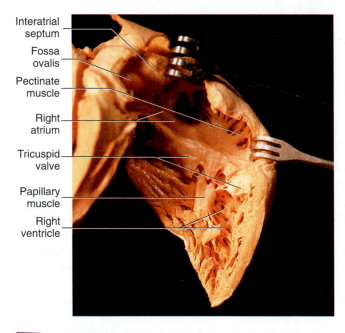

Interatrial septum
Fossa ovalis
Pectinate muscle
Right atrium
Tricuspid valve
Papillary muscle
Right ventricle

FIGURE 21.4
Details of the Right Atrium

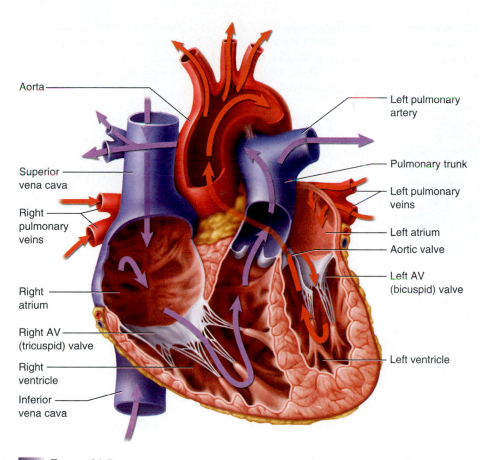

Aorta

Superior
vena cava

Right
pulmonary
veins

Right
atrium

Right AV
(tricuspid) valve

Right
ventricle

Inferior
vena cava

Left pulmonary
artery

Pulmonary trunk

Left pulmonary
veins

Left atrium

Aortic valve

Left AV
(bicuspid) valve

Left ventricle

FIGURE 21.5

Blood Flow through the Heart Purple arrows indicate deoxygenated blood; red arrows, oxygenated blood.

EXTERIOR OF THE HEART

Examine the heart model and notice that the heart has a pointed end, or **apex,** and a blunt end, or **base.** The apex of the heart is inferior, and the great vessels leaving the heart are located at the base (therefore, in the case of the heart, the base is superior to the apex). Compare the model with figure 21.6 to see how the **aorta** curves to the left in an anterior view of the heart and is posterior to the **pulmonary trunk.**

Locate the anterior features of the heart. The **left ventricle** extends to the apex of the heart and is delineated from the **right ventricle** by the **interventricular sulcus,** or **groove.** Some of the **coronary arteries** and **cardiac veins** are in this groove. These vessels will be discussed later in this exercise. The right ventricle occu-

pies less area than the left ventricle. Note the two earlike flaps that occur on the anterior, superior region of the heart. These structures are the **auricles,** which are part of the atria.

If you examine the heart from the posterior side you will see the atria more clearly. At the junction of the **right atrium** and the right ventricle is the **atrioventricular sulcus,** or **groove.** The **coronary sinus,** a large venous chamber that carries blood from the cardiac veins to the right atrium, is located in this sulcus. Locate the **superior vena cava** and the **inferior vena cava,** two vessels that also return blood to the right atrium. Locate the **pulmonary veins,** which carry blood from the lungs to the left atrium. Compare the heart model with figure 21.7.

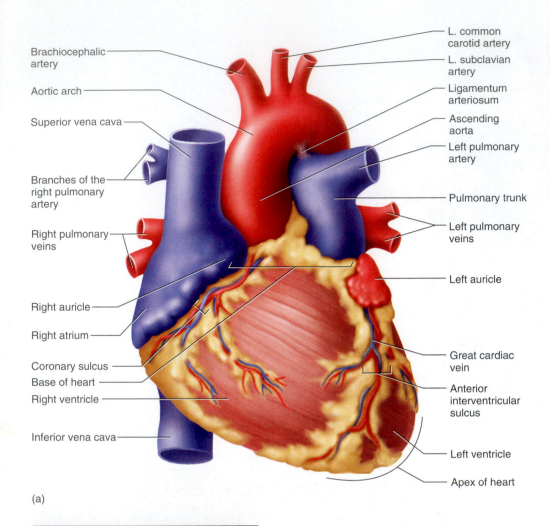

Brachiocephalic artery

Aortic arch

Superior vena cava

Branches of the right pulmonary artery

Right pulmonary veins

Right auricle

Right atrium

Coronary sulcus

Base of heart

Right ventricle

Inferior vena cava

L. common carotid artery

L. subclavian artery

Ligamentum arteriosum

Ascending aorta

Left pulmonary artery

Pulmonary trunk

Left pulmonary veins

Left auricle

Great cardiac vein

Anterior interventricular sulcus

Left ventricle

Apex of heart

(a)

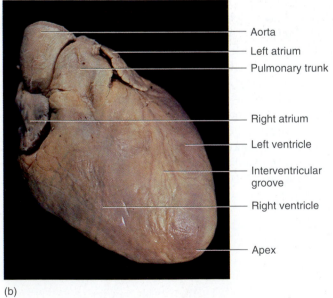

Aorta

Left atrium

Pulmonary trunk

Right atrium

Left ventricle

Interventricular groove

Right ventricle

Apex

(b)

FIGURE 21.6

Surface Anatomy of the Heart, Anterior View (a) Diagram; (b) photograph.

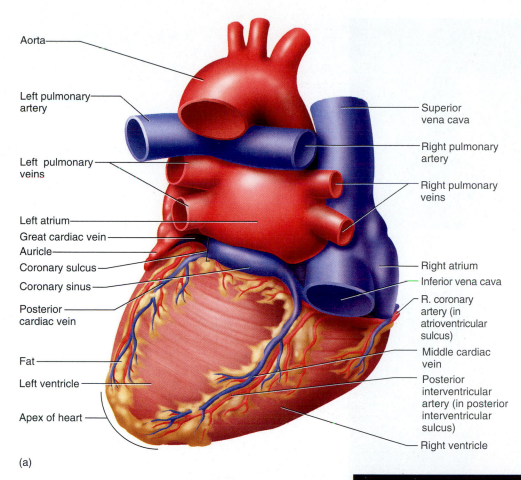

Aorta

Left pulmonary artery

Left pulmonary veins

Left atrium

Great cardiac vein

Auricle

Coronary sulcus

Coronary sinus

Posterior cardiac vein

Fat

Left ventricle

Apex of heart

Superior vena cava

Right pulmonary artery

Right pulmonary veins

Right atrium

Inferior vena cava

R. coronary artery (in atrioventricular sulcus)

Middle cardiac vein

Posterior interventricular artery (in posterior interventricular sulcus)

Right ventricle

(a)

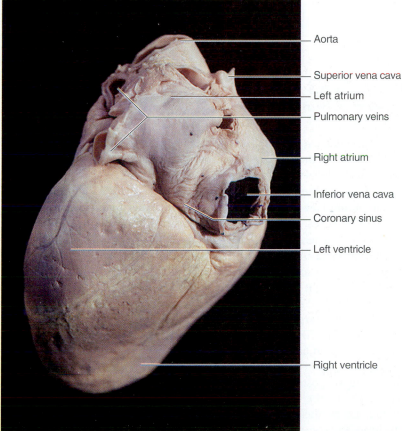

Aorta

Superior vena cava

Left atrium

Pulmonary veins

Right atrium

Inferior vena cava

Coronary sinus

Left ventricle

Right ventricle

FIGURE 21.7

Surface Anatomy of the Heart, Posterior View
(a) Diagram; (b) photograph.

(b)

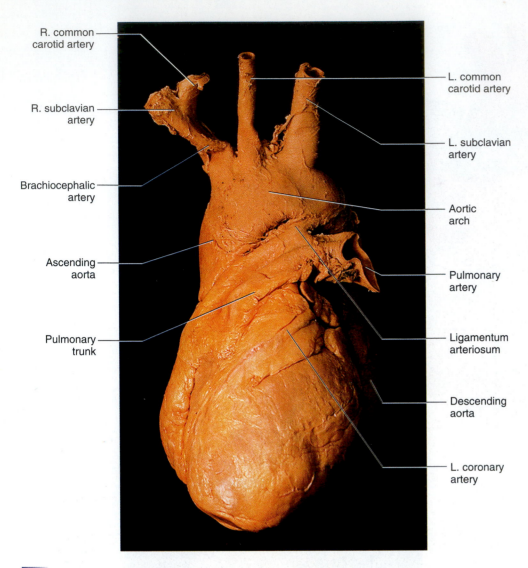

R. common
carotid artery

R. subclavian
artery

Brachiocephalic
artery

Ascending
aorta

Pulmonary
trunk

L. common
carotid artery

L. subclavian
artery

Aortic
arch

Pulmonary
artery

Ligamentum
arteriosum

Descending
aorta

L. coronary
artery

FIGURE 21.8
Vessels of the Heart, Anterior View

The major vessels of the heart are illustrated in figures 21.6 to 21.9. Locate the **pulmonary trunk, pulmonary arteries, ligamentum arteriosum** (what remains of fetal blood vessel that shunts blood between the pulmonary trunk and aortic arch), **ascending aorta, pulmonary veins, superior vena cava, inferior vena cava, coronary arteries,** and **cardiac veins.** The heart tissue is nourished by coronary arteries. The **left coronary artery** arises from the **ascending aorta** and then branches into the **anterior interventricular artery** (left anterior descending artery) and the **circumflex branch.** The **right coronary artery** also arises from the ascending aorta and branches to form the **posterior interventricular branch** and the **right marginal branch.** These major arteries of the heart supply blood to the myocardium. On the return flow from the heart muscle the **great cardiac vein** follows the depression of the interventricular groove and the atrioventricular groove to the **coronary sinus.** On the posterior, right side of the heart the **middle cardiac vein** leads to the coronary sinus, which empties into the right atrium. Locate these vessels on the external surface of the model of the heart and compare them with figure 21.9.

Dissection of the Sheep Heart

Caution Be careful when handling preserved materials. Ask your instructor for the proper procedure for working with preserving fluid and for handling and disposal of the specimen. Do not dispose of animal material in the sinks. Place them in an appropriate waste container.

The sheep heart is similar to the human heart and usually is readily available as a dissection specimen. Dissection of anatomic material is valuable in that you can examine structures that are represented more accurately in preserved material than in models. Also, the preserved material has greater flexibility and is more easily manipulated. There are some differences between sheep hearts and human hearts, especially in the position of the **superior** and **inferior venae cavae.** In sheep these are called the anterior and posterior venae cavae, but they are referred to using the human terminology.

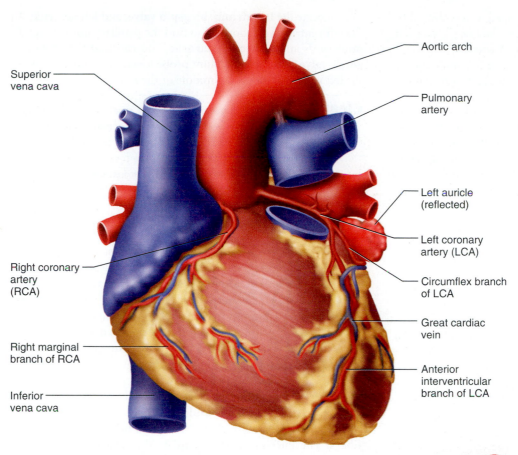

Superior
vena cava

Aortic arch

Pulmonary
artery

Left auricle
(reflected)

Left coronary
artery (LCA)

Right coronary
artery
(RCA)

Circumflex branch
of LCA

Great cardiac
vein

Right marginal
branch of RCA

Anterior
interventricular
branch of LCA

Inferior
vena cava

(a)

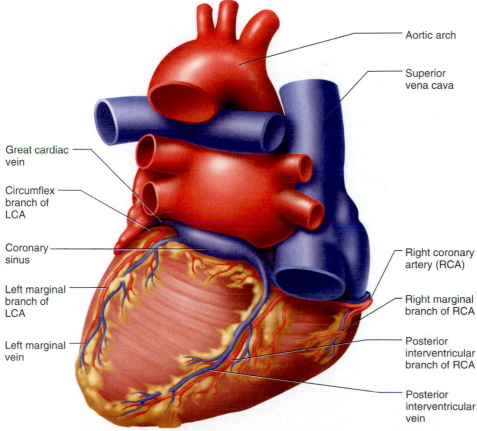

Aortic arch

Superior
vena cava

Great cardiac
vein

Circumflex
branch of
LCA

Coronary
sinus

Left marginal
branch of
LCA

Left marginal
vein

Right coronary
artery (RCA)

Right marginal
branch of RCA

Posterior
interventricular
branch of RCA

Posterior
interventricular
vein

FIGURE 21.9

Vessels of the Heart (a) Anterior view;
(b) posterior view.

(b)

If your sheep heart has not been dissected, you will need to open the heart. If your sheep or other mammalian heart has been previously dissected, you can skip the next paragraph.

Place the heart under running water for a few moments to rinse off the preserving fluid. Examine the external features of the heart. Determine if you are looking at the anterior or posterior surface. Note the fat layer on the heart. The amount of fat on the human or sheep heart is variable. Locate the **left ventricle,** the **right ventricle,** the **interventricular sulcus (groove),** the **right atrium,** and the **left atrium.** Note the **auricles** that extend on the anterior surface of the atria.

Using a sharp scalpel or razor blade, make an incision along the right side of the heart (lateral side) from the apex of the heart to the lateral side of the right atrium. If you are unsure about how to proceed during any part of the dissection, ask your instructor for directions. Make another long cut from the lateral side of the left atrium through the lateral side of the left ventricle. You will have made a coronal section of the heart if you cut through the **interventricular septum.** Once you have opened the heart, compare the structures of the sheep heart with the human heart in figure 21.3.

You can locate the vessels of the heart by inserting a blunt metal probe into the vessels and determining which chamber the vessel goes to or comes from. Place the heart in anatomic position and insert the probe into the large, anterior vessel that exits toward the specimen's left side. The blunt end of the probe should enter into the **right ventricle.** The vessel you have placed the probe into is the **pulmonary trunk.** The pulmonary trunk may still have the **pulmonary arteries** attached. Locate the large vessel directly posterior to the pulmonary trunk (fig. 21.6.) This is the **ascending aorta.** If the vessels are cut farther away from the heart, you can see the **aortic arch.** Insert the probe into this vessel and into the left ventricle.

Turn the heart to the posterior surface and locate the **superior vena cava** and **inferior vena cava.** Insert the probe into the superior and inferior vena cavae, pushing the probe into the **right atrium.** If you find only one large opening in the atrium, you may have cut through either the superior or inferior vena cava during your initial dissection. The probe can be felt through the wall more easily here than in a ventricle because the atrial walls are thinner than those of the ventricles. On the left side either the **pulmonary veins** appear as four separate veins or you may just see a large hole on each side of the left atrium if the vessels were cut close to the atrial wall. Locate the same structures in the sheep heart as you found on the posterior side of the heart model and compare them with figure 21.6.

Cut into the right atrium and use your blunt probe to locate the opening of the **coronary sinus** in the posterior, inferior portion of the atrium. It is small and somewhat difficult to find. Examine the opening between the right atrium and the right ventricle to locate the **tricuspid valve.** You may have dissected through one of the cusps as you opened the heart. Locate the major features of the right ventricle. Find the **tendinous cords** and the **papillary muscles.** You can find the **pulmonary trunk** by inserting a blunt probe into the superior portion of the right ventricle. Make an incision in the pulmonary trunk near the right ventricle to expose the three thin cusps of the **pulmonary valve.** Note how the cusps press against the wall of the pulmonary trunk when the probe is pushed against them in a superior direction. These cusps close when blood begins to flow back into the right ventricle as the ventricle relaxes.

Locate the **left atrium, bicuspid valve,** and **left ventricle.** In the left ventricle you should also find the papillary muscles, tendinous cords and **trabeculae carneae.** You can find the **aorta** and **aortic valve** by inserting a blunt probe toward the superior end of the left ventricle toward the middle of the heart.

 Clean Up When you have finished your study of the sheep heart, make sure you clean your dissection equipment with soap and water. Be careful with sharp blades. Place the sheep heart either back in the preserving fluid or in the appropriate waste container as directed by your instructor.

Conduction in the Heart

Locate the structures of the heart's conduction system in figure 21.10. The initiation of the electrical impulse in the heart begins at the **sinoatrial (SA) node,** which is commonly known as the **pacemaker.** Conduction from the sinoatrial node travels across the atria, causing the muscles of the atria to contract. The impulse that spreads out across the atria reaches the **atrioventricular (AV) node.** The impulse has a slight delay (0.1 second) in the node before being conducted farther. This delay allows the atrial cardiac muscle to contract prior to ventricular firing.

The electrical impulse then travels from the AV node to the **atrioventricular bundle (bundle of His),** to the **right** and **left bundle branches,** and finally to the **Purkinje (conduction) fibers.** The Purkinje fibers stimulate the cardiac muscle of the ventricles to contract. The ventricles are thus stimulated from the apex toward the base, and the contraction proceeds from the inferior end of the ventricles toward the atria. The conduction system of the heart consists of these specialized muscle cells that initiate the heartbeat.

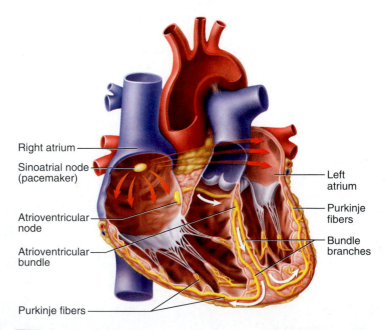

FIGURE 21.10

Conduction System of the Heart Conduction follows the path indicated by arrows.

Name _____ Date _____

1. The heart is located between the lungs in an area known as the _____.

2. What arrangement of cardiac muscle fibers makes the heart efficient?

3. What is the innermost layer of the heart wall called?

4. Which cell type makes up most of the myocardium?

5. Is the apex of the heart superior or inferior to the rest of the heart?

6. What is the name of the depression on the anterior surface of the heart that is between the two ventricles?

7. Are auricles extensions of the atria or the ventricles?

8. What separates the left atrium from the right atrium?

9. What is the name of the thin spot between the atria?

10. The bicuspid valve is located between what two chambers of the heart?

11. Name the structure between the left atrioventricular valve and the papillary muscle.

12. What adaptation do you see with the walls of the left ventricle being thicker than those of the right ventricle?

13. What is the function of the aortic valve?

14. What is another name for the tricuspid valve?

15. What three vessels take blood to the right atrium?

16. The great cardiac vein and the middle cardiac vein lead to what vessel?

17. What blood vessels nourish the heart tissue?

18. Label the following illustration using the terms provided.

right atrium   left atrium

interatrial septum   apex

right ventricle (wall)   tendinous cords

left ventricle (wall)   interventricular septum

aorta   bicuspid valve

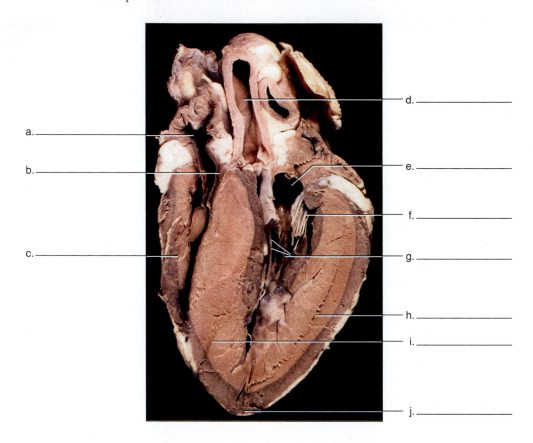

19. The sinoatrial node has a common name. What is it?

20 After the AV node depolarizes, what structures conduct the impulse to the myocardium of the ventricles?

Introduction to the Blood Vessels and Arteries of the Upper Body

Cardiovascular System

INTRODUCTION

The heart is the mechanical pump of the cardiovascular system, and the blood vessels are the conduits that carry oxygen, nutrients, and other materials to the cells, as well as remove wastes from the tissue fluid near the cells. These vascular conduits consist of numerous vessels, including arteries, arterioles, capillaries, venules, and veins. **Arteries** are blood vessels that carry blood *away from* the heart. Most arteries carry oxygenated blood, but there are a few exceptions to this. Arteries have thicker walls than veins, and the thickness of arterial walls reflects the higher blood pressure found in them. Arteries are frequently named for the region of the body they pass through. The *brachial* artery is found in the region of the arm, while the *femoral* artery is found in the thigh region. Arteries become progressively smaller and turn into **arterioles.** Arterioles become even smaller and become **capillaries,** which are the sites of exchange between the blood and the cells of the body. **Venules** return blood to the **veins,** which carry blood back to the heart.

In this exercise you learn the basic structure of blood vessels and locate the arteries of the upper body.

LEARNING OBJECTIVES

At the end of this exercise you should be able to

1. draw a cross section of the wall of a generalized blood vessel showing the three layers;
2. distinguish between arteries and veins;
3. differentiate between conducting arteries and distributing arteries;
4. describe the sequence of major arteries that branch from the aortic arch;
5. list the arteries of the upper extremity;
6. trace the arteries that supply blood to the head;
7. describe the major organs that receive blood from the upper arteries.

MATERIALS

Microscopes

Prepared slides of arteries and veins

Models of the blood vessels of the body

Charts and illustrations of the arterial system

Microscope slides of arteries and veins in cross section

PROCEDURE

Overview of Blood Vessels

Obtain a prepared microscope slide of a cross section of artery and vein. Both arteries and veins have walls that consist of three layers. The outer layer is known as the **tunica externa (adventia)** and consists of a connective tissue sheath. The middle layer, or **tunica media,** is composed of smooth muscle in both arteries and veins. The tunica media is thicker in arteries and there may be pronounced elastic fibers in the wall of the tunica media in arteries. The wall of a blood vessel requires oxygen, which is supplied by the **vaso vasorum.** The innermost layer (near the blood) is the **tunica interna (intima)** and consists of a thin layer of connective tissue and a thin layer of simple squamous epithelium, known as **endothelium.** Endothelium is the layer closest to the blood. In arteries there is an **inner elastic membrane,** which is a thin layer of elastic tissue. These layers are seen in figure 22.1. Examine a prepared slide of an artery and vein and locate these layers.

A feature of veins not found in arteries is valves, which are described further in laboratory exercise 24. You can easily distinguish veins from arteries in a prepared slide where the vessels are cut in cross section. Veins have a large lumen (though the vein is frequently collapsed in prepared sections) and a thin tunica media relative to its overall size. You may also find nerves in the prepared slide. These are also circular structures but they are solid, not hollow. Compare your slide with figure 22.1.

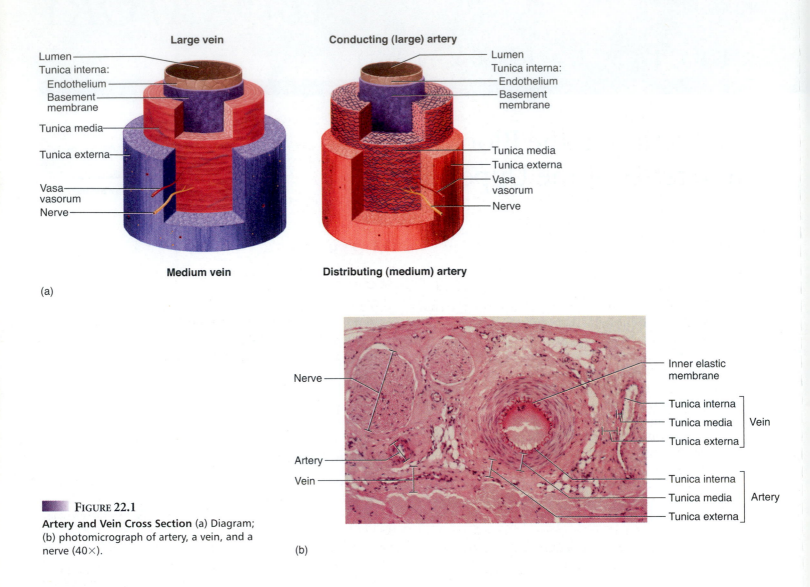

FIGURE 22.1

Artery and Vein Cross Section (a) Diagram; (b) photomicrograph of artery, a vein, and a nerve (40×).

Types of Arteries

Conducting (elastic) arteries are larger arteries close to the heart. Their appearance is made distinctive by the presence of significant amounts of elastic tissue in the tunica media. **Distributing (muscular) arteries** are found farther away from the heart and have more smooth muscle in the tunica media.

Specific Arteries

Examine the models, charts, and illustrations of the cardiovascular system in the lab and locate the major arteries of the upper body. Read the following descriptions of the arteries. Name the vessels that take blood to an artery and those that receive blood from the artery in question. An overview of the major arteries of the body is shown in figure 22.2. You should refer to this illustration for this exercise and for the vessels in laboratory exercise 23.

AORTIC ARCH ARTERIES

Locate the heart and find the large **aorta** that exits from the left ventricle. This is the **ascending aorta,** and it is a large vessel about the size of a garden hose. The ascending aorta is relatively thick-walled due to the high pressure of the blood coming from the heart. The first two arteries that arise from the ascending aorta are the **coronary arteries,** which were covered in detail in laboratory exercise 21. The ascending aorta curves to the left side of the body and forms the **aortic arch** (fig. 22.3). In humans, there are three main arteries that receive blood from the aortic arch. The first major artery to receive blood is on the right side of the body and is called the **brachiocephalic artery (brachiocephalic trunk).** This artery shortly divides into arteries that feed the right side of the head (the **right common carotid artery**) and the right upper extremity (the **right subclavian artery**), as seen in figure 22.3. The aortic arch has two other arteries. On the left side is the **left common carotid artery,** which takes blood to the left side of the head, and the **left subclavian artery,** which takes blood to the left upper extremity.

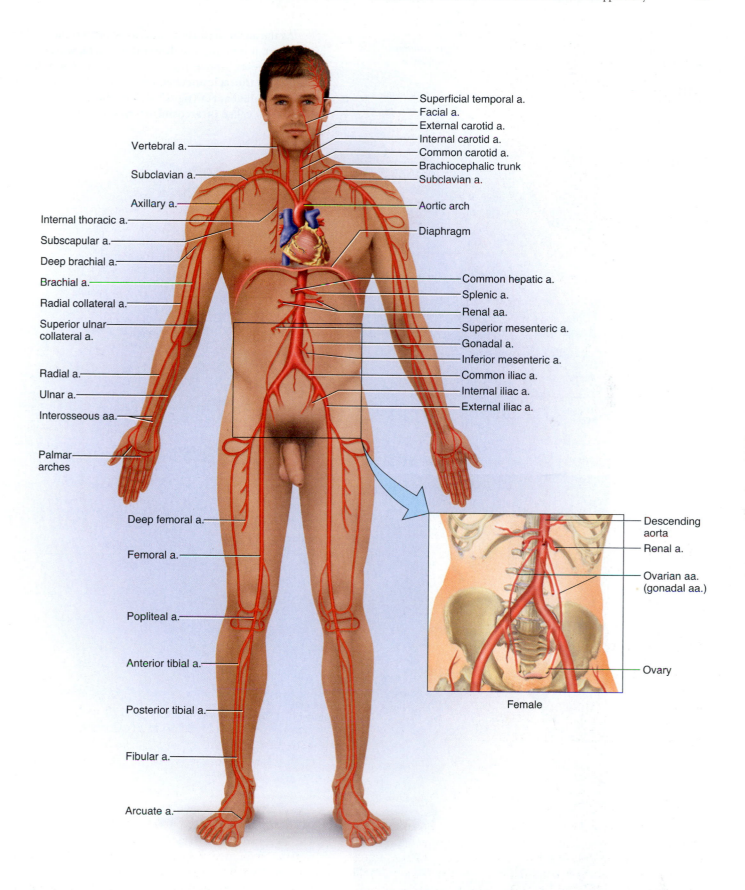

Superficial temporal a.
Facial a.
External carotid a.
Internal carotid a.
Common carotid a.
Brachiocephalic trunk
Subclavian a.
Aortic arch
Diaphragm
Common hepatic a.
Splenic a.
Renal aa.
Superior mesenteric a.
Gonadal a.
Inferior mesenteric a.
Common iliac a.
Internal iliac a.
External iliac a.

Vertebral a.
Subclavian a.
Axillary a.
Internal thoracic a.
Subscapular a.
Deep brachial a.
Brachial a.
Radial collateral a.
Superior ulnar collateral a.
Radial a.
Ulnar a.
Interosseous aa.
Palmar arches
Deep femoral a.
Femoral a.
Popliteal a.
Anterior tibial a.
Posterior tibial a.
Fibular a.
Arcuate a.

Descending aorta
Renal a.
Ovarian aa. (gonadal aa.)
Ovary

Female

FIGURE 22.2

Major Arteries of the Body (a. = artery and aa. = arteries)

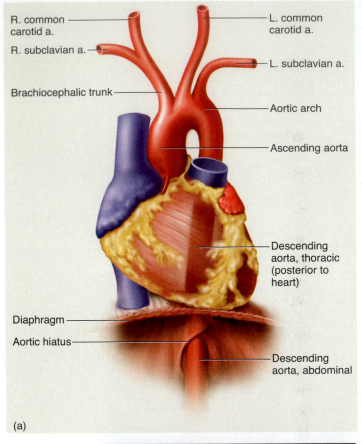

R. common carotid a.

R. subclavian a.

Brachiocephalic trunk

L. common carotid a.

L. subclavian a.

Aortic arch

Ascending aorta

Descending aorta, thoracic (posterior to heart)

Diaphragm

Aortic hiatus

Descending aorta, abdominal

(a)

As the aortic arch turns inferiorly behind the posterior part of the heart, it becomes the **descending aorta,** which is composed of two segments. Above the diaphragm the descending aorta is known as the **thoracic aorta** and below the diaphragm it is known as the **abdominal aorta** (fig. 22.4). The thoracic aorta has numerous branches, called **intercostal arteries,** that run between the ribs (fig. 22.5).

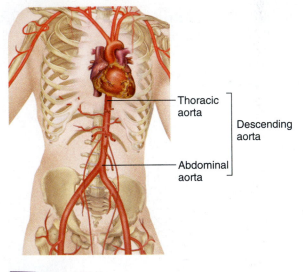

Thoracic aorta

Descending aorta

Abdominal aorta

FIGURE 22.4
Thoracic and Abdominal Aorta

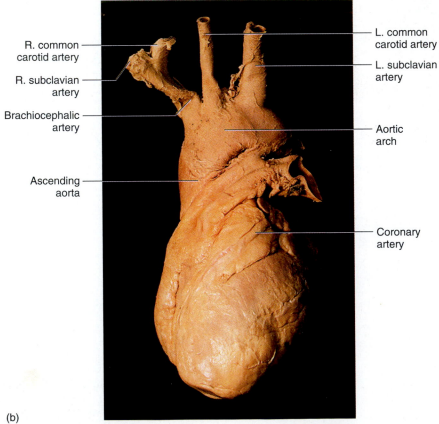

R. common carotid artery

R. subclavian artery

Brachiocephalic artery

Ascending aorta

L. common carotid artery

L. subclavian artery

Aortic arch

Coronary artery

(b)

FIGURE 22.3
Arteries of the Aortic Arch, Anterior View (a) Diagram; (b) photograph.

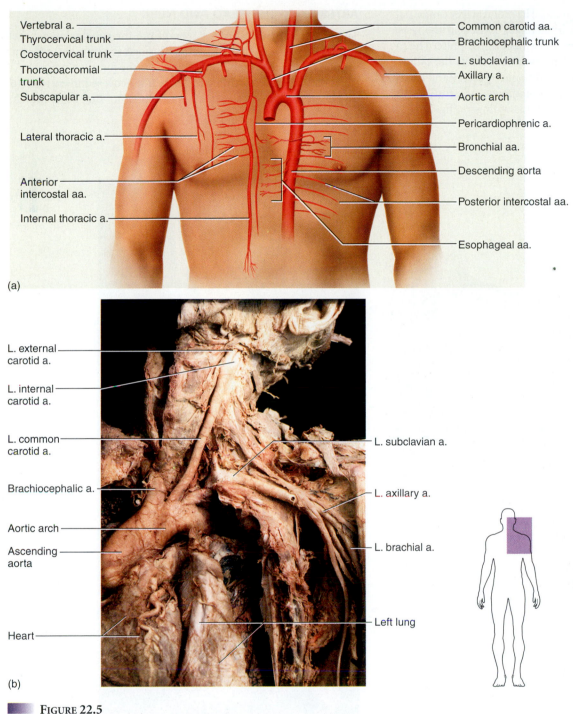

Vertebral a.
Thyrocervical trunk
Costocervical trunk
Thoracoacromial trunk
Subscapular a.
Lateral thoracic a.
Anterior intercostal aa.
Internal thoracic a.

Common carotid aa.
Brachiocephalic trunk
L. subclavian a.
Axillary a.
Aortic arch
Pericardiophrenic a.
Bronchial aa.
Descending aorta
Posterior intercostal aa.
Esophageal aa.

(a)

L. external carotid a.
L. internal carotid a.
L. common carotid a.
Brachiocephalic a.
Aortic arch
Ascending aorta
Heart

L. subclavian a.
L. axillary a.
L. brachial a.
Left lung

(b)

FIGURE 22.5

Branches of the Subclavian Artery (a) Diagram; (b) photograph of cadaver.

ARTERIES THAT FEED THE UPPER LIMBS

If you return to the brachiocephalic artery, you can see that it is a short tube that takes blood from the aortic arch and then branches into the **right common carotid artery** and the **right subclavian artery.** The right subclavian artery has one branch that takes blood to the brain, called the vertebral artery, while another branch is the **right axillary artery.** The **left subclavian artery** also has a vertebral artery branch that takes blood to the brain and another branch that is the **left axillary artery** (figs. 22.2 and 22.5).

The axillary artery turns into the **brachial artery,** which has a pulse that can be palpated by finding a groove on the medial, distal region of the arm between the biceps brachii and brachialis muscles. This location is the site for the placement of the diaphragm of a stethoscope during blood pressure measurement. The brachial artery supplies blood to the muscles of the arm and the humerus. The brachial artery bifurcates (splits in two) to form the **radial artery** on the lateral side of the antebrachium and the **ulnar artery** on the medial side of the antebrachium. These arteries supply blood to the forearm and hand. The radial artery can be palpated just lateral to the flexor carpi radialis muscle at the wrist, which is a common site for the measurement of the pulse. The radial artery and ulnar artery become united again as the **palmar arch arteries.** There is a **superficial palmar arch artery** and a **deep palmar arch artery.** They join, or anastomose, and send out the small **digital arteries** that supply blood to the fingers (fig. 22.6).

ARTERIES OF THE HEAD AND NECK

The **vertebral arteries** receive blood from the **subclavian arteries** and take it to the brain by traveling through the transverse foramina of the cervical vertebrae and then into the foramen magnum of the skull. The arterial supply to the brain is covered in exercise 16. The other main arteries that take blood to the head are the

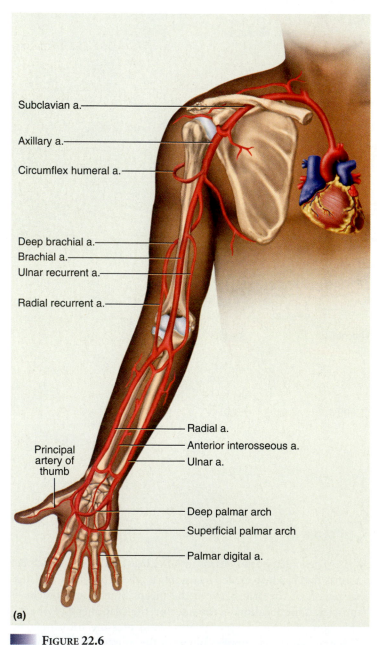

Subclavian a.
Axillary a.
Circumflex humeral a.
Deep brachial a.
Brachial a.
Ulnar recurrent a.
Radial recurrent a.
Radial a.
Anterior interosseous a.
Principal artery of thumb
Ulnar a.
Deep palmar arch
Superficial palmar arch
Palmar digital a.

(a)

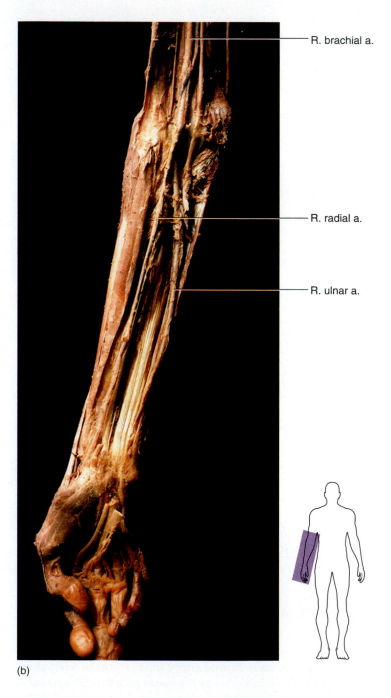

R. brachial a.
R. radial a.
R. ulnar a.

(b)

FIGURE 22.6

Arteries of the Upper Extremity (a) Diagram; (b) photograph of a cadaver.

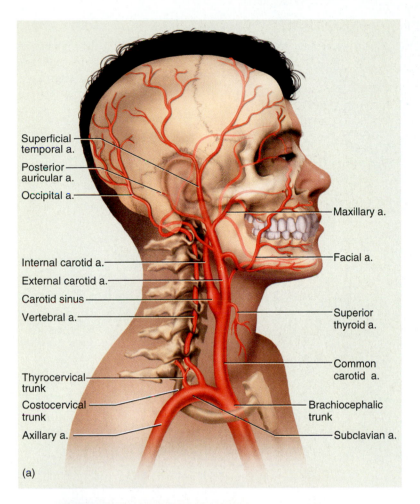

Superficial temporal a.
Posterior auricular a.
Occipital a.
Maxillary a.
Internal carotid a.
External carotid a.
Carotid sinus
Vertebral a.
Facial a.
Superior thyroid a.
Common carotid a.
Thyrocervical trunk
Costocervical trunk
Axillary a.
Brachiocephalic trunk
Subclavian a.

(a)

common carotid arteries, which occur on the anterior sides of the neck. Each common carotid artery branches just below the angle of the mandible to form the **external carotid artery,** which takes blood to the face, and the **internal carotid artery,** which passes through the carotid canal and takes blood to the brain. The external carotid artery on each side has several branches, including the **facial artery,** the **superficial temporal artery,** the **maxillary artery,** and the **occipital artery** (fig. 22.7). Once you have studied all the arteries described in this exercise, select an artery for your lab partner to name. Quiz each other on the charts or models available in lab.

Cat Anatomy

If you are using cats, turn to section 6, "Blood Vessel Anatomy of the Cat," on page 427 of this lab manual.

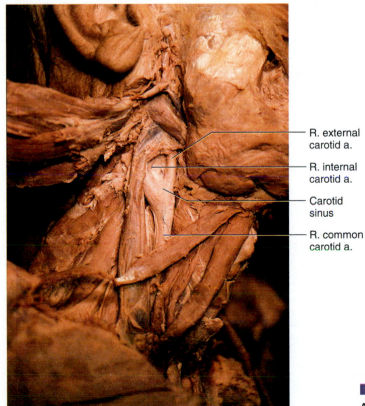

R. external carotid a.
R. internal carotid a.
Carotid sinus
R. common carotid a.

(b)

FIGURE 22.7
Arteries of the Head (a) Diagram; (b) photograph of a cadaver.

Introduction to the Blood Vessels and Arteries of the Upper Body

Name _____ Date _____

1. Label the following illustration with the major arteries of the body using the terms provided. Try to complete the illustration first and then review the material in this exercise to determine your accuracy.

abdominal aorta   right radial artery   left common carotid artery   right ulnar artery

right subclavian artery   right axillary artery   aortic arch   right brachial artery

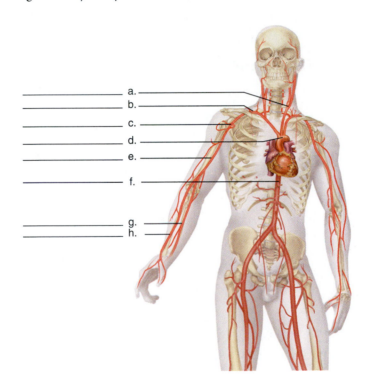

_____ a. _____

_____ b. _____

_____ c. _____

_____ d. _____

_____ e. _____

_____ f. _____

_____ g. _____
_____ h. _____

2. Blood from the left subclavian artery flows into what vessel as it moves toward the left arm?

3. Blood in the radial artery comes from what blood vessel?

4. An aneurysm is a weakened, expanded portion of an artery. Ruptured aneurysms can lead to rapid blood loss. Describe the significance of an aortic aneurysm versus a digital artery aneurysm.

5. The pulmonary arteries carry deoxygenated blood from the heart to the lungs. Umbilical arteries carry a mixture of oxygenated and deoxygenated blood. Why are these blood vessels called arteries?

6. What is the name of the outermost layer of a blood vessel?

7. What kind of blood vessels have valves?

8. Blood from the common carotid artery next travels to what two vessels?

9. Where does blood in the right subclavian artery come from?

10. The internal carotid artery takes blood to what organ?

11. The descending aorta receives blood from what vessel?

12. Blood in the right common carotid artery receives blood from what vessel?

13 Name three blood vessels that exit from the aortic arch.

Arteries of the Lower Body

Cardiovascular System

INTRODUCTION

In this exercise you continue your study of arteries. The arteries of the lower body receive blood directly or indirectly from the descending aorta. As with the arteries of the upper extremities, many of these vessels are named for their location, such as the *iliac* artery, the *femoral* artery, and the *mesenteric* arteries, or they are named for the organs they serve, such as the common *hepatic* (*hepatic* = liver) artery and the *splenic* artery.

LEARNING OBJECTIVES

At the end of this exercise you should be able to

1. describe the sequence of major arteries that originate from the abdominal aorta or its derivatives;
2. list the arteries of the lower extremities;
3. describe the major organs that receive blood from the lower arteries;
4. name the vessels that take blood to a particular artery and the vessels that receive blood from a particular artery.

MATERIALS

Models of the blood vessels of the body

Charts and illustrations of arterial system

Cadaver (if available)

Microscope

Prepared slide of arteriosclerosis

PROCEDURE

Examine the models and charts in the lab and the accompanying illustrations to locate the major arteries in the abdomen and lower body. Read the following descriptions of the vessels and locate the individual arteries. You should refer to figure 22.2 for an overview of the major arteries of the body.

Abdominal Arteries

The **abdominal aorta** is the portion of the **descending aorta** inferior to the diaphragm. The first major branch of the abdominal aorta is the **celiac trunk** (also known as the **celiac artery**). The celiac trunk splits into three separate arteries: the **splenic artery,** taking blood to the spleen, pancreas, and part of the stomach; the **left gastric artery,** taking blood to the stomach and esophagus; and the **common hepatic artery,** taking blood to the liver, stomach, duodenum, and pancreas. Locate the celiac trunk and the branches of the celiac in figure 23.1.

Just below the celiac trunk is the **superior mesenteric artery.** This vessel takes blood from the abdominal aorta and continues through the mesentery until it reaches the small intestine and proximal portions of the large intestine, including the cecum, the ascending colon, and part of the transverse colon. The major branches of the superior mesenteric artery are the **intestinal arteries,** the **ileocolic artery,** and the right and middle **colic arteries.** These are illustrated in figure 23.2.

The next vessels to branch from the aorta are the paired **suprarenal arteries,** which take blood to the adrenal glands. Below the suprarenal arteries are the **left** and **right renal arteries.** These arteries take blood to the kidneys. Locate these vessels in the lab and in figures 23.2 and 23.3.

The **gonadal arteries** branch inferior to the renal arteries and descend to either the testes or the ovaries. The **inferior mesenteric artery** is the next vessel to branch from the aorta, and it takes blood to the lower portion of the large intestine, including part of the transverse colon, the descending colon, the sigmoid colon, and the rectum. These can be located in figures 22.2 and 23.3.

The abdominal aorta terminates by dividing into the two **common iliac arteries.** The common iliac arteries take blood to the internal and **external iliac arteries.** The **internal iliac artery** takes blood to the pelvic region, including branches that feed the rectum,

FIGURE 23.1
Celiac trunk and its branches

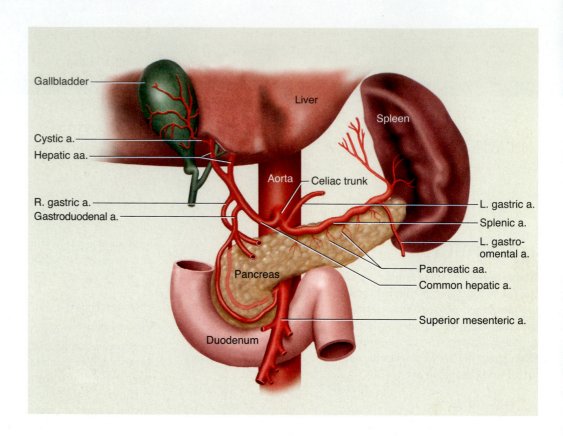

FIGURE 23.2
Middle Abdominal Arteries

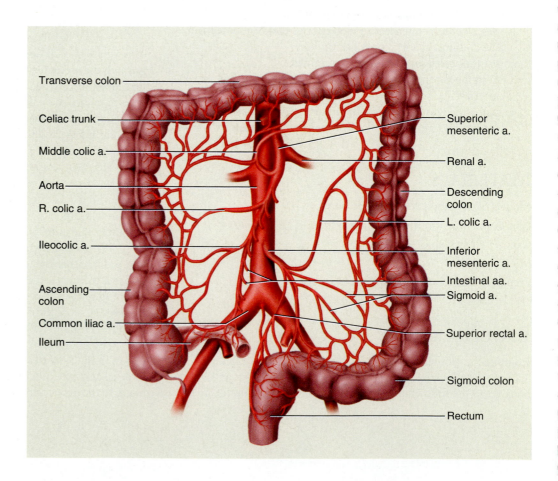

pelvic floor, external genitalia, groin muscles, hip muscles, uterus, ovary, and vagina. These arteries are illustrated in figure 23.3.

Arteries of the Lower Limb

Each external iliac artery branches from a common iliac artery and exits the body wall near the **inguinal canal.** Once the external iliac artery leaves the body cavity, it continues into the thigh as the **femoral artery.** The femoral artery has a superficial branch that feeds the thigh and a deeper branch, called the **deep femoral artery,** that takes blood to the muscles of the thigh, the knee, and the femur. The femoral artery continues posterior to the knee as the **popliteal artery** and then divides into the **posterior tibial artery** and **anterior tibial artery,** which take blood to the knee and leg. Another branch of the popliteal artery is the **fibular artery,** which supplies the muscles on the lateral side of the leg. The foot is supplied with blood from several arteries, including the **plantar arteries,** the **dorsal pedal artery,** and the **digital arteries.** Locate these arteries in figure 23.4.

Arteriosclerosis

Arteriosclerosis is a condition known commonly as hardening of the arteries. Examine a microscope slide or an illustration of arteriosclerosis and note the development of **cholesterol plaque** under the endothelial layer. Draw what you see in the space provided.

Drawing of an artery with arteriosclerosis:

Cat Anatomy

If you are using cats, turn to section 6, "Blood vessel Anatomy of the Cat," on page 427 of this lab manual.

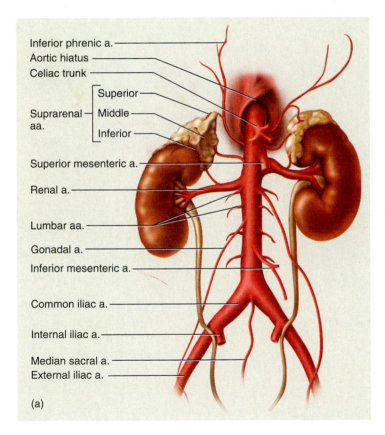

(a)

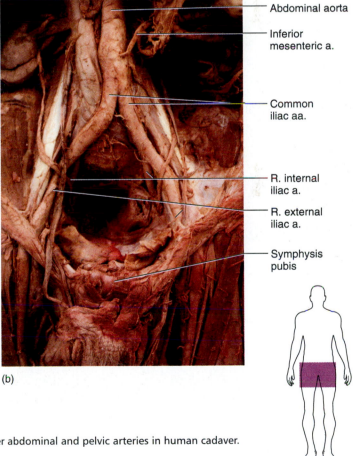

(b)

FIGURE 23.3

Abdominal Arteries (a) Diagram of abdominal arteries; (b) photograph of lower abdominal and pelvic arteries in human cadaver.

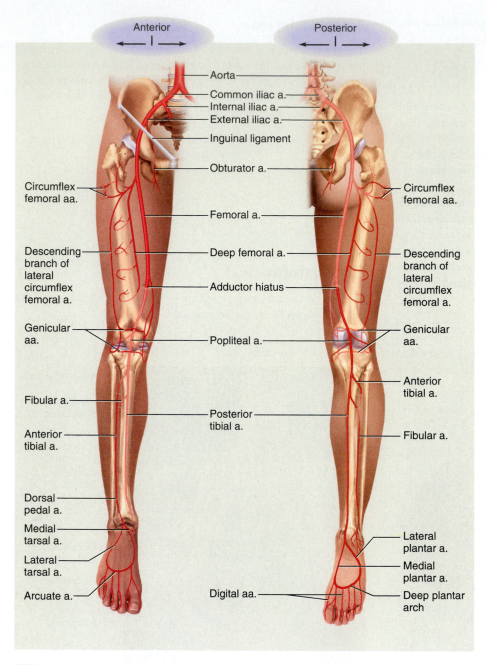

FIGURE 23.4
Arteries of the Lower Extremity

Name _____ Date _____

1. Name the section of descending aorta inferior to the diaphragm.

2. Blood from the celiac artery flows into three different blood vessels. What are these vessels?

3. Blood from the superior mesenteric artery feeds which major abdominal organs?

4. What vessels take blood to the kidneys?

5. The ovaries or testes receive blood from which arteries?

6. Blood in the inferior mesenteric artery travels to what organs?

7. Where does blood in the external iliac artery come from?

8. What artery takes blood directly to the femoral artery?

9. Blood from the popliteal artery comes directly from what artery?

10. What is arteriosclerosis?

11. In what part of the arterial wall does cholesterol plaque develop?

12. Label the following illustration using the terms provided.

celiac trunk

anterior tibial artery

internal iliac artery

superior mesenteric artery

gonadal artery

popliteal artery

femoral artery

common iliac artery

inferior mesenteric artery

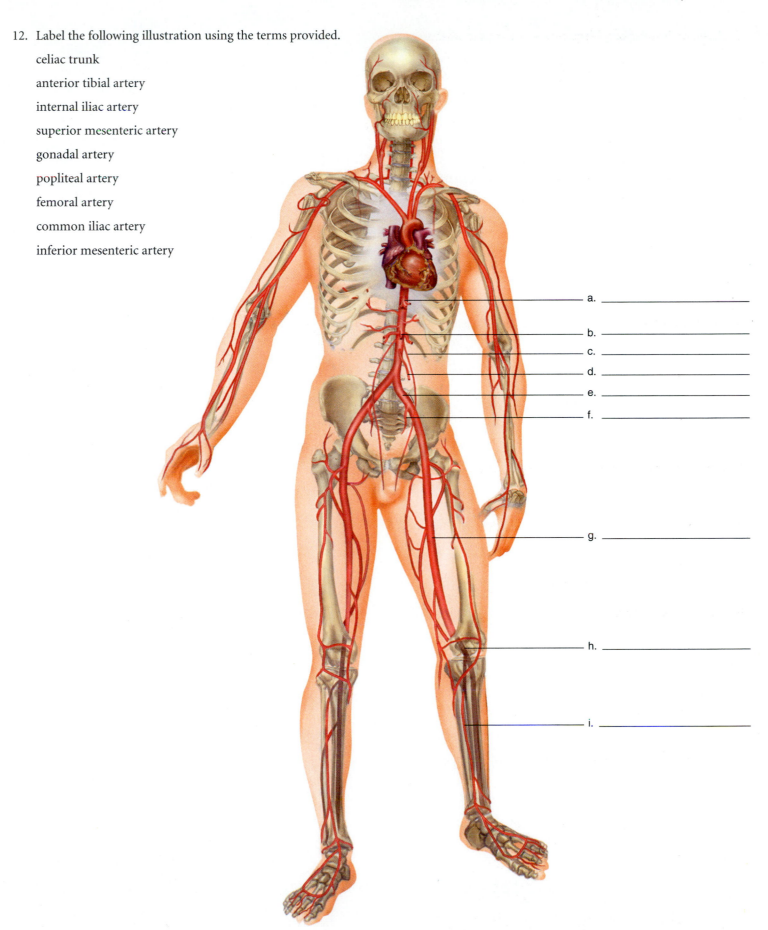

a. _____

b. _____

c. _____

d. _____

e. _____

f. _____

g. _____

h. _____

i. _____

Veins and Special Circulations

**Cardiovascular System,
Lymphatic System**

INTRODUCTION

Veins are blood vessels that carry blood toward the heart. They can be superficial (just under the skin) or deep. The deep veins of the body frequently travel alongside the major arteries and take on the arterial names (for example, the femoral vein), while superficial veins have names specific to themselves (for example, the great saphenous vein). Veins have thinner walls than arteries and contain valves for a one-way flow of blood to the heart.

A review of the differences in the walls of arteries and veins is in laboratory exercise 22. In this exercise you begin the study of veins by examining charts or models of the entire venous system and then study the veins of specific regions of the body. Studying the functions of veins and the lymphatic system is vital to an understanding of the circulatory pattern. Fluid in the circulatory route may travel by different pathways. Normally blood is pumped from the **heart** through **arteries** and distributed through the **arterioles** to the **capillaries.** At the capillary level, nutrients, water, and oxygen are exchanged with the cells of the body. The blood returns via **venules** to **veins** and back to the heart under relatively low pressure.

Blood cells and plasma protein typically stay in the vessels, yet fluid from the plasma leaks from the capillaries and bathes the cells of the body. This fluid flows between the cells of the body and is known as **interstitial fluid.** It provides nutrients to the cells and receives dissolved wastes from the cells along with cellular debris. Most of this fluid flows back into the capillaries but some is picked up by **lymph capillaries,** which return the fluid, now known as **lymph,** to the **collecting vessels.** Collecting vessels take the lymph back to the venous system. In this way the remaining interstitial fluid is returned to the cardiovascular system. **Lymphatic vessels** refer to any type of lymph vessel (lymph capillary, collecting duct, etc.).

In addition to the lymph system returning fluid to the cardiovascular system, it protects the body and is instrumental in the absorption of lipids from the digestive system. Lipids in the small intestine are converted to **chylomicrons** (phospholipids and other molecules) in the cells that line the digestive tract. These chylomicrons are conducted into the lymphatic system, which then transports the material through lymphatic vessels to the cardiovascular system.

As the fluid flows through the lymphatic vessels, lymph nodes clean the lymph of cellular debris and foreign material (bacteria, viruses) that may have entered the lymphatic system.

LEARNING OBJECTIVES

At the end of this exercise you should be able to
1. identify major veins in models or a cadaver;
2. describe the sequence of major veins that drain the upper limbs;
3. trace the blood flow from the brain to the heart;
4. distinguish a portal system from normal venous return flow;
5. describe the major digestive organs that supply blood to the hepatic portal system;
6. trace the path of blood flow in fetal circulation;
7. identify the valves in lymph vessels;
8. locate the major histologic features of a lymph node;
9. demonstrate on a model the location of the tonsils and other lymph organs;
10. describe the function of the spleen and thymus;
11. identify selected lymph vessels.

MATERIALS

Charts, diagrams, and models of blood vessels and the lymph
   system
Microscopes
Microscope slides of lymphatic vessels with valves
Torso models
Cadaver (if available)

PROCEDURE

Veins of the Body

Examine the models, charts, and illustrations in the lab and locate the major veins of the body. Read the following descriptions of the veins and find them as they are represented in lab. As you locate a specific vein, name the vessel that takes blood to the vein and those that receive blood from the vein. Veins in this exercise are studied in the direction of their flow from the cells of the body to the heart. In this way they resemble tributaries of rivers as smaller veins flow into larger veins.

An overview of the major veins of the body is shown in figure 24.1. Compare that figure with the following list of some of the major veins of the body. Place a check mark next to the name of the vein when you locate it.

_____ Internal jugular vein

_____ External jugular vein

_____ Brachiocephalic vein

_____ Superior vena cava

_____ Axillary vein

_____ Cephalic vein

_____ Basilic vein

_____ Inferior vena cava

_____ Common iliac vein

_____ External iliac vein

_____ Internal iliac vein

_____ Femoral vein

_____ Great saphenous vein

_____ Tibial veins

VEINS OF THE UPPER LIMBS

Examine the models and charts in the lab and locate the veins of the upper limbs. The fingers are drained by the small **digital veins,** which lead to the **palmar arch veins.** The major superficial veins of each upper limb are the **basilic vein,** which is on the anterior, medial side of the forearm and arm, and the **cephalic vein,** which is on the anterior, lateral side of the forearm and arm. The two vessels have many anastomosing branches (cross-connections) between them. One of the significant anastomosing veins is the **median cubital vein,** which crosses the anterior cubital fossa and is a common site for the withdrawal of blood. Locate these superficial veins in figure 24.2a.

The deep veins of the forearm are the **radial veins** and the **ulnar veins,** which can be found traveling near the arteries of the same name. The deep veins of the arm are the **brachial veins,** which are next to the brachial artery in the proximal portion of the arm. The brachial veins are formed by the union of the radial and ulnar veins and merge superiorly with the **basilic vein** to form the short **axillary vein.** The axillary vein connects with the cephalic vein to form the **subclavian vein.** The drainage of the upper limbs is carried to the heart by the left and right subclavian veins, which take the blood via the **brachiocephalic veins** to the **superior vena cava** and finally to the right atrium of the heart. Locate the deep veins of the upper limbs in figure 24.2.

VEINS OF THE HEAD AND NECK

Examine the models and charts in the lab and locate the various veins that drain blood from the head. The drainage of the brain occurs as veins take blood from the brain and pass through the venous sinuses in the subdural spaces. The blood from the venous sinuses flows into the **internal jugular veins** and **facial veins.**

Another vessel that takes blood from the brain is the **vertebral vein,** which, like the vertebral arteries, travels through the transverse foramina of the cervical vertebrae. The vertebral veins take blood to the subclavian veins, which, in turn, flow to the brachiocephalic veins. These vessels are illustrated in figure 24.3.

Each internal jugular vein passes through a jugular foramen of the skull, taking blood along the lateral aspect of the neck as it moves toward the brachiocephalic vein. The brachiocephalic vein is formed by the union of the internal jugular vein and the subclavian vein. The superficial regions of the head (musculature and skin of the scalp and face) are drained by the **external jugular veins.** The external jugular veins join up with the subclavian veins prior to reaching the brachiocephalic veins. The left and right brachiocephalic veins take blood to the superior vena cava. Locate these major vessels of the head and neck in figure 24.3.

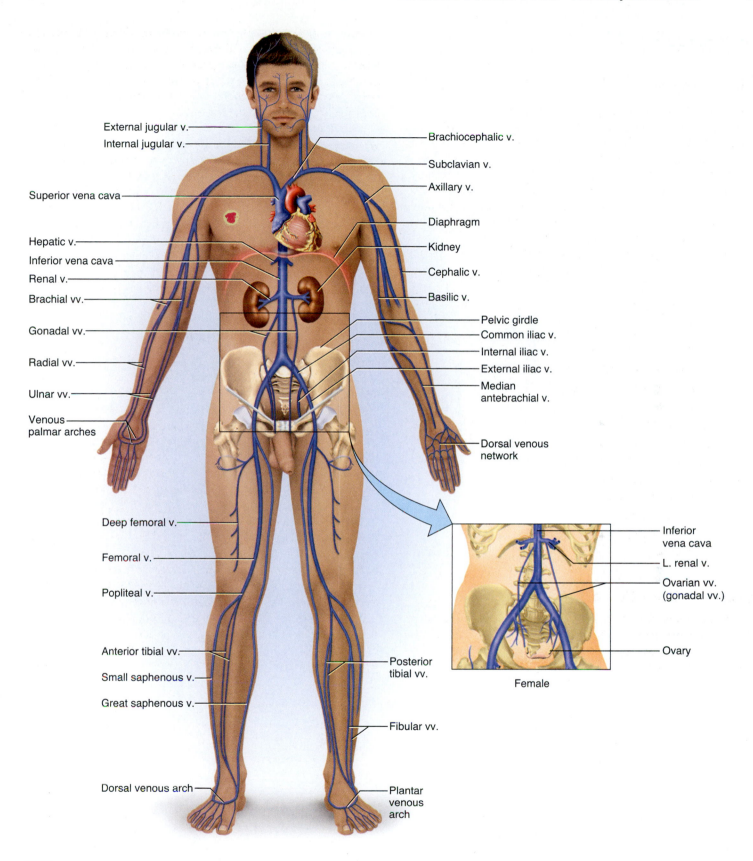

External jugular v.
Internal jugular v.
Superior vena cava
Hepatic v.
Inferior vena cava
Renal v.
Brachial vv.
Gonadal vv.
Radial vv.
Ulnar vv.
Venous palmar arches

Brachiocephalic v.
Subclavian v.
Axillary v.
Diaphragm
Kidney
Cephalic v.
Basilic v.
Pelvic girdle
Common iliac v.
Internal iliac v.
External iliac v.
Median antebrachial v.
Dorsal venous network

Deep femoral v.
Femoral v.
Popliteal v.
Anterior tibial vv.
Small saphenous v.
Great saphenous v.
Dorsal venous arch

Posterior tibial vv.
Fibular vv.
Plantar venous arch

Inferior vena cava
L. renal v.
Ovarian vv. (gonadal vv.)
Ovary

Female

FIGURE 24.1
Major Systemic Veins (v. = vein and vv. = veins)

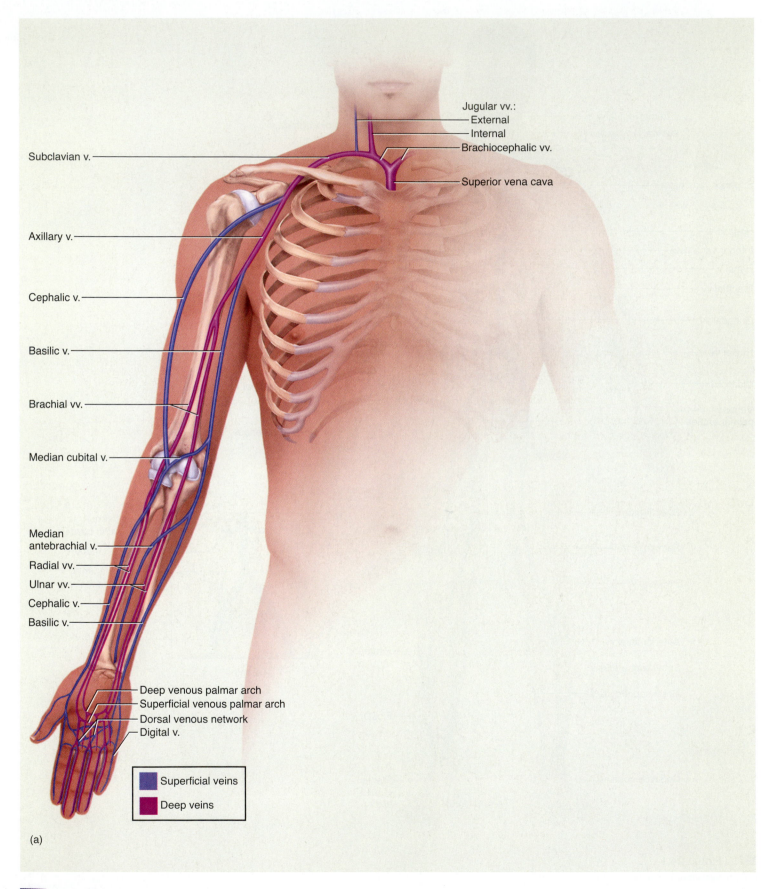

Subclavian v.

Axillary v.

Cephalic v.

Basilic v.

Brachial vv.

Median cubital v.

Median
antebrachial v.

Radial vv.

Ulnar vv.

Cephalic v.

Basilic v.

Jugular vv.:
External
Internal
Brachiocephalic vv.

Superior vena cava

Deep venous palmar arch
Superficial venous palmar arch
Dorsal venous network
Digital v.

Superficial veins

Deep veins

(a)

FIGURE 24.2

Veins of the Upper Extremity Diagram (a) Superficial and deep veins.

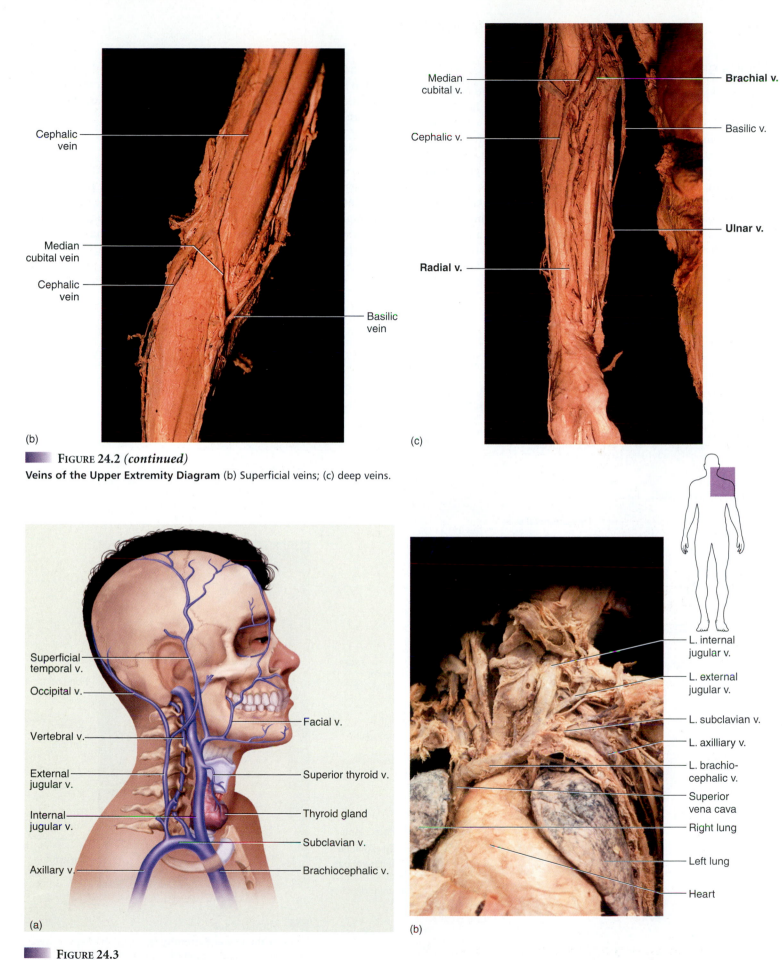

(b)

Cephalic vein

Median cubital vein

Cephalic vein

Basilic vein

(c)

Median cubital v.

Cephalic v.

Radial v.

Brachial v.

Basilic v.

Ulnar v.

FIGURE 24.2 (continued)
Veins of the Upper Extremity Diagram (b) Superficial veins; (c) deep veins.

Superficial temporal v.

Occipital v.

Vertebral v.

External jugular v.

Internal jugular v.

Axillary v.

Facial v.

Superior thyroid v.

Thyroid gland

Subclavian v.

Brachiocephalic v.

(a)

L. internal jugular v.

L. external jugular v.

L. subclavian v.

L. axilliary v.

L. brachio-cephalic v.

Superior vena cava

Right lung

Left lung

Heart

(b)

FIGURE 24.3
Veins of the Head and Neck (a) Diagram of right side; (b) photograph of left side of cadaver.

VEINS OF THE LOWER LIMB

The blood vessels that drain the lower limbs operate under relatively low pressure and must take blood back to the heart against gravity. The anterior and posterior tibial veins, along with the fibular vein, and the small saphenous vein receive blood from the foot and leg and eventually take blood to the femoral vein.

The longest vessel in the human body is the **great saphenous vein,** which can be found just deep to the skin on the medial aspect of the lower limb at the level of the medial malleolus and traversing the lower limb to the proximal thigh. This vessel frequently is imbedded deep in adipose tissue below the skin, yet it is considered a superficial vein. Two other vessels in the thigh are the **femoral vein** and the **deep femoral vein.** The femoral vein travels alongside the femoral artery. As the great saphenous vein

reaches the inguinal region, it joins with the femoral vein, which takes blood from the thigh region and passes under the inguinal ligament. Examine these vessels in the lab and compare them with figure 24.4.

VEINS OF THE PELVIS

Once the femoral vein crosses under the ligament it becomes the external iliac vein. The **external iliac vein** joins with another vein, the **internal iliac vein,** which drains the region of the pelvis. Their union forms the **common iliac vein.** The common iliac veins from both sides of the body unite and form the **inferior vena cava,** which travels superiorly along the right side of the vertebrae, taking blood to the right atrium of the heart. Locate these veins and compare them with figure 24.5.

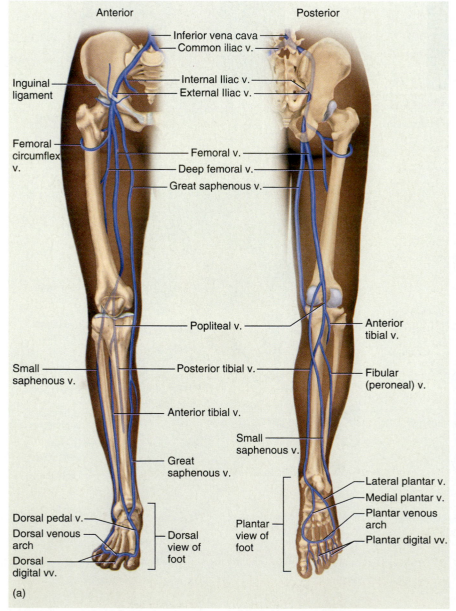

(a)

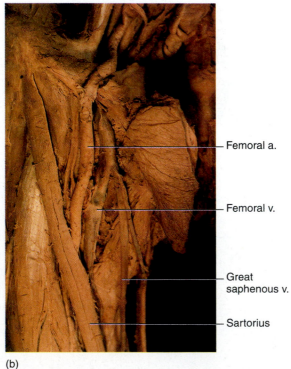

(b)

FIGURE 24.4

Veins of the Lower Extremity (a) Diagram of entire lower extremity; (b) photograph of upper right thigh

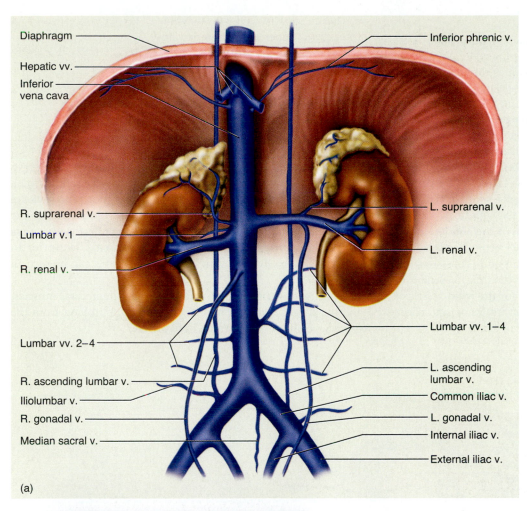

(a)

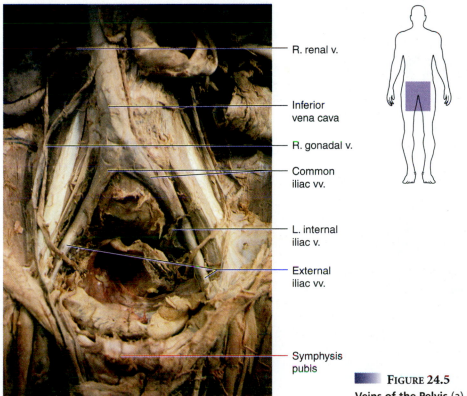

(b)

FIGURE 24.5
Veins of the Pelvis (a) Diagram; (b) photograph.

VEINS OF THE TRUNK

Veins of the abdominal region drain the major organs of the digestive tract and other abdominal organs, such as the spleen. Frequently, these veins are named for the organs from which they receive blood (e.g., the splenic vein takes blood from the spleen). The abdominal veins are a unique group of blood vessels, some of which constitute the hepatic portal system. The portal system pattern is different from the normal venous blood flow. Given the normal venous flow, blood from an organ begins in the capillary bed of that organ and then travels through venules to veins and then to the heart.

In a **portal system,** a series of vessels takes blood from the *capillary beds* of an organ (or organs) through a series of veins and then to another *capillary bed.* In the case of the **hepatic portal system** the blood flows from the capillary beds of the abdominal organs through numerous veins to the capillary bed of the liver. The liver receives blood from the digestive organs and processes it before sending it through the **hepatic vein** to the inferior vena cava and toward the heart. Examine figure 24.6 for the main vessels of the hepatic portal system, identify the following veins, and check them off when you find them on models or in charts in lab.

_____ Inferior mesenteric vein
_____ Superior mesenteric vein
_____ Splenic vein
_____ Right gastro-omental vein
_____ Hepatic portal vein

VEINS OF THE THORAX, ABDOMEN, AND PELVIS

Some abdominal veins take blood directly to the inferior vena cava and some pass through the liver before reaching the inferior vena cava. Those that take blood directly to the inferior vena cava are the **renal veins,** the **suprarenal veins,** and the **lumbar veins.** The **right gonadal vein** takes blood directly to the inferior vena cava but the **left gonadal vein** takes blood to the renal vein prior to flowing into the inferior vena cava. These vessels can be seen in figure 24.7.

Numerous **intercostal veins** drain the intercostal muscles, and the **azygos vein** and **hemiazygos vein** drain blood from the thoracic region. These are seen in figure 24.7.

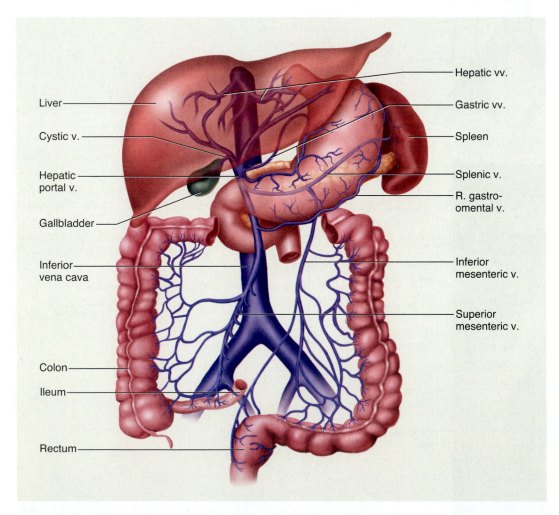

Liver
Cystic v.
Hepatic portal v.
Gallbladder
Inferior vena cava
Colon
Ileum
Rectum

Hepatic vv.
Gastric vv.
Spleen
Splenic v.
R. gastro-omental v.
Inferior mesenteric v.
Superior mesenteric v.

FIGURE 24.6
Hepatic Portal System

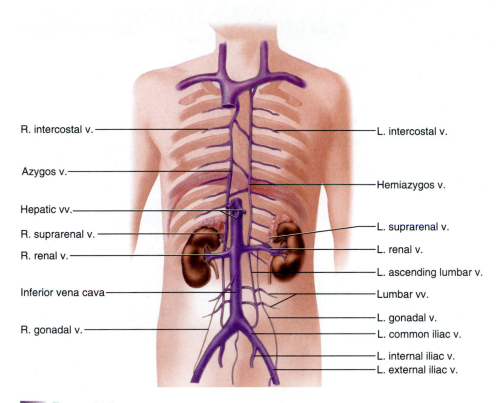

R. intercostal v.

Azygos v.

Hepatic vv.

R. suprarenal v.

R. renal v.

Inferior vena cava

R. gonadal v.

L. intercostal v.

Hemiazygos v.

L. suprarenal v.

L. renal v.

L. ascending lumbar v.

Lumbar vv.

L. gonadal v.

L. common iliac v.

L. internal iliac v.

L. external iliac v.

FIGURE 24.7
Veins of the thorax, abdomen, and pelvis

Fetal Circulation

The pathway of fetal blood is somewhat different from that of the adult in that the lungs are nonfunctional in the fetus. Oxygen and nutrients move from the maternal side of the placenta to the fetal bloodstream, while carbon dioxide and metabolic wastes move from the fetal bloodstream to the placenta. Examine figure 24.8 and note how the numbers in the description are placed in the diagram. From the placenta[1] the blood flows through the **umbilical vein,**[2] which is located in the **umbilical cord.** The blood from the umbilical vein travels through the **ductus venosus,**[3] which serves as a shunt to the inferior vena cava[4] of the fetus. The maternal blood from the umbilical vein, which is relatively high in oxygen and nutrients, mixes with the deoxygenated, nutrient-poor fetal blood from the inferior vena cava and thus the fetus receives a mixture of blood.

The blood from the inferior vena cava travels to the right atrium of the heart. While in the right atrium the blood travels to the right ventricle (which pumps blood to the lungs) and through a hole in the right atrium called the **foramen ovales.**[5] Since the lungs are nonfunctional in the fetus, the foramen ovale serves as a bypass route away from the lungs and to the chambers of the heart that will pump blood to the body. Blood in the right ventricle is pumped to the pulmonary trunk, where another shunt vessel, the **ductus arteriosus,**[6] carries blood to the aortic arch bypassing the

lungs. The lungs do receive some blood, but it is for the nourishment of the lung tissue, not for gas exchange.

Blood from the heart exits the left ventricle and passes through the aorta to the systemic arteries. Blood travels down the internal iliac arteries[7] to the pelvic organs and lower limbs and some moves into the **umbilical arteries,**[8] which carry blood to the placenta. At birth, the pressure changes in the newborn's lungs and heart cause the closing of a flap of tissue over the foramen ovale, leaving a thin spot in the **interatrial septum** known as the **fossa ovalis.** Lack of closure of this foramen can lead to a condition known as "blue baby."

Lymph System

The lymph system is difficult to study in preserved specimens because it collapses at death. The lymph system is composed of **lymphatic capillaries, lymphatic vessels, lymph nodes, lymph organs,** and **lymph tissue.** An overview of the system is illustrated in figure 24.9. Examine charts and models in lab and compare them with this figure.

Lymph originates as interstitial fluid that bathes the cells of the body. This fluid flows from blood capillaries and carries oxygen, nutrients, and other dissolved materials to the cells. Some of the interstitial fluid does not return to the blood capillary bed but enters the lymph capillaries by way of small, valvelike slits in the lymph capillary wall. This process is illustrated in figure 24.10.

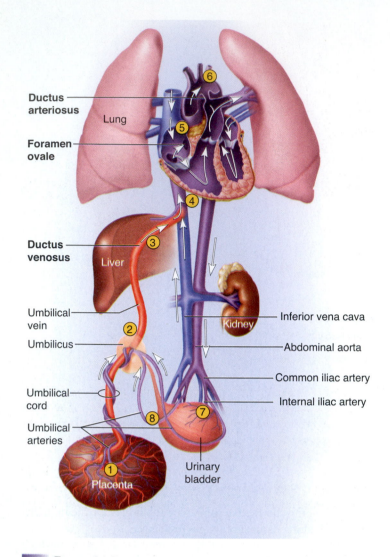

Ductus arteriosus

Lung

Foramen ovale

Ductus venosus

Liver

Umbilical vein

Umbilicus

Umbilical cord

Umbilical arteries

Placenta

Kidney

Inferior vena cava

Abdominal aorta

Common iliac artery

Internal iliac artery

Urinary bladder

FIGURE 24.8
Fetal Circulation in the Human

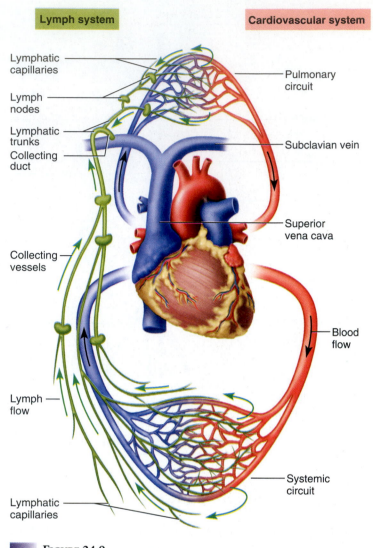

Lymph system

Cardiovascular system

Lymphatic capillaries

Lymph nodes

Lymphatic trunks

Collecting duct

Collecting vessels

Lymph flow

Lymphatic capillaries

Pulmonary circuit

Subclavian vein

Superior vena cava

Blood flow

Systemic circuit

FIGURE 24.9
Flow of Interstitial Fluid and Lymph

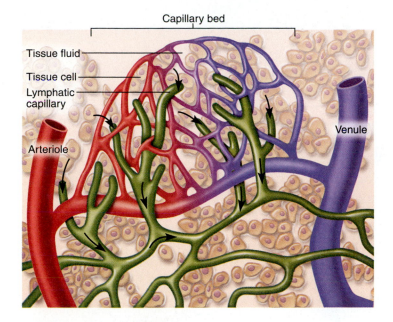

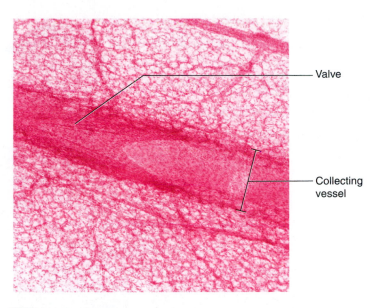

FIGURE 24.11
Collecting Vessel with Valve

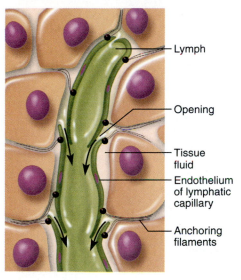

FIGURE 24.10
Lymphatic Capillary

LYMPHATIC VESSELS

Once the lymph is in the lymph capillaries it travels through the **collecting vessels.** As with the capillaries there is a one-way flow that occurs in the collecting vessels due to the presence of valves. Examine a prepared slide of a collecting vessel and note the presence of the **valve.** Compare your slide with figure 24.11.

Use figure 24.12 to examine the drainage pattern of lymph in the body. The **thoracic duct** drains most of the body, taking lymph to the left subclavian vein, where the fluid is returned to the car-diovascular system. Identify the **cisterna chyli,** which is an enlarged portion of the thoracic duct in the abdominal region. The **right lymphatic duct** drains the right side of the head and neck, the right thoracic region, and the right upper extremity. The right lymphatic duct returns lymph to the right subclavian vein.

Note that the collecting vessels lead to regions of the body where lymph nodes are found. Nodes are clustered in the groin (inguinal region), axilla, antecubital fossa, popliteal region, neck, thorax, and abdomen.

LYMPH NODE

Look at a slide of a **lymph node.** The lymph node is enclosed by a sheath of tissue called the **capsule.** Locate the dark purple **lymphatic nodules** in the lymph node. The cortex is the outer region of the node that has numerous lymphatic nodules. In the **medulla, medullary cords** (reticular fibers and immune cells, such as macrophages and plasma cells), and **sinuses** cleanse the lymph of foreign particles and cellular debris. **Afferent lymphatic vessels** take lymph to the node, and **efferent lymphatic vessels** take lymph away from the node. Compare your slide with figure 24.13.

LYMPH ORGANS

Examine models and charts in the lab and compare the lymph organs with those represented in figure 24.12. The major lymph organs and tissue of the lymph system consist of the tonsils, thymus, spleen, lymph nodes along the collecting vessels, and Peyer's patches in the intestine.

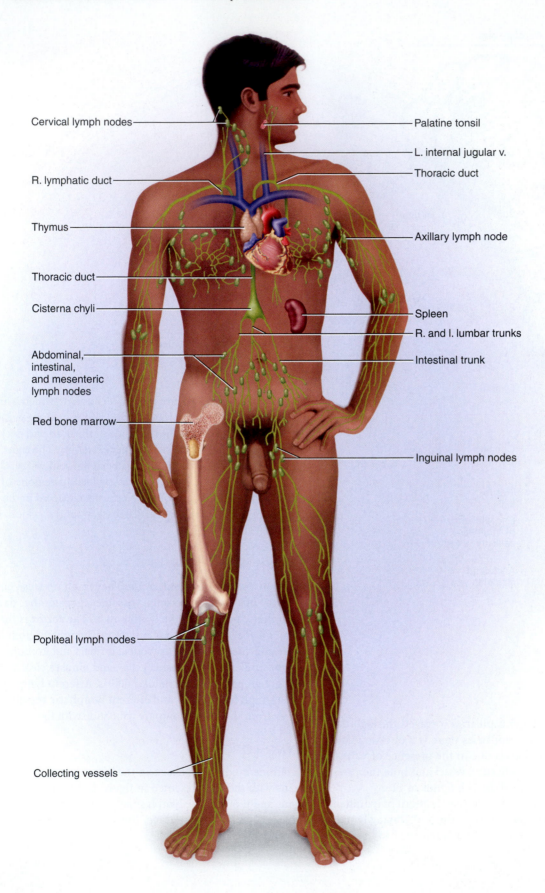

Cervical lymph nodes

Palatine tonsil

L. internal jugular v.

R. lymphatic duct

Thoracic duct

Thymus

Axillary lymph node

Thoracic duct

Cisterna chyli

Spleen

R. and l. lumbar trunks

Abdominal, intestinal, and mesenteric lymph nodes

Intestinal trunk

Red bone marrow

Inguinal lymph nodes

Popliteal lymph nodes

Collecting vessels

FIGURE 24.12

Lymph Organs and Collecting Vessels of the Body

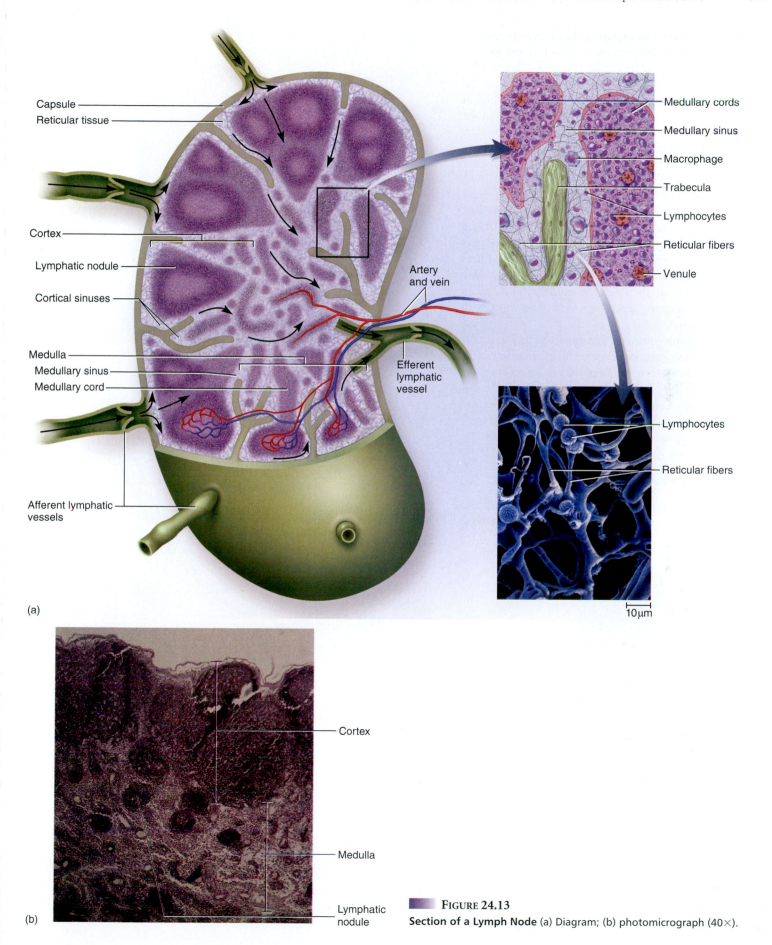

Capsule

Reticular tissue

Cortex

Lymphatic nodule

Cortical sinuses

Medulla

Medullary sinus

Medullary cord

Afferent lymphatic vessels

Artery and vein

Efferent lymphatic vessel

Medullary cords

Medullary sinus

Macrophage

Trabecula

Lymphocytes

Reticular fibers

Venule

Lymphocytes

Reticular fibers

10μm

(a)

Cortex

Medulla

Lymphatic nodule

(b)

FIGURE 24.13

Section of a Lymph Node (a) Diagram; (b) photomicrograph (40×).

Tonsils are lymph organs in the oral cavity and nasopharynx. Look at illustrations in your text and models in the lab and locate the three pair of tonsils. These are the **palatine tonsils** along the sides of the oral cavity near the oropharynx, the **lingual tonsils** at the posterior portion of the tongue, and the **pharyngeal tonsils** (adenoids) in the nasopharynx. Compare these with figure 24.14. The tonsils have **tonsillar crypts** lined with **lymphatic nodules** that serve as a first line of defense against foreign matter that may be inhaled or swallowed.

The **spleen** is an important lymph organ that filters blood, removing aging erythrocytes and foreign particles. The spleen is located to the left of the stomach. It is a highly vascular organ that filters blood and produces lymphocytes. The spleen contains **red pulp,** which filters blood, and **white pulp,** which is involved in producing lymphocytes. Locate the spleen in models and charts in the lab, or in cadaver specimens, and compare it with figures 24.12 and 24.15.

The **thymus** overlays the vessels superior to the heart. It is an important site in determining **immune competence.** Lymphocytes travel from the bone marrow to the thymus, where they undergo maturation essential to immune responses. Once these cells mature they are known as **T cells.** Examine the charts and models, or cadaver specimens in the lab, and compare them with figure 24.16.

Cat Anatomy

If you are using cats, turn to section 6, "Blood Vessels and Lymph System of the Cat" on page 427 of this lab manual.

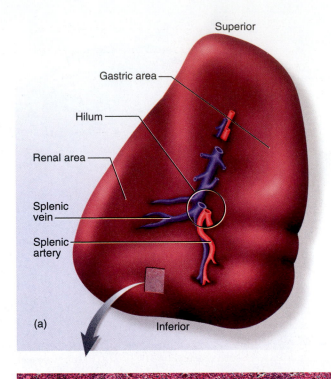

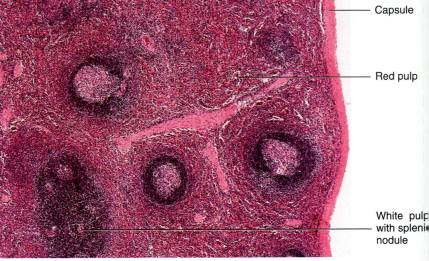

(b)

FIGURE 24.15

Spleen (a) Medial surface; (b) histology with red and white pulp.

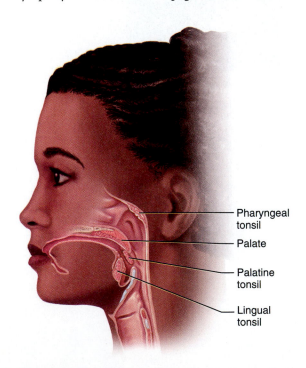

FIGURE 24.14

Tonsils

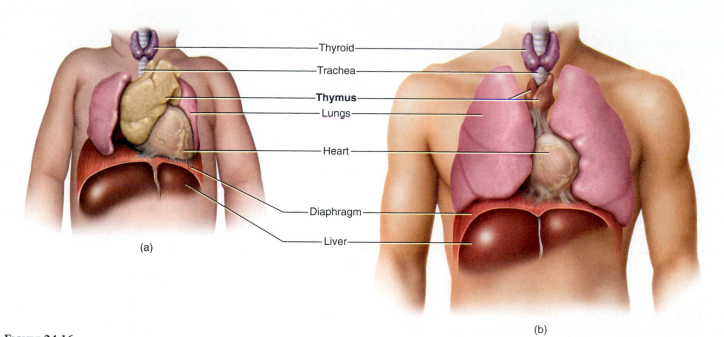

Thyroid

Trachea

Thymus

Lungs

Heart

Diaphragm

Liver

(a)

(b)

FIGURE 24.16

Thymus (a) Young individual; (b) mature adult.

Name _____ Date _____

1. Label the following illustration using the terms provided.

   brachial vein

   brachiocephalic vein

   basilic vein

   superior vena cava

   great saphenous vein

   cephalic vein

   femoral vein

   internal iliac vein

   internal jugular vein

   inferior vena cava

   common iliac vein

   external jugular vein

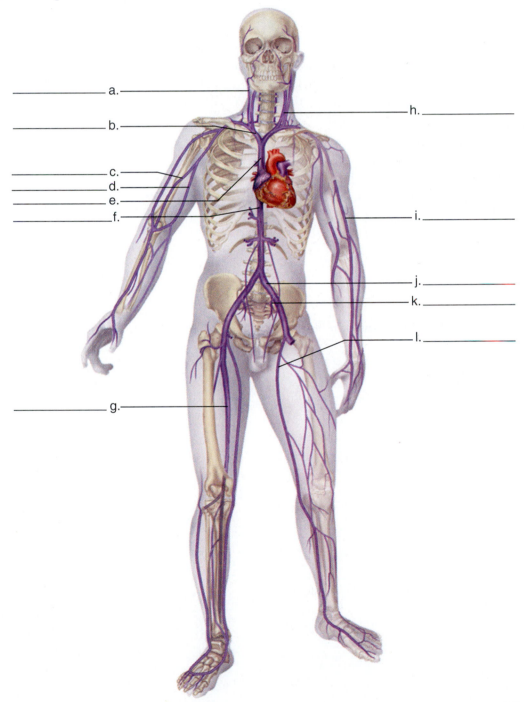

_____ a. _____

_____ b. _____

_____ c. _____

_____ d. _____

_____ e. _____

_____ f. _____

_____ g. _____

h. _____

i. _____

j. _____

k. _____

l. _____

2. Which veins (superficial/deep) have names that do not correlate with arteries?

3. The internal jugular vein takes blood from what area?

4. What veins pass through the transverse vertebral foramina?

5. The external jugular veins drain what area?

6. The brachiocephalic veins take blood to what vessel?

7. Is the radial vein a superficial or deep vein?

8. Where is the median cubital vein found?

9. What vessel receives blood from the ulnar vein?

10. What region of the body houses the cephalic vein?

11. Blood from the right axillary vein travels next to what vessel?

12. What vessels take blood to the left femoral vein?

13. In what region of the body is the great saphenous vein?

14. Where does blood flow after it leaves the femoral vein?

15. A common iliac vein receives blood from two vessels. What are these two vessels?

16. What is the functional nature of a portal system, and how is it different from normal venous return flow?

17. What major vessels take blood to the hepatic portal vein?

18. In the fetal heart what is the name of the shunt between the pulmonary trunk and the aortic arch?

19. Name the opening between the atria in the fetal heart.

20. What kind of vessels carry lymph from the lymph capillaries to the veins?

21. Once interstitial fluid enters the collecting vessels, what is it called?

22. What is the name of the inner region of a lymph node?

23. What kind of vessel takes lymph to a lymph node?

24. The adenoids are also known as _____ tonsils.

25. Which tonsils are found on the sides of the oral cavity?

26. What tonsils are located at the back of the tongue?

27. Blood is filtered by which lymph organ in the adult?

28. What part of the spleen is involved in producing lymphocytes?

29. In the analysis of breast cancer, lymph nodes of the axillary region are removed and a biopsy is performed. The removal of the nodes is done to determine if cancer has spread from the breast to other regions of the body. What effect would the removal of lymph nodes have on the drainage of the pectoral region?

30 Label the illustration of the lymph system using the terms provided.

cistera chyli   thymus gland   lymph nodes

collecting vessels   spleen   thoracic duct

right lymphatic duct

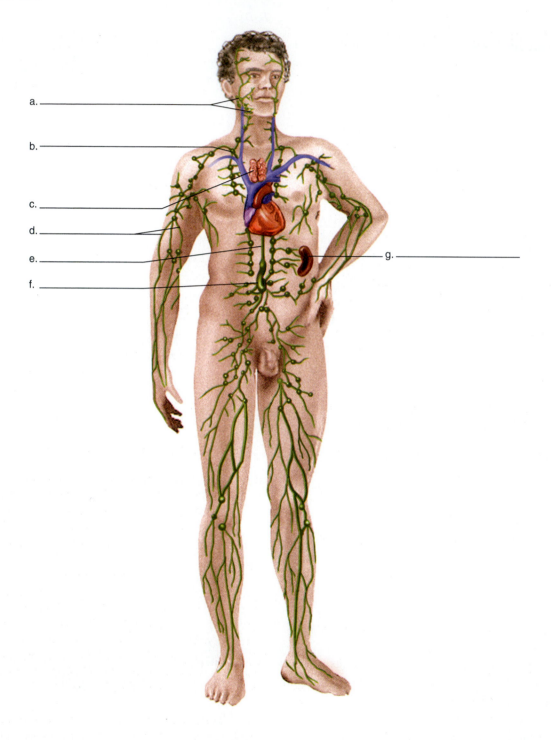

The Respiratory System

Respiratory System

INTRODUCTION

The respiratory system exchanges oxygen and carbon dioxide with the atmosphere. Atmospheric oxygen moves into the lungs and diffuses into the cardiovascular system. It subsequently reaches the individual cells of the body while the metabolic waste product, carbon dioxide, is released from the intercellular environment and travels via the blood to the lungs, where it is released by exhalation. The respiratory system is therefore integrated with the other body systems. The **conducting division** involves the airflow, while the **respiratory division** is the part of the system where gas exchange occurs. As you study the anatomy of the system be aware of the role that other systems, such as the cardiovascular system, play in respiration.

LEARNING OBJECTIVES

At the end of this exercise you should be able to
1. list the organs and significant structures of the respiratory system;
2. define the role of the respiratory system in terms of the overall function of the body;
3. explain the physical reason for the tremendous surface area of the lungs;
4. identify the cartilages of the larynx;
5. distinguish among a bronchus, bronchiole, and respiratory bronchiole.

MATERIALS

Lung models or detailed torso model, including a midsagittal section of head

Model of larynx

Microscopes

Prepared microscope slides of lung tissue

Prepared microscope slides of "smoker's lung"

Charts and illustrations of the respiratory system

PROCEDURE

Look at the charts and models of the respiratory system and compare them with figure 25.1. Locate the following structures:

_____ Nose

_____ Nasal cavity

_____ Pharynx

_____ Larynx

_____ Trachea

_____ Anterior naris

_____ Pleural cavity

_____ Bronchus

_____ Lungs

Nose and Nasal Cartilages

Examine a midsagittal section of a model or chart of the head and look for the **nose, nasal cartilages, anterior (external) nares (nostrils)** and **nasal septum.** The nasal septum separates the nasal cavities and is composed of the perpendicular plate of the ethmoid bone, the vomer, and the nasal cartilage. Examine these features in figures 25.1 and 25.2.

The entrance of the anterior nares is protected by guard hairs. The region of the nose, just posterior to the anterior nares, is the **nasal vestibule,** which is lined with stratified squamous epithelium. Behind the vestibule is the **nasal cavity,** which is lined with a **mucous membrane** that moistens air entering the respiratory system. The membrane consists of **respiratory epithelium,** which is composed of **pseudostratified ciliated columnar epithelium** with **goblet cells.** This mucous membrane overlays a superficial venous plexus that warms the air.

The lateral walls of the cavity have three protrusions that push into the nasal cavity. These are the **nasal conchae.** They cause

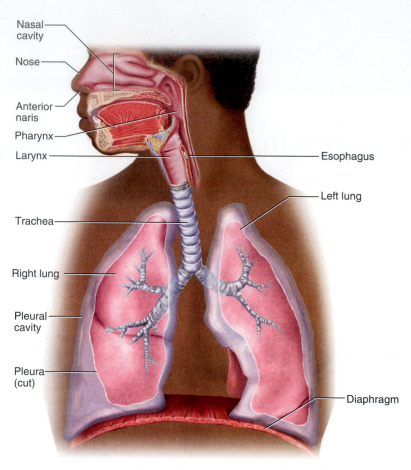

FIGURE 25.1

Overview of the Anatomy of the Respiratory System

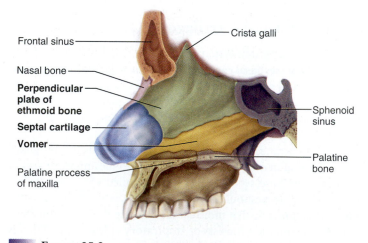

FIGURE 25.2

Structures That Make Up the Nasal Septum

the inhaled air to swirl in the nasal cavity and come into contact with the mucous membrane. This moistens the air. Locate the **superior, middle,** and **inferior conchae** in the nasal cavity. The nasal cavity terminates where two openings, the **posterior nares (internal nares)** (singular, **naris**), **choanae** lead to the **pharynx.** Locate these structures on the materials in lab and compare them with figure 25.3.

Pharynx

The pharynx can be divided into three regions based on location. The uppermost area is the **nasopharynx,** which is directly posterior to the nasal cavity. The nasopharynx has two openings on the lateral walls, which are the openings of the **auditory,** or **eustachian, tubes.** These equalize pressure in the middle ear. As the pharynx descends behind the oral cavity it becomes the **oropharynx.** The oropharynx is a common passageway for food, liquid, and air. The **uvula** is a small, pendulous structure that partially separates the **oral cavity** from the oropharynx. The uvula flips upward during swallowing, reducing the chance of fluids entering the nasopharynx. The most inferior portion of the pharynx is the **laryngopharynx,** which is located posterior to the larynx. Find the regions of the pharynx in figure 25.3.

Larynx

The **larynx** is commonly known as the "voice box" because it is an important organ for sound production in humans. It controls the pitch of the voice, while the shape of the oral cavity and the placement and size of the paranasal sinuses are responsible for the sonority, or sound quality, of the voice. The larynx is at the level of the fourth through sixth cervical vertebrae and consists of a num-

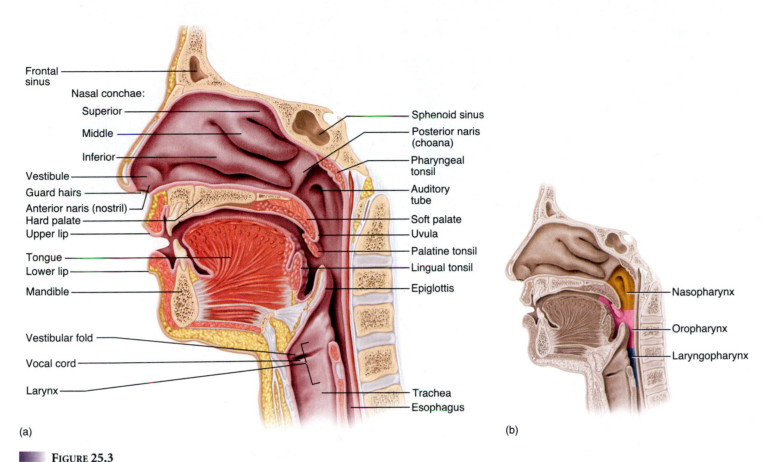

FIGURE 25.3

Upper Respiratory Tract (a) Midsagittal section of the head with nasal septum removed; (b) three regions of the pharynx.

ber of cartilages. The most prominent cartilage in the larynx is the **thyroid cartilage,** which is a shield-shaped structure made of hyaline cartilage.

The thyroid cartilage is more prominent in males because it increases in size under the influence of testosterone. Its common name is "Adams apple." Inferior to the thyroid cartilage is the **cricoid cartilage.** The cricoid cartilage is also composed of hyaline cartilage, and it is a thin band of cartilage when seen from the anterior aspect but increases in size at its posterior surface. Superior to the cricoid cartilage in the posterior wall of the larynx are the paired **arytenoid cartilages.** These cartilages attach to the posterior end of the **true vocal folds (vocal cords).** Movement of the arytenoid cartilages pulls on the true vocal folds, causing them to tighten, thus increasing the pitch of the voice. This occurs by the contraction of **intrinsic muscles** attached to the arytenoid cartilages from the back, while the vocal cords are held stationary by the thyroid cartilage in the front. Above the true vocal folds are the **vestibular folds (false vocal cords)** (fig. 25.4).

At the very posterior, superior edge of the larynx are the **corniculate** and **cuneiform cartilages.** These are also made of hyaline cartilage. The most superior cartilage of the larynx is the **epiglottis,** which is composed of elastic cartilage. During swallowing, the epiglottis protects the opening of the larynx, which is known as the **glottis.** In the swallowing reflex, muscles pull the epiglottis down over the glottis. This is not a perfect system, as anyone knows who has started swallowing a liquid and responded by laughing at a joke. Inhalation at the beginning of the laugh causes fluid to move into the larynx and trachea, irritating the respiratory lining. This causes another reflex called the **cough reflex,** which propels the liquid out of the respiratory system. Locate the structures of the larynx on models or charts in the lab and in figure 25.4.

Trachea and Bronchi

The trachea is commonly known as the "windpipe" because it conducts air from the larynx to the lungs. The **trachea** is a straight tube whose lumen is kept open by **tracheal (C-shaped) cartilages.** Examine these cartilages by running your fingers gently down the outside of your throat. Palpate the cartilage rings below the larynx. The tracheal cartilages are composed of hyaline cartilage. At the most inferior portion of the trachea is a center point known as the **carina** (*carina* = keel). The carina divides the incoming air into the bronchi. Locate the features of the trachea in figures 25.5 and 25.6. The trachea is also lined with respiratory epithelium. Obtain a prepared slide of the trachea and find the tracheal cartilage, respiratory epithelium, and **posterior tracheal membrane** (which includes the trachealis muscle) (fig. 25.6). The membrane allows the esophagus to expand into the trachea during swallowing.

The trachea splits into two tubes, which enter the lungs. These tubes are the **main bronchi.** Each lung receives air from a main bronchus, which contains hyaline cartilage and is lined with

Anterior

Posterior

Median

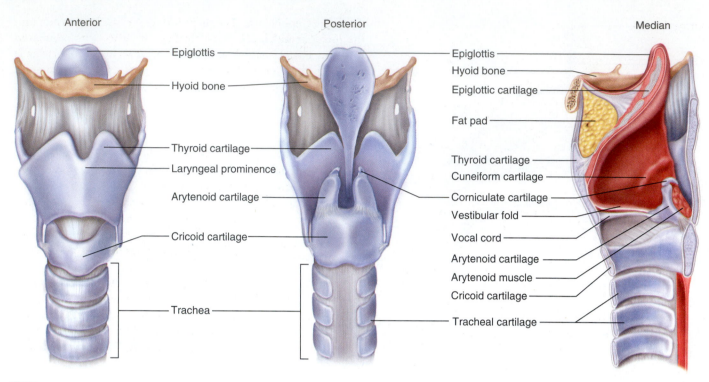

Epiglottis

Hyoid bone

Thyroid cartilage

Laryngeal prominence

Arytenoid cartilage

Cricoid cartilage

Trachea

Epiglottis

Hyoid bone

Epiglottic cartilage

Fat pad

Thyroid cartilage

Cuneiform cartilage

Corniculate cartilage

Vestibular fold

Vocal cord

Arytenoid cartilage

Arytenoid muscle

Cricoid cartilage

Tracheal cartilage

FIGURE 25.4

Larynx

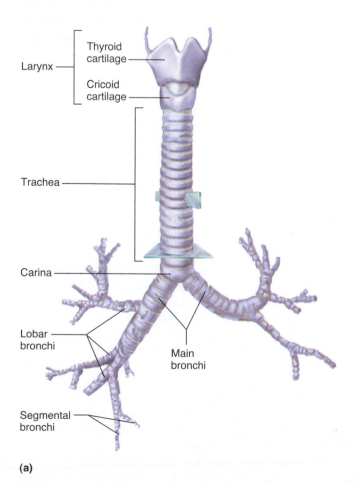

Larynx

Thyroid cartilage

Cricoid cartilage

Trachea

Carina

Lobar bronchi

Main bronchi

Segmental bronchi

(a)

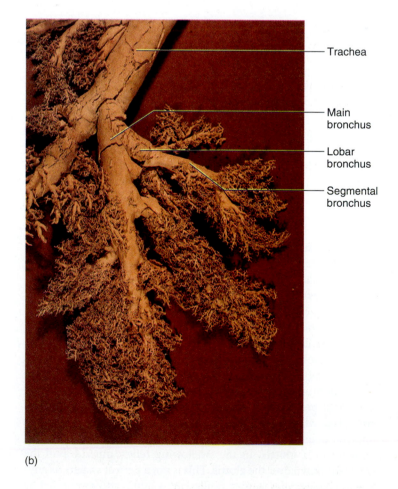

Trachea

Main bronchus

Lobar bronchus

Segmental bronchus

(b)

FIGURE 25.5

Larynx, Trachea, and Bronchial Tree (a) Diagram anterior view; (b) cast of bronchial tree.

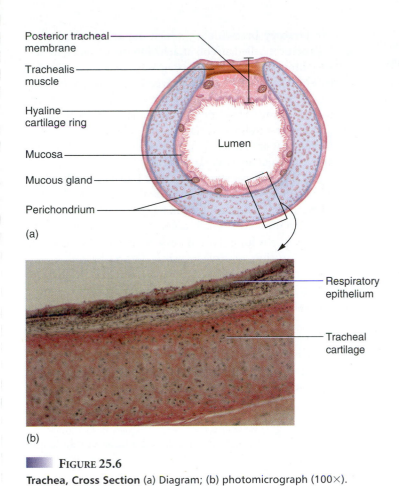

Posterior tracheal membrane

Trachealis muscle

Hyaline cartilage ring

Mucosa

Mucous gland

Perichondrium

Lumen

(a)

Respiratory epithelium

Tracheal cartilage

(b)

FIGURE 25.6
Trachea, Cross Section (a) Diagram; (b) photomicrograph (100×).

respiratory epithelium. The main bronchi of the lung divide into the **lobar bronchi,** and these further divide to form **segmental bronchi.** The extensive branching of the bronchi produces what is known as the bronchial tree (fig. 25.5). Each segmental bronchus takes air to its specific bronchopulmonary segment.

Lungs

There are two **lungs** in humans; the right lung has **three lobes** known as the **superior, middle,** and **inferior lobes,** separated by a superior **horizontal fissure** and an inferior **oblique fissure.** The left lung has **two lobes** with an **oblique fissure** and the indentation left by the heart known as the **cardiac impression.** The lobes of the left lung are the **superior** and **inferior lobes.** Look at models or charts in the lab and identify the major features, as shown in figure 25.7.

The lungs are located in the **pleural cavities** on each side of the mediastinum. The **parietal pleura** is the outer membrane on the chest cavity wall, and the membrane that is adhered to the surface of the lungs is the **visceral pleura.** The space between the membranes is known as the pleural cavity. These are shown in figure 25.8.

HISTOLOGY OF THE LUNG

The bronchi continue to divide until they become **bronchioles,** small respiratory tubules with smooth muscle in their walls. Obtain a prepared slide of lung and scan first under low power and then under higher powers. The bronchioles in the lung further divide

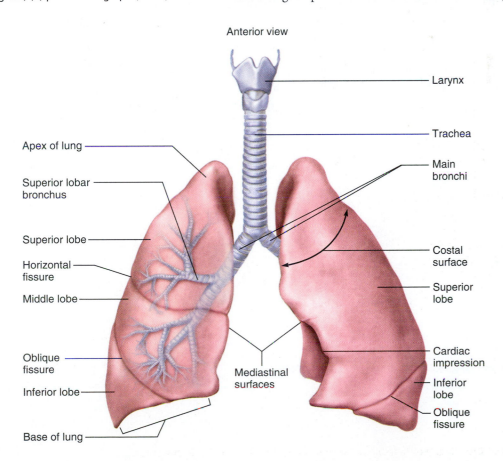

Anterior view

Apex of lung

Superior lobar bronchus

Superior lobe

Horizontal fissure

Middle lobe

Oblique fissure

Inferior lobe

Base of lung

Larynx

Trachea

Main bronchi

Costal surface

Superior lobe

Cardiac impression

Inferior lobe

Oblique fissure

Mediastinal surfaces

FIGURE 25.7
Thoracic Region with Lungs

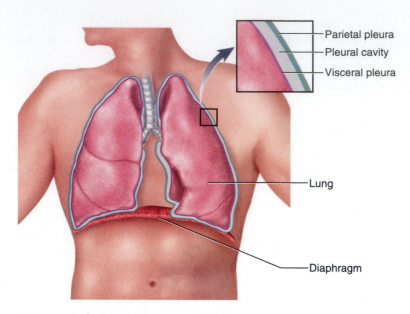

FIGURE 25.8
Pleural Membranes and Cavity

Parietal pleura
Pleural cavity
Visceral pleura

Lung

Diaphragm

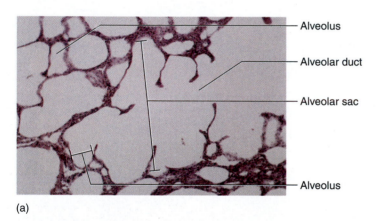

Alveolus

Alveolar duct

Alveolar sac

Alveolus

(a)

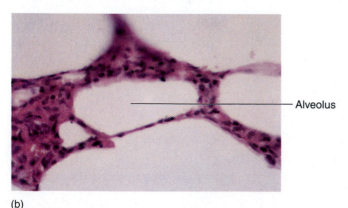

Alveolus

(b)

FIGURE 25.9
Histology of the Lung (a) Alveolar sac (100×); (b) alveolus (400×).

into **respiratory bronchioles,** which are so named due to small structures called alveoli attached to their walls. The respiratory bronchioles lead to passageways known as **alveolar ducts,** which branch into alveoli. **Alveoli** are air sacs in the lung that exchange oxygen and carbon dioxide with the blood capillaries of the lungs. Examine your slide and locate the bronchi, bronchioles, respiratory bronchioles, alveolar ducts, and alveoli. Alveoli that are clustered around an alveolar duct are collectively known as an **alveolar sac** (fig. 25.9).

The alveoli have **squamous type I alveolar cells** which are simple squamous epithelium as are the capillaries that surround the alveoli. These are seen in figure 25.9. The division of the lung into numerous small sacs tremendously increases the surface area of the lung. This increase is vital for the rapid and extensive diffusion of oxygen across the respiratory membranes. You may see other cell shapes that occur in the prepared lung sections. Some of these cells are **great type 2 alveolar cells.** They decrease the surface tension of the lung by the secretion of **surfactant.**

Examine a prepared slide of smoker's lung. Note the dark material in the lung tissue and the general destruction of the alveoli. Breakdown of the alveoli leads to a disease known as **emphysema.**

PATHWAY OF AIR

Air moves from the anterior nares to the nasal vestibule and into the nasal cavity. From there the warmed and moistened air passes through the pharyax and into the larynx, trachea, and bronchial tree. The air moves into the alveolar ducts and finally to the alveolus. Exhalation is essentially the reverse process.

Cat Anatomy

If you are using cats, turn to section 7, "Respiratory Anatomy of the Cat," on page 438 of this lab manual.

Name _____ Date _____

1. What is the common name for the anterior nares?

2. The apex of the nose is composed of cartilage. What functional adaptation does cartilage have over bone in making up the external framework of the nose?

3. The nasal cavity is divided in two by what structure?

4. Three structures make up the nasal septum. What are they?

5. What is the function of respiratory epithelium and the venous plexuses in the nasal cavity?

6. From the nasal cavity, what is the name of the opening into the nasopharynx?

7. What is the name of the space behind the oral cavity and above the laryngopharynx?

8. What is the name of the structure that prevents fluid in the oral cavity from entering the nasopharynx during swallowing?

9. What is the name of the large cartilage of the anterior larynx?

10. What is the structure that covers the glottis, protecting the larynx from fluid when swallowing?

11. Which lung has just two lobes?

12. What membrane attaches directly to the external surface of the lungs?

13. The trachea branches into two tubes that go to the lungs. What are these tubes called?

14. Where is the bronchial tree found?

15. What small structure in the lung is the site of exchange of oxygen with the blood capillaries?

16. The surface area of the lungs in humans is about 70 square meters. How can this be if the lungs are located in the small space of the thoracic cavity? What role do alveoli play in the nature of surface area?

17. Emphysema is a destruction of the alveoli of the lungs. What effect does this have on the surface area of the lungs?

18. Fill in the following illustration of the human respiratory system using the terms provided.

main bronchus   superior lobe   middle lobe

trachea   nasal cavity   cricoid cartilage

nasopharynx   inferior lobe   epiglottis

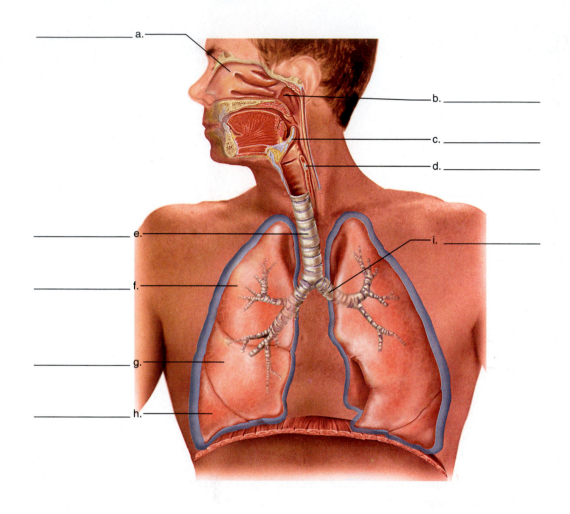

The Digestive System

Digestive System

INTRODUCTION

The digestive system can be divided into two major parts—the **digestive tract** and the **accessory organs.** The digestive tract is a long tube that runs from the mouth to the anus; it comes into contact with food or the breakdown products of digestion. The digestive tract is also known as the **alimentary canal.** The portion of the alimentary canal from the stomach to the anus is the **gastrointestinal (GI) tract.** Some of the organs of the digestive tract are the esophagus, stomach, small intestine, large intestine, and anus. The accessory organs are important in that they secrete many substances necessary for digestion, yet these organs do not come into direct contact with ingested material. Examples of accessory organs are the salivary glands, liver, gallbladder, and pancreas.

The functions of the digestive system include the ingestion of food, the physical breakdown of food, the chemical breakdown of food, nutrient storage, nutrient and water absorption, vitamin synthesis, and the elimination of indigestible material. In this exercise you examine the anatomy of the digestive system, correlating the structure of the digestive organs with their functions.

LEARNING OBJECTIVES

At the end of this exercise you should be able to
1. list, in sequence, the major organs of the digestive tract;
2. describe the basic function of the accessory digestive organs;
3. note the specific anatomic features of each major digestive organ;
4. describe the layers of the wall of the digestive tract;
5. describe the major functions of the stomach, small intestine, and large intestine.

MATERIALS

Models, charts, or illustrations of the digestive system
Hand mirror

Skull, human teeth, or cast of teeth
Cadaver (if available)
Microscopes
Microscope slides:
    Esophagus
    Stomach
    Small intestine
    Large intestine
    Liver

PROCEDURE

Overview of the Digestive System

Begin this exercise by examining a torso model or charts in the lab and compare them with figure 26.1. Locate the major digestive organs and, when you find them, place a check mark next to the following terms.

_____ Mouth

_____ Pharynx

_____ Esophagus

_____ Stomach

_____ Small intestine

_____ Large intestine (colon)

_____ Liver

_____ Pancreas

_____ Anus

_____ Gallbladder

_____ Salivary glands (parotid, submandibular, and sublingual glands)

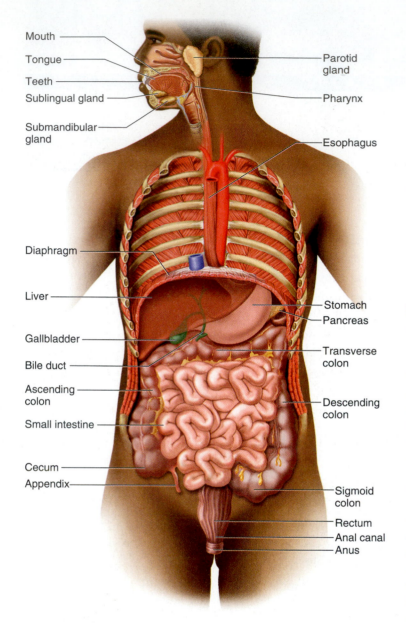

FIGURE 26.1
Overview of the Digestive System

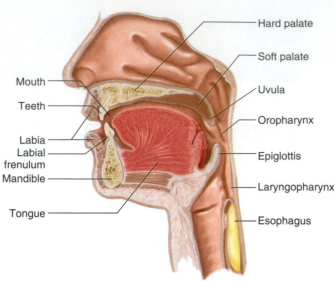

FIGURE 26.2
Mouth, Midsagittal Section

Organs of the Digestive Tract

Begin your study of the digestive tract with the **mouth (oral cavity).** Examine a midsagittal section of the head, as represented in figure 26.2, and locate the major anatomic features.

At the beginning of the digestive tract is the mouth, which begins as an opening surrounded by lips, or **labia.** The **labial frenulum** is a membranous structure that keeps the lip adhered to the gums, or **gingivae.** The mouth is a space anterior to the **oropharynx,** and medial to the cheeks. The hard and soft palates form the roof of the mouth, and the mylohyoid muscle is the inferior border. The **hard palate** is composed of the palatine processes of the maxillae and palatine bones. The **soft palate** is composed of connective tissue and a mucous membrane. At the posterior por-

tion of the mouth is the **uvula** (small grape), a structure that is suspended from the posterior edge of the soft palate. The uvula helps prevent food or liquid from moving into the nasal cavity during swallowing. The oral cavity is lined with **nonkeratinized stratified squamous epithelium,** which protects the underlying tissue from abrasion.

In a midsagittal section the tongue appears as a broad, fan-shaped muscular structure. One of the major muscles of the tongue is the **genioglossus.** As a digestive organ, the tongue moves food toward the teeth for chewing and acts as a piston to propel food to the **oropharynx** for swallowing. The tongue is held down to the floor of the mouth by a thin mucous membrane called the **lingual frenulum.** On the superior surface of the tongue are four types of **papillae,** or raised areas, on the tongue. These are the **foliate, fungiform, filiform,** and **circumvallate** papillae. Papillae increase the frictional surface of the tongue. The tongue also has taste receptors located in **taste buds.** Taste buds are found on the tongue along the sides of the papillae. The sense of taste is covered in laboratory exercise 18. Using a mirror, examine your tongue and locate the papillae. The mouth is important in digestion for the physical breakdown of food. This process is driven by powerful muscles called the **muscles of mastication.** The **masseter** and the **temporalis muscles** are involved in the closing of the jaws, and the **pterygoid muscles** are important in the sideways grinding action of the molar and premolar teeth.

STRUCTURE OF TEETH

Examine models of teeth, dental casts, or real teeth on display in the lab and compare them with figure 26.3. A tooth consists of a **crown,** a **neck,** and a **root.** The crown is the exposed part of the tooth; the neck is a constricted portion of the tooth is normally at the surface of the gingivae; and the root is embedded in the jaw. Examine a model or an illustration of a longitudinal section of a tooth and find the outer **enamel,** an extremely hard material. Inside this layer is the **dentin,** which is made of bonelike material. The innermost

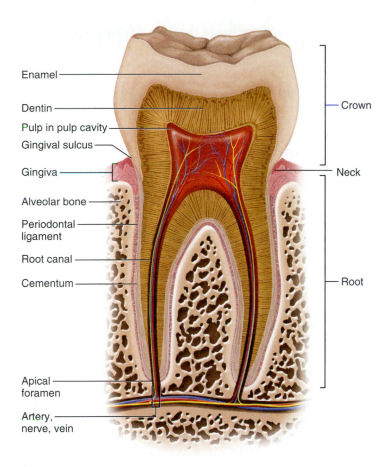

Enamel

Dentin

Pulp in pulp cavity

Gingival sulcus

Gingiva

Alveolar bone

Periodontal ligament

Root canal

Cementum

Apical foramen

Artery, nerve, vein

Crown

Neck

Root

FIGURE 26.3
Tooth, Longitudinal Section

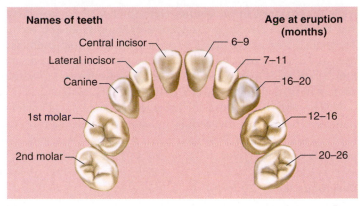

Names of teeth

Age at eruption (months)

Central incisor — 6–9
Lateral incisor — 7–11
Canine — 16–20
1st molar — 12–16
2nd molar — 20–26

(a)

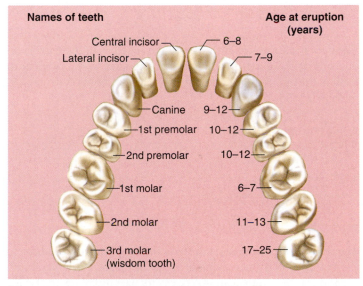

Names of teeth

Age at eruption (years)

Central incisor — 6–8
Lateral incisor — 7–9
Canine   9–12
1st premolar   10–12
2nd premolar   10–12
1st molar   6–7
2nd molar   11–13
3rd molar (wisdom tooth)   17–25

(b)

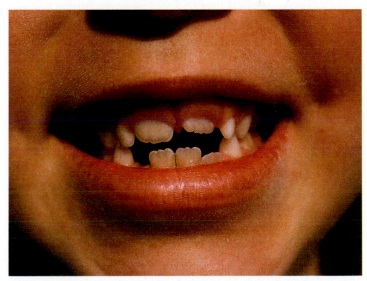

(c)

FIGURE 26.4
Dentition (a) Upper teeth of a child. Notice the absence of premolar teeth. (b) Upper teeth of an adult. (c) Replacement of deciduous teeth with permanent teeth.

portion of the tooth is the **pulp cavity,** which leads to the **root canal,** a passageway for nerves and blood vessels into the tooth. The nerves and blood vessels enter the tooth through the **apical foramen** at the tip of the root of the tooth. The teeth are set in depressions in the mandible or maxilla called **alveolar sockets** and are anchored to the bone by periodontal ligaments.

There are four types of teeth in the adult mouth:

- **Incisors** are flat, bladelike front teeth that nip food. There are eight incisors in the adult mouth.

- **Canines,** or **cuspids,** are pointed teeth just lateral to the incisors that shear food. There are four cuspids in the adult, and they can be identified as the teeth that have just one cusp, or point.

- **Premolars,** or **bicuspids,** are lateral to the cuspids and grind food. There are eight premolars in the adult mouth, and they can be identified by their two cusps.

- **Molars** are the most posterior. Like the premolars, these teeth grind food. There are 12 molars in the adult mouth (including the third molars, or **wisdom teeth**). Molars typically have three to five cusps.

Examine a human skull or dental cast and identify the characteristics of the four types of teeth (fig. 26.4). Humans have two sets of teeth: the **primary,** or **deciduous, teeth** (milk teeth) appear first, and these are replaced by the **secondary,** or **permanent, teeth.** There are 20 deciduous teeth. There are no deciduous premolar teeth and there are only 8 deciduous molar teeth. In adults there are 8 premolar teeth and 12 molar teeth. Compare figure 26.4*a*, the adult pattern, to figure 26.4*b* and *c*, the deciduous teeth.

The pattern of tooth structure is represented by a dental formula, which describes the teeth by quadrants. The dental formula for the primary teeth is illustrated here: I = incisor, C = cuspid, P = premolar, and M = molar.

Deciduous Teeth (20 Total)

Dental formula	I	C	P	M	One side (quadrant)
	2	1	0	2	Top
	2	1	0	2	Bottom

The upper numbers refer to the number of teeth in the maxilla on one side of the head. The lower numbers refer to the number of teeth in the mandible on one side of the head. The adult dental formula is as follows:

Adult (Permanent) Teeth (32 Total)

	I	C	P	M
	2	1	2	3
	2	1	2	3

PHARYNX

The space posterior to the mouth is the **oropharynx.** Superior to the oropharynx is the **nasopharynx,** which leads to the nasal cavity, and inferior to the oropharynx is the **laryngopharynx,** which leads to the larynx and the esophagus. The oropharynx is composed of nonkeratinized stratified squamous epithelium and is a common passageway for food, liquids, and air. Muscles around the wall of the oropharynx are the **pharyngeal constrictor muscles** and are involved in swallowing. Food is moved by the tongue to the region of the pharynx where it is propelled into the **esophagus.** Locate the oropharynx and the esophagus in figures 26.1 and 26.2.

ESOPHAGUS

The esophagus conducts food from the laryngopharynx, through the diaphragm, and into the stomach. The esophagus has an inner layer of stratified squamous epithelium near the lumen (the space where ingested material travels throughout the alimentary canal) and an outer connective tissue layer called the **adventitia.** Unlike the trachea, which is an open tube, the esophagus is a closed tube that begins approximately at the level of the sixth cervical vertebra. As a lump of food, or **bolus,** enters the esophagus, **skeletal muscle** begins to move it toward the stomach. The middle portion of the esophagus is composed of both skeletal and smooth muscle, while the lower portion of the esophagus is made of **smooth muscle.** In the lower region of the esophagus the smooth muscle contracts, moving the bolus by a process known as **peristalsis.** Locate the esophagus in charts and models in lab and compare it with figure 26.1. The lower portion of the esophagus has an **esophageal sphincter,** which prevents the backflow of stomach acids. Esophageal reflex (heartburn) occurs if the stomach contents pass through the esophageal sphincter and irritate the esophageal lining.

ABDOMINAL PORTIONS OF THE DIGESTIVE TRACT

Membranes   The inner structure of the body has frequently been referred to as a "tube within a tube." The body wall forms the outer tube, and the digestive tract, including the stomach, small intestine, and large intestine, forms the inner tube. Specialized serous membranes cover the various organs and line the inner wall of the **coelom** (**body cavity**). The membrane lining the outer surface of the digestive tract is called the **visceral peritoneum** (**serosa**) and continues as a double-folded membrane called the **mesentery,** which attaches the tract to the back of the body wall. The mesentery is continuous with the membrane on the inner side of the body wall, where it is called the **parietal peritoneum.** Locate these three membranes in figure 26.5.

Layers of the Wall of the Digestive Tract   Obtain a slide of a cross section of the digestive tract and look for the layers that make up the wall of the digestive tract. In general, the stomach, small intestine, and large intestine have the same layers from the lumen to the coelom. The innermost layer is the **mucosa,** which consists of a **mucous membrane** closest to the lumen; a connective tissue layer called the **lamina propria;** and an outer muscular layer called the **muscularis mucosa.**

The next layer is called the **submucosa,** which is made mostly of **connective tissue** and contains numerous blood vessels. This is followed by the **muscularis** (**muscularis externa**), which is typically made of two or three layers of smooth muscle. The muscularis propels material through the digestive tract and mixes ingested material with digestive juices. The outermost layer is called the **serosa,** or **visceral peritoneum,** and this layer is closest to the coelom. Examine a microscope slide of the digestive tract (small intestine) and locate the layers as represented in figure 26.6.

Stomach   Examine a model of the stomach or charts in the lab and compare these with figure 26.7. The **stomach** is located on the superior middle and left portion of the abdomen and receives its contents from the esophagus. The food that enters the stomach is stored and mixed with enzymes and hydrochloric acid (HCl) to form a soupy material called **chyme.** The stomach can have a pH as low as 1 or 2. Chyme remains in the stomach as the acids denature proteins and enzymes reduce proteins to shorter fragments. The acid of the stomach also has an antibacterial action, as most microbes do not grow well in conditions of low pH.

Locate the upper portion of the stomach called the **cardiac region,** or **cardia.** A part of the cardiac region extends superiorly as a domed section called the **fundus,** or **fundic region.** The main part of the stomach is called the **body,** and the terminal portion of the stomach, closest to the small intestine, is called the **pyloric region.** The pyloric region has an expanded area called the **antrum** and a narrowed region called the **pyloric canal** that leads to the duodenum.

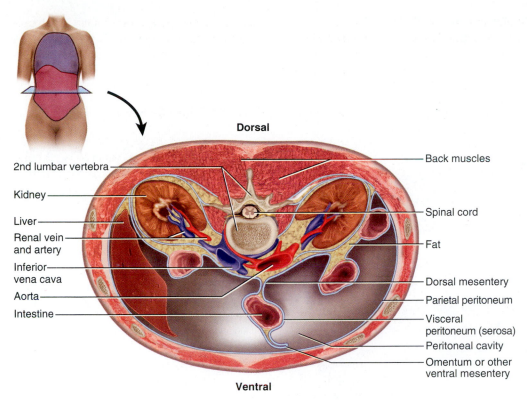

Dorsal

2nd lumbar vertebra

Kidney

Liver

Renal vein and artery

Inferior vena cava

Aorta

Intestine

Back muscles

Spinal cord

Fat

Dorsal mesentery

Parietal peritoneum

Visceral peritoneum (serosa)

Peritoneal cavity

Omentum or other ventral mesentery

Ventral

FIGURE 26.5
Membranes of the Digestive Tract

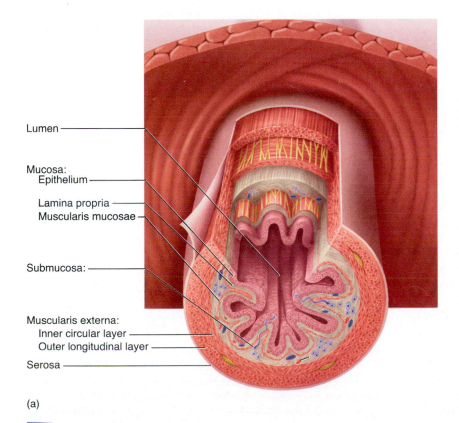

Lumen

Mucosa:
  Epithelium

  Lamina propria
  Muscularis mucosae

Submucosa:

Muscularis externa:
  Inner circular layer
  Outer longitudinal layer

Serosa

(a)

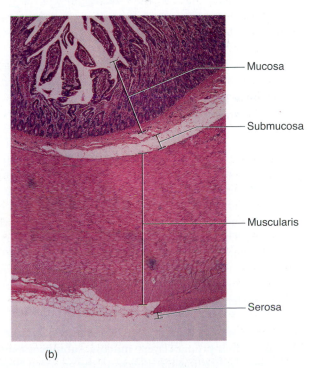

Mucosa

Submucosa

Muscularis

Serosa

(b)

FIGURE 26.6
Digestive Tract, Cross Section (a) Diagram; (b) photomicrograph (100×).

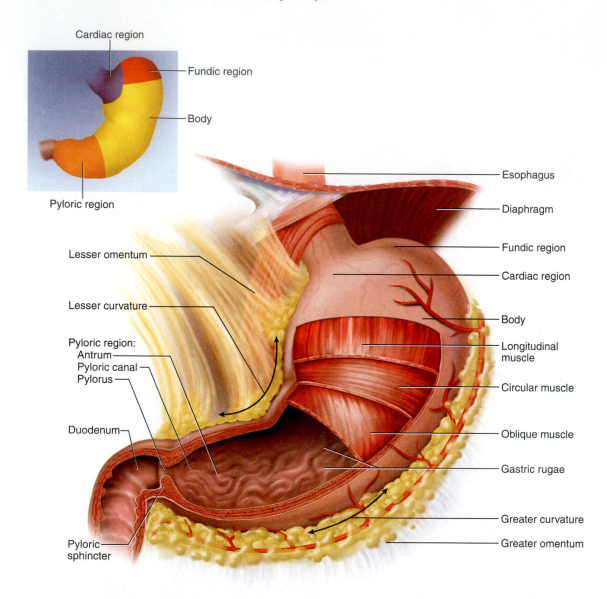

Gross Anatomy of the Stomach

The left side of the stomach is arched and forms the **greater curvature,** while the right side of the stomach is a smaller arc, forming the **lesser curvature.** The inner surface of the stomach has a series of folds called **rugae,** which allow for significant expansion of the stomach. The contents are held in the stomach by two sphincters. The upper one, the **esophageal sphincter,** prevents stomach contents from moving into the esophagus. The lower sphincter is called the **pyloric sphincter;** it prevents the premature release of stomach contents into the small intestine. Locate the pyloric sphincter at the terminal portion of the stomach.

Stomach Histology   Examine a prepared slide of the stomach. Identify the four primary layers: **mucosa, submucosa, muscularis,** and **serosa.** Notice how the mucosa in the prepared slide has a series of indentations. These depressions are **gastric glands,** in the inner lining of the stomach. The **mucous membrane** is composed predominantly of **simple columnar epithelium,** along with several specialized cells.

Surface mucous cells are located in the membrane; they secrete mucus, which protects the stomach lining from erosion by stomach acid and proteolytic (protein-digesting) enzymes. Other specialized cells that you might find in the mucosa are **chief cells,** which secrete **pepsinogen** (the inactive state of a proteolytic enzyme). Chief cells contain blue-staining granules in some prepared slides. Other cells are **parietal cells,** which secrete **HCl.** They typically contain orange-staining granules. When pepsinogen comes into contact with HCl, pepsinogen is activated as **pepsin.**

Deeper to the mucous membrane, locate the **lamina propria,** which is a connective tissue layer. The **muscularis mucosa** is even farther away from the lumen; it moves the mucous membrane. The **submucosa** is the next layer and is typically lighter in color in prepared slides.

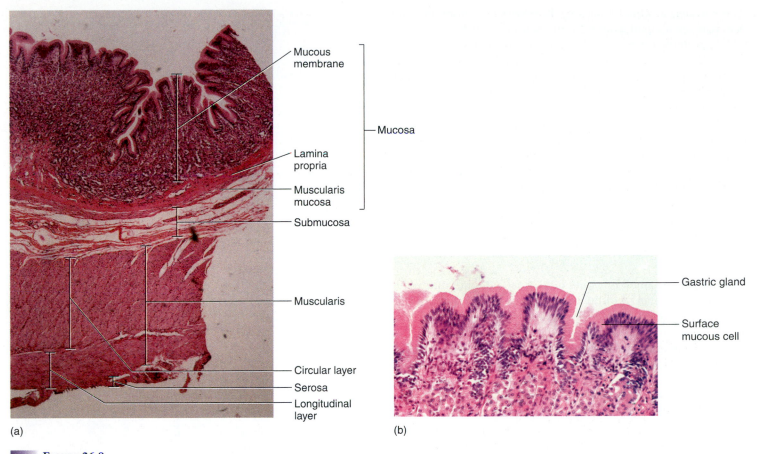

(a)

(b)

FIGURE 26.8

Histology of the Stomach (a) Overview (40×); (b) mucosa (100×).

The next layer of the stomach is the **muscularis.** In some parts of the stomach there are three layers of the muscularis—an inner **oblique layer,** a middle **circular layer,** and an outer **longitudinal layer.** The muscularis moves chyme from the stomach through the pyloric sphincter and into the small intestine.

The outermost layer is the **serosa,** and it is composed of a thin layer of connective tissue and **simple squamous epithelium.** Locate these structures and compare them with figure 26.8.

Small Intestine   The **small intestine** is approximately 5 m (17 ft) long in living humans yet can be longer in cadaveric specimens due to the stretching of the smooth muscle. Movement through the small intestine occurs by peristalsis, which is smooth muscle contraction. It is called the small intestine because it is small in diameter. The small intestine is typically 3 to 4 cm (1.5 in.) in diameter when empty. The primary function of the small intestine is nutrient absorption.

Locate the three major regions of the small intestine on a model or chart and compare them with figure 26.9. The first part of the small intestine is the **duodenum,** which forms a **C**-shaped curve attached to the pyloric region of the stomach. The duodenum begins in the body cavity and then moves behind the parietal peritoneum and returns to the body cavity. The duodenum is

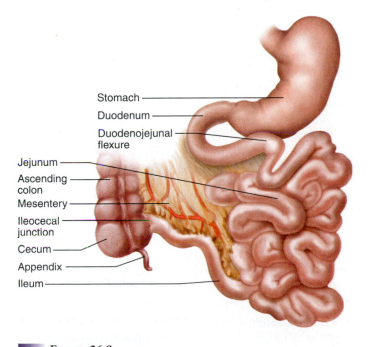

FIGURE 26.9

Gross Anatomy of the Small Intestine

approximately 25 cm (10 in.) long. It receives fluid from both the **pancreas** and the **gallbladder.** The junction between the duodenum and the jejunum is known as the **duodenojejunal flexure.**

The second portion of the small intestine is the **jejunum,** which is approximately 2 m (6.5 ft) long. The terminal portion of the small intestine is the **ileum,** which is approximately 3 m (10 ft) long. The sphincter between the small intestine and large intestine is called the **ileocecal valve.** This valve keeps material in the large intestine from reentering the small intestine. Locate the small intestine and associated structures in figure 26.9.

The surface area of the small intestine is tremendous. **Circular folds (plicae circulares)** are macroscopic folds that churn material as it passes through the small intestine and increase the surface area. **Villi** (fingerlike structures) and **microvilli** (cellular processes) increase the surface area of the small intestine even more.

Histology of the Small Intestine   The small intestine, due to the presence of villi, can be distinguished from both the stomach and the large intestine (fig. 26.10). Each villus contains **blood vessels,** which transport sugars and amino acids from the intestine to the liver. In addition, the villi contain **lacteals,** which transport fatty acids via lymphatics to the venous system. The villi give the lining of the small intestine a velvety appearance to the naked eye. The inner lining of the small intestine consists of **simple columnar epithelium** with **goblet cells.** You may distinguish the three sections of the small intestine by noting that the duodenum has **duodenal glands** in the portion of the mucosa away from the lumen. The jejunum and the ileum lack these glands. The ileum is distinguished by the presence of **aggregated lymphatic nodules,** or **Peyer's patches,** present in the mucosa. These lymphatic nodules produce lymphocytes, which protect the body from the bacterial flora in the lumen of the small intestine. Compare the prepared slides of the small intestine with figure 26.11.

Large Intestine   The **large intestine,** so named because it is large in diameter, approximately 7 cm (3 in.); it is 1.4 m (4.5 ft) long. The function of the large intestine is absorption of water and formation of feces. The mucosa of the large intestine is made of **simple columnar epithelium** with a large number of **goblet cells.** There are no villi present, nor aggregated lymphatic nodules, yet the wall of the large intestine has solitary lymphatic nodules. Look at a model or chart of the large intestine and note the major regions in the following list. Compare them with figure 26.12.

- **Cecum:** first part of the large intestine. The cecum is a pouchlike area that articulates with the small intestine at the ileocecal junction. The appendix (described later) is attached to the cecum.

- **Ascending colon:** found on the right side of the body. It becomes the transverse colon at the **right colic (hepatic) flexure.**

- **Transverse colon:** traverses the body from right to left. It leads to the descending colon at the **left colic (splenic) flexure.**

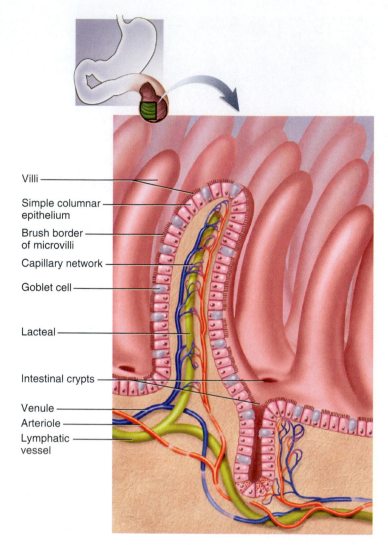

FIGURE 26.10
Villi of the Small Intestine

Labels: Villi, Simple columnar epithelium, Brush border of microvilli, Capillary network, Goblet cell, Lacteal, Intestinal crypts, Venule, Arteriole, Lymphatic vessel

- **Descending colon:** passes inferiorly on the left side of the body and joins with the sigmoid colon

- **Sigmoid colon:** S-shaped segment of the large intestine in the left inguinal region

- **Rectum:** straight section of colon in the pelvic cavity. The rectum has superficial veins in its wall called **hemorrhoidal veins.** When these enlarge they produce hemorrhoids.

The large intestine has some unique structures. The longitudinal muscle of the large intestine is thickened along the length of the colon into three bands called **taeniae coli.** The large intestine also has pouches or puckers called **haustra** (singular, **haustrum**). Another unique feature of the large intestine is fat lobules along the outer wall called **epiploic appendages.** Locate these structures in figure 26.12.

Fecal material passes through the large intestine by peristalsis and is stored in the rectum and sigmoid colon. Defecation occurs as mass peristalsis causes a bowel movement.

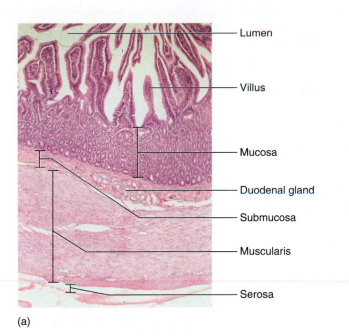

— Lumen

— Villus

— Mucosa

— Duodenal gland

— Submucosa

— Muscularis

— Serosa

(a)

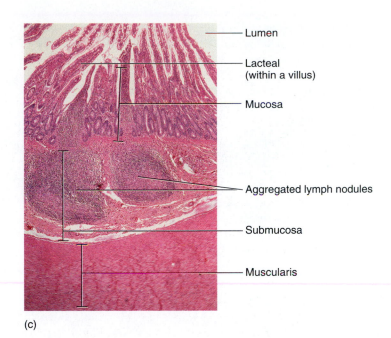

— Lumen

— Lacteal (within a villus)

— Mucosa

— Aggregated lymph nodules

— Submucosa

— Muscularis

(c)

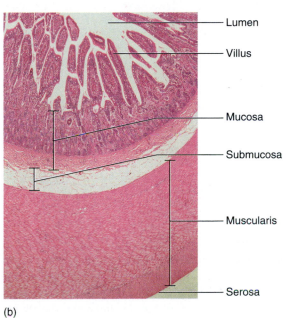

— Lumen

— Villus

— Mucosa

— Submucosa

— Muscularis

— Serosa

(b)

FIGURE 26.11

Three Sections of Small Intestine (40×) (a) Duodenum; (b) jejunum; (c) ileum.

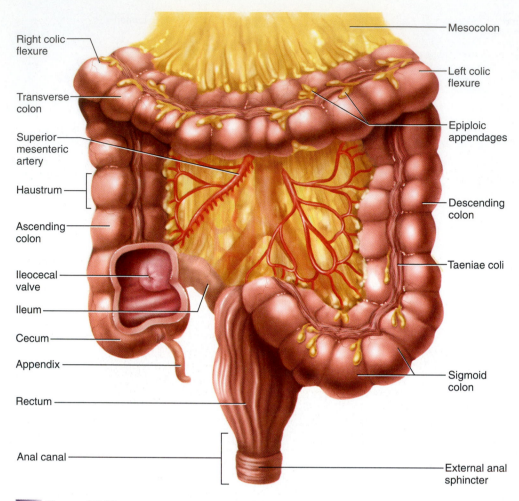

Labels on figure:
Right colic flexure
Transverse colon
Superior mesenteric artery
Haustrum
Ascending colon
Ileocecal valve
Ileum
Cecum
Appendix
Rectum
Anal canal
Mesocolon
Left colic flexure
Epiploic appendages
Descending colon
Taeniae coli
Sigmoid colon
External anal sphincter

FIGURE 26.12
Gross Anatomy of the Large Intestine and Anal Canal

Histology of the Large Intestine   Examine a prepared slide of the large intestine and compare it with figure 26.13. Locate the mucosa, submucosa, muscularis, and serosa in your slide. The large intestine is distinguished from the small intestine by the absence of villi and it is distinguished from the stomach by the presence of large numbers of goblet cells. Examine your slide for these characteristics.

Anal Canal   The **anal canal** is not part of the large intestine but is a short tube that leads to an external opening, the **anus.** Locate the anal canal in figure 26.12.

ACCESSORY STRUCTURES

The accessory structures are frequently associated with segments of the digestive tract. The first of these structures, the **salivary glands,** are located in the head and secrete saliva into the oral cavity. Saliva is a watery secretion that contains **mucus,** a protein lubricant, and **salivary amylase,** a starch-digesting enzyme.

There are three pairs of salivary glands. The first of these, the **parotid glands,** are located superficial to the masseter muscles just anterior to the ears. Each gland secretes saliva through a **parotid duct,** a tube that traverses the buccal (cheek) region and enters the mouth just posterior to the upper second molar. The second pair, the **submandibular glands,** are located just medial to the mandible on each side of the face. The submandibular glands secrete saliva into the mouth by a single duct on each side of the oral cavity inferior to the tongue. The third pair, the **sublingual glands,** are located inferior to the tongue and these glands open into the mouth by several ducts. Locate these salivary glands on a model of the head and in figure 26.14.

Appendix   The **appendix** is about the size of the little finger and is attached to the cecum (at the region of the ileocecal valve). The appendix has lymph nodules but its function is not apparent. Locate the appendix on a torso model or chart in the lab and compare it with figures 26.12 and 26.15.

Omenta   The **lesser omentum** is an extension of the peritoneum that forms a double fold of tissue between the stomach and the

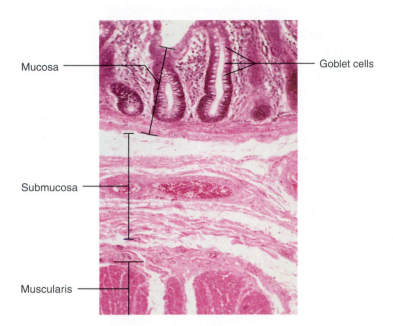

FIGURE 26.13
Histology of the Large Intestine (100×)

Mucosa

Submucosa

Muscularis

Goblet cells

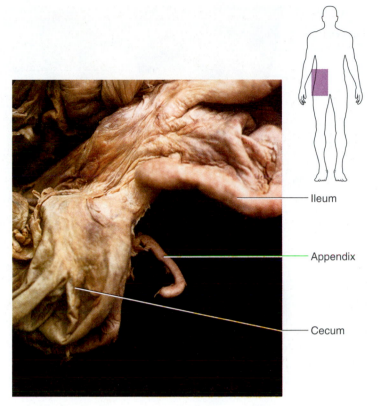

FIGURE 26.15
Appendix

Ileum

Appendix

Cecum

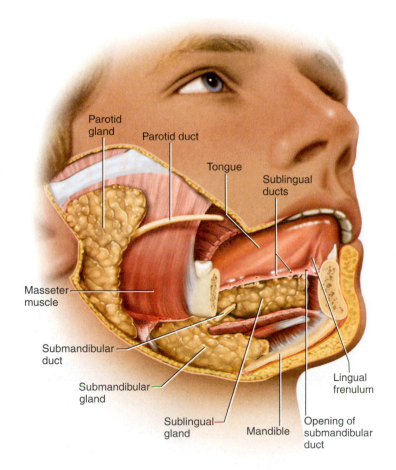

FIGURE 26.14
Salivary Glands

Parotid gland

Parotid duct

Tongue

Sublingual ducts

Masseter muscle

Submandibular duct

Submandibular gland

Sublingual gland

Mandible

Opening of submandibular duct

Lingual frenulum

liver. The **greater omentum** is a section of peritoneum that attaches to the inferior margin of the stomach and drapes over the intestines like a fatty curtain. In the large intestine the membrane that attaches the colon to the posterior body wall is called the **mesocolon.** Locate the lesser and greater omenta in figure 26.16.

Liver   The **liver** is a complex organ with numerous functions, some of which are digestive but many of which are not. The liver processes digestive material from the vessels returning blood from the intestines and has a role in either moving nutrients into the bloodstream or storing them in the liver tissue. The liver also produces blood plasma proteins, detoxifies harmful chemicals that have been produced by the body or introduced into the body, and produces bile.

Examine a model or chart of the liver and note its relatively large size. The liver is located mainly on the right side of the body and is divided into four lobes—the **right, left, quadrate,** and **caudate lobes.** The right lobe of the liver is the largest lobe; the left is also fairly large. The quadrate lobe is located in the middle portion of the liver, adjacent to the gallbladder and anterior to the caudate lobe. Traversing through the liver is the **falciform ligament,** which attaches the liver to the inferior side of the diaphragm. Locate the **gallbladder,** which is on the inferior aspect of the liver, and compare the gallbladder and liver structures with figure 26.17.

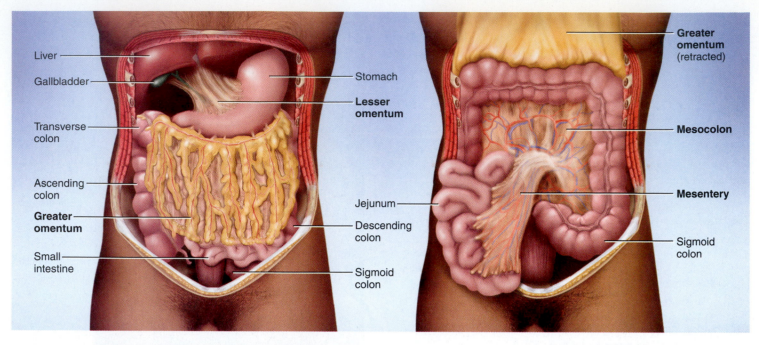

FIGURE 26.16
Omenta, Mesentery, and Mesocolon

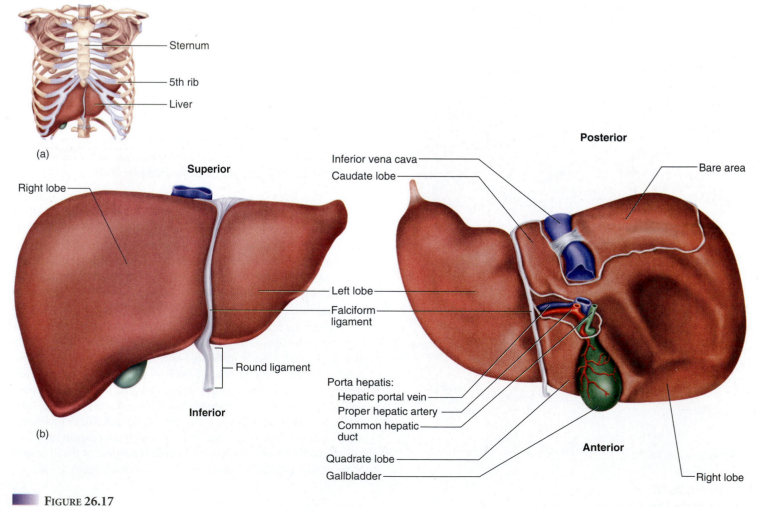

FIGURE 26.17
Liver (a) Overview in body; (b) anterior view; (c) inferior view.

Liver Histology   Examine a prepared slide of liver tissue. Note the hexagonal structures in the specimen. These are the **liver lobules.** Each lobule has a blood vessel in the middle called the **central vein.** Vessels that carry blood to the central vein are the **liver sinusoids,** and these are lined with a double row of cells called **hepatocytes.** Hepatocytes carry out the various functions of the liver. The tissue of the liver is extremely vascular and functions as a great sponge. Oxygenated blood from the hepatic artery and deoxygenated blood from the hepatic portal vein mix in the liver. **Kupffer cells** throughout the liver tissue function as phagocytic cells. Locate the lobules, central vein, sinusoids, and hepatocytes in a prepared slide using figure 26.18 to aid you.

Pancreas   The **pancreas** is located inferior to the stomach on the left side of the body. It has both endocrine and exocrine functions. The hormonal function of the pancreas is covered in laboratory exercise 19. The pancreas consists of a **tail** near the spleen, an elongated **body,** and a rounded **head** near the duodenum. Digestive enzymes and buffers that neutralize stomach acids pass from the tissue of the pancreas into the **pancreatic duct** and then into the duodenum. Locate the pancreatic structures, as represented in figure 26.19.

Gallbladder   The **gallbladder** releases bile, which emulsifies lipids, into the duodenum. The lipids break into smaller droplets, which increase the surface area for digestion. The gallbladder is located just inferior to the liver. The liver is the site of bile production. Bile is secreted from hepatocytes into **bile canaliculi,** which take the bile to **bile ductules.** The canaliculi flow between the plates of the liver lobules and the bile ductules run perpendicular to the bile canaliculi. The bile eventually flows from the bile ductules into the **left** and **right hepatic ducts.** These join to form the **common hepatic duct.** A short tube, the **cystic duct,** takes bile to and from the gallbladder, and the **bile duct** is located at the union of the common hepatic duct and the cystic duct. Sometimes the bile duct empties into the duodenum as a singular tube, but it usually joins, the pancreatic duct to form the **hepatopancreatic ampulla** (ampulla of Vater). At the duodenum the **hepatopancreatic sphincter** is normally closed. Bile fills the ducts and flows into the gallbladder. When the stomach empties, the hepatopancreatic sphincter opens, the gall bladder constricts, and bile and pancreatic juice enter the duodenum. Locate these structures on a model and compare them with figure 26.19.

Cat Anatomy

If you are using cats, turn to section 8, "Digestive Anatory of the Cat" on page 440 of this lab manual.

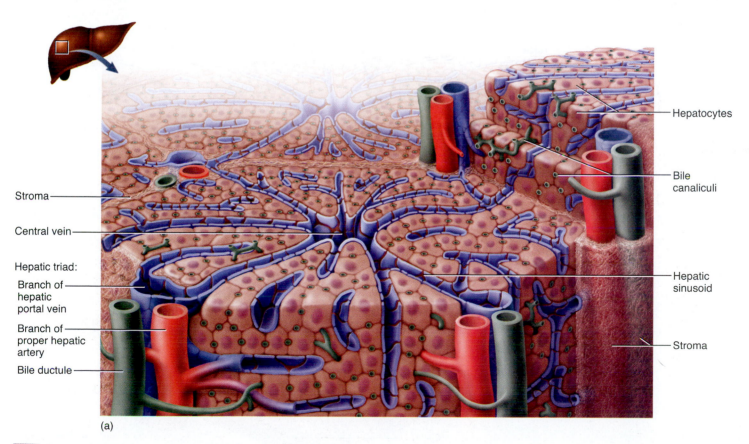

(a)

FIGURE 26.18
Histology of the Liver (a) Diagram.

FIGURE 26.18 *(Continued)*

Histology of the Liver
(b) photomicrograph (100×).

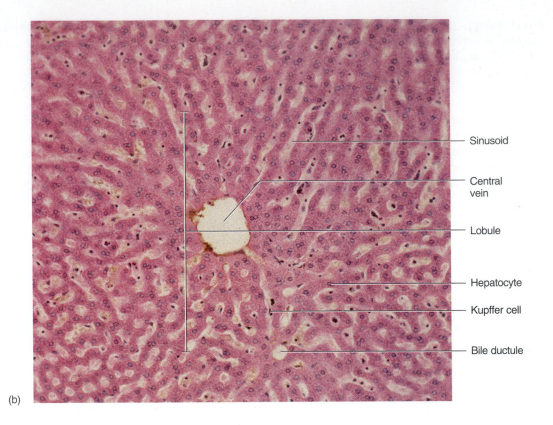

Sinusoid

Central vein

Lobule

Hepatocyte

Kupffer cell

Bile ductule

(b)

FIGURE 26.19

Liver, Gallbladder, and Pancreas

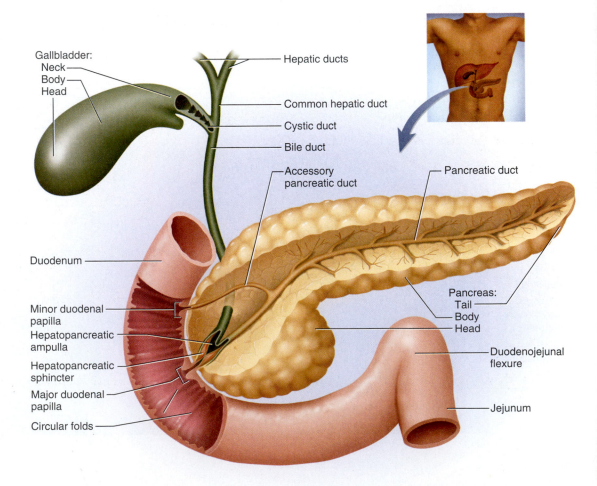

Gallbladder:
  Neck
  Body
  Head

Hepatic ducts

Common hepatic duct

Cystic duct

Bile duct

Accessory pancreatic duct

Pancreatic duct

Duodenum

Minor duodenal papilla

Hepatopancreatic ampulla

Hepatopancreatic sphincter

Major duodenal papilla

Circular folds

Pancreas:
  Tail
  Body
  Head

Duodenojejunal flexure

Jejunum

Name _____ Date _____

1. The pancreas belongs to what part of the digestive system (alimentary canal/accessory structure)?

2. The descending colon belongs to what part of the digestive system (alimentary canal/accessory structure)?

3. Name the layer lining the outer surface of the stomach and small intestine.

4. What is partially digested food in the stomach called?

5. What is the name of the portion of the stomach closest to the small intestine?

6. What cell type constitutes the inner lining of the mucosa (near the lumen) of the small intestine?

7. How long is the large intestine?

8. What is the middle length of the small intestine called?

9. In what specific structure do you find lacteals in the digestive tract?

10. What membrane holds the tongue to the floor of the oral cavity?

11. What region of the tooth is found above the neck?

12. What is the layer of a tooth superficial to the dentin?

13. Which adult teeth are directly posterior to the canine teeth?

14. What kind of muscle lines the small intestine?

15. What is the name for a segment, or pouch, in the large intestine?

16. Name the salivary gland directly anterior to the ear.

17. The lesser omentum is found between what two organs?

18. Where is bile stored?

19. Stomach acidity is approximately pH 2. How does this affect bacterial growth in the digestive tract?

20. Trace the flow of bile from the liver to the duodenum, listing all the structures that come into contact sequentially with the bile on its journey.

21. How does the large intestine differ from the small intestine in terms of length?

22. How does the large intestine differ from the small intestine in terms of diameter?

23. Name two functions of the pancreas.

24. Describe the pathway that food takes from the mouth to the anus, listing each structure that food comes into contact with.

25. Label the following illustration using the terms provided.

ascending colon	appendix	sigmoid colon
rectum	tongue	small intestine
mouth	duodenum	esophagus
stomach	parotid gland	liver

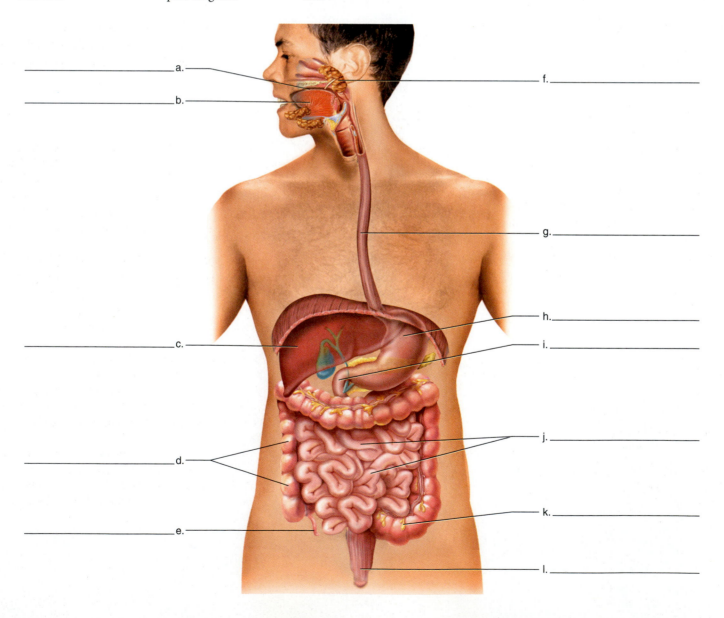

a. _____

b. _____

c. _____

d. _____

e. _____

f. _____

g. _____

h. _____

i. _____

j. _____

k. _____

l. _____

LABORATORY

The Urinary System

Urinary System

INTRODUCTION

The urinary system filters dissolved material from the blood, regulates ions and fluid volume, concentrates and stores waste products, and reabsorbs metabolically important substances, returning them to the circulatory system. The system consists of two kidneys, two ureters, a single urinary bladder, and a single urethra.

In this exercise you examine the gross and microscopic anatomy of the urinary system, study the major organs as represented in humans, and dissect a mammalian kidney if available.

LEARNING OBJECTIVES

At the end of this exercise you should be able to
1. identify the major organs of the urinary system;
2. describe the blood flow through the kidney;
3. describe the flow of filtrate through the kidney;
4. name the major parts of the nephron;
5. trace the flow of urine from the kidney to the exterior of the body;
6. distinguish among the parts of the nephron in histologic sections.

MATERIALS

Models and charts of the urinary system

Models and illustrations of the kidney and nephron system

Microscopes

Microscope slides of kidney and bladder

Samples of renal calculi (if available)

Preserved specimens of sheep or other mammalian kidney

Dissection trays and materials

Scalpels or razor blades

Forceps

Blunt probes

Latex gloves

Waste container

PROCEDURE

You can begin the study of the urinary system by locating its principal organs. Look at figure 27.1 and compare it with material available in the lab. Find the **kidneys,** the **ureters,** the **urinary bladder,** and the **urethra.**

Kidneys

The kidneys are **retroperitoneal** (posterior to the parietal peritoneum) and are embedded in **renal fat pads (perirenal fat).** These adipose pads cushion the kidneys, which are found mostly below the protection of the rib cage. The kidneys are located adjacent to the vertebral column approximately at the level of T12 to L3. The right kidney is slightly more inferior than the left. This is illustrated in figure 27.1.

Examine a model of the kidney and compare it with figure 27.2. Locate the outer **renal capsule,** a tough connective tissue layer, the outer **cortex,** and the inner **medulla** of the kidney. The kidney has a depression on the medial side where the **renal artery** enters the kidney and the **renal vein** and the **ureter** exit the kidney. This depression is called the **hilum.**

If you examine a coronal section of the kidney, you can find the **renal pyramids,** found inside the renal medulla. Each renal pyramid ends in a blunt point called the **renal papilla** (fig. 27.2). Urine drips from many papillae toward the middle of the kidney. The renal pyramids are separated by **renal columns,** which are extensions of the cortex.

The urine drips into the **minor calyces** (singular, **calyx**), which enclose the renal papillae. Minor calyces act somewhat as funnels that collect fluid and lead to the **major calyces.** These, in turn, conduct urine into the large **renal pelvis.** The renal pelvis is

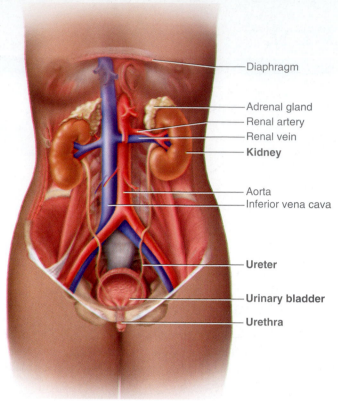

Major Organs of the Urinary System

- Diaphragm
- Adrenal gland
- Renal artery
- Renal vein
- **Kidney**
- Aorta
- Inferior vena cava
- **Ureter**
- **Urinary bladder**
- **Urethra**

▐ FIGURE 27.2

Gross Anatomy of the Kidney, Entire and Coronal Section Diagram
(a) coronal section of the kidney and details of renal pyramid.

found in a space known as the **renal sinus.** The renal pelvis is like a glove in a coat pocket. The pocket is the renal sinus, and the membranous glove that occupies the space is the renal pelvis. Locate these structures in figure 27.2. The renal pelvis is connected to the ureter at the medial side of the kidney.

BLOOD FLOW THROUGH THE KIDNEY

There are two fluid flows in the kidney. One is the flow of the filtrate and the other is the flow of blood. The blood enters the kidney by way of the **renal artery** to the **interlobar arteries** and into the **arcuate arteries.** From the arcuate arteries the blood then moves to the **interlobular arteries.** These blood vessels can be seen in figure 27.3. From one of the interlobular arteries, blood flows into an **afferent arteriole,** into the **glomerulus** (a tuft of capillaries), and to the **efferent arteriole.** The blood then enters the **peritubular capillaries** and returns via the **interlobular veins,** the **arcuate veins,** the **interlobar veins** to the **renal vein,** and finally to the inferior vena cava. The separation between the cortex and the medulla occurs at the level of the arcuate veins (fig. 27.3).

MICROSCOPIC EXAMINATION OF THE KIDNEY

Before you examine the sections of kidney under the microscope, become familiar with the structure of the **nephron.** Look at figure 27.4 and locate the **renal corpuscle, proximal convoluted tubule, nephron loop (loop of Henle),** and **distal convoluted tubule.** These four structures make up the nephron.

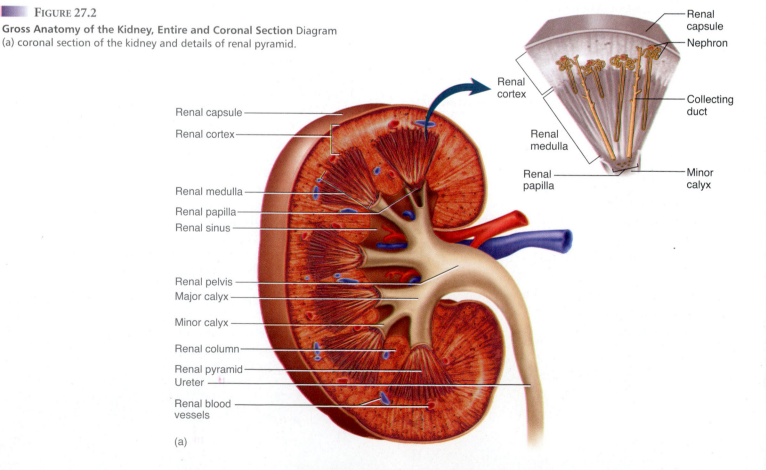

- Renal capsule
- Nephron
- Renal cortex
- Renal medulla
- Renal papilla
- Collecting duct
- Minor calyx

- Renal capsule
- Renal cortex
- Renal medulla
- Renal papilla
- Renal sinus
- Renal pelvis
- Major calyx
- Minor calyx
- Renal column
- Renal pyramid
- Ureter
- Renal blood vessels

(a)

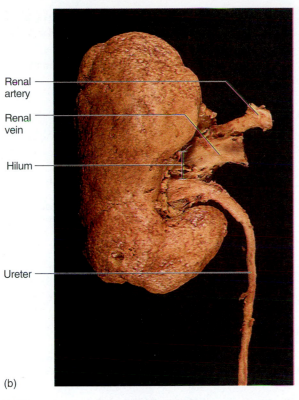

Renal
artery

Renal
vein

Hilum

Ureter

(b)

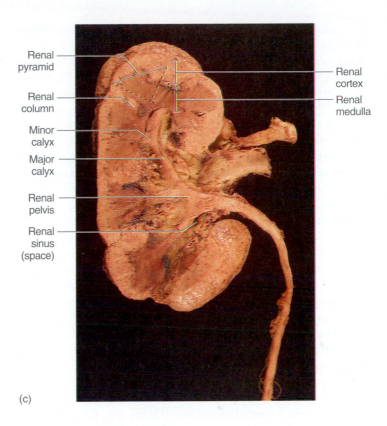

Renal
pyramid

Renal
column

Minor
calyx

Major
calyx

Renal
pelvis

Renal
sinus
(space)

Renal
cortex

Renal
medulla

(c)

FIGURE 27.2 (*Continued*)

Gross Anatomy of the Kidney, Entire and Coronal Section Photograph (b) entire kidney; (c) coronal section.

FIGURE 27.3

Blood Flow through the Kidney (a) Major arteries and veins of the kidney; (b) diagram of blood flow. Arteries are in red, veins are in blue.

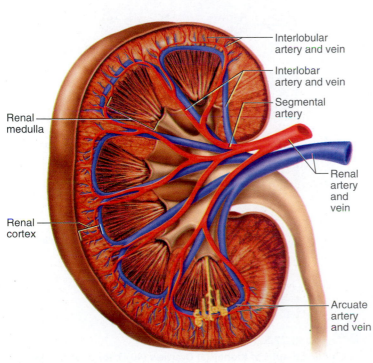

Interlobular
artery and vein

Interlobar
artery and vein

Segmental
artery

Renal
artery
and
vein

Arcuate
artery
and vein

Renal
medulla

Renal
cortex

(a)

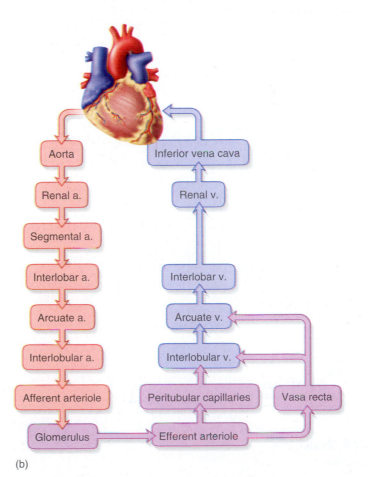

Aorta

Renal a.

Segmental a.

Interlobar a.

Arcuate a.

Interlobular a.

Afferent arteriole

Glomerulus

Efferent arteriole

Inferior vena cava

Renal v.

Interlobar v.

Arcuate v.

Interlobular v.

Peritubular capillaries

Vasa recta

(b)

Blood travels to the nephron via the afferent arteriole. When the blood reaches the **glomerulus,** a cluster of capillaries in the **glomerular (Bowman's) capsule,** the plasma is filtered by blood pressure forcing fluid across the capillary membranes. This fluid, which is present in the nephron, is called filtrate. The glomerulus along with the glomerular capsule, are known as the **renal corpuscle.**

As the filtrate flows through the nephron, water, glucose, and many ions are returned to the blood. The urea is concentrated as it passes through the entire nephron and into the **collecting duct,** a tube that receives the end product of the nephrons. Locate the glomerulus, glomerular capsule, proximal convoluted tubule, nephron loop (loop of Henle), distal convoluted tubule, and collecting duct in figure 27.4 and on models or charts in the lab.

Examine a kidney slide under low power. You should see the **cortex** of the kidney, which has a number of round spheres scattered throughout. These are the **glomeruli** and they are composed of capillary tufts. The other part of the kidney slide should have fairly open parallel spaces. These spaces, the **collecting ducts,** are located in the **medulla.** Compare the slide with figure 27.5a.

Examine the slide under higher magnification and locate the glomerulus and the glomerular capsule (fig. 27.5b). The capsule is composed of simple squamous epithelium and specialized cells called **podocytes.** Examine the outer edge of the capsule around the glomerulus. If you move the slide around in the cortex, you should find the proximal convoluted tubules with the **brush border** or **microvilli** on the inner edge of the tubule. The inner surface of the tubule appears fuzzy. The microvilli increase the surface area for reabsorption in the proximal convoluted tubule.

The distal convoluted tubules do not have brush borders; therefore, the inner surface of a tubule does not appear fuzzy. The cells of the distal convoluted tubules generally have darker nuclei and cytoplasm that is relatively clear when compared with the cells of the proximal convoluted tubules (fig. 27.5b). Examine the

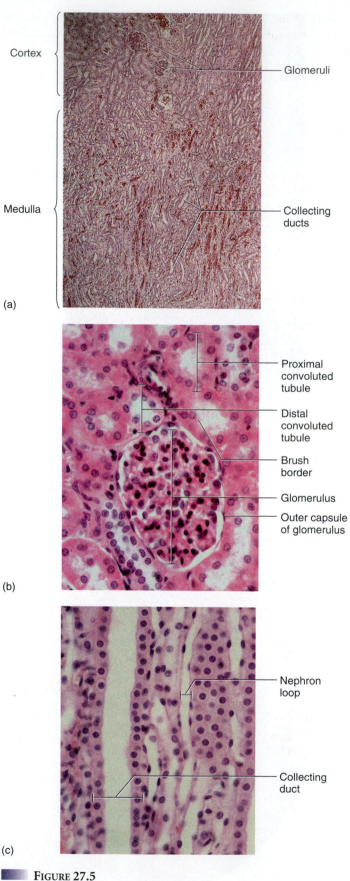

(a)

(b)

(c)

FIGURE 27.5

Photomicrographs of Kidney (a) Overview (40×); (b) cortex (400×); (c) medulla (400×).

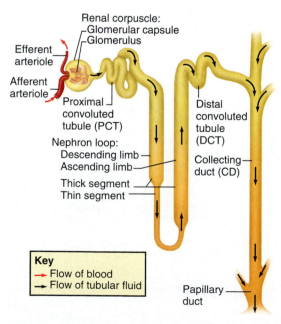

FIGURE 27.4

Nephron

medulla of the kidney under high magnification and locate the thin-walled nephron loop and the larger-diameter collecting ducts (fig. 27.5c).

RENAL CALCULI

Examine renal calculi (kidney stones) if they are available in lab.

Ureters

The **ureters** are long, thin tubes that conduct urine from the kidneys to the urinary bladder. The ureters have **transitional epithelium** as an inner lining and smooth muscle in their outer wall. Urine is moved from the kidney by **peristalsis** to the urinary bladder. Examine the models in the lab and compare them with figure 27.1.

Urinary Bladder

The **urinary bladder** is located anterior to the parietal peritoneum and is thus described as being **anteperitoneal.** Locate the urinary bladder in the torso model and see that it is found just posterior to the symphysis pubis. You can also see the position of the urinary bladder in figure 27.6. On the posterior wall of the urinary bladder is a triangular region known as the trigone. The **trigone** is defined by the entrances of the ureters and the inferior exit of the urethra, as seen in figure 27.7.

HISTOLOGY OF THE BLADDER

The urinary bladder has an inner lining of **transitional epithelium.** This epithelium can withstand a significant amount of stretching (distension) when the bladder fills with urine. Examine a slide of transitional epithelium under the microscope and compare it with figure 27.8. Look at the inner surface of the prepared section for the epithelial layer. Note that the cells are shaped somewhat like teardrops. Transitional epithelium can be distinguished from stratified squamous epithelium in that the cells of transitional epithelium from an empty bladder do not flatten at the surface of the tissue.

Urethra

The terminal organ of the urinary system is the **urethra.** The urethra is approximately 3 to 4 cm long in females. It passes from the urinary bladder to the **external urethral orifice,** located anterior to the vagina and posterior to the clitoris (figs. 27.6 and 27.7). The urethra is about 20 cm long in males. It begins at the urinary bladder, passes through the body wall, and then exits through the penis to the external urethral orifice at the tip of the glans penis (figs. 27.6 and 27.7).

Dissection of the Sheep Kidney

Take a sheep kidney and dissection equipment back to your table. Examine the outer **capsule** of the kidney. You may see some tubes coming from an indendation in the kidney. The indentation is the

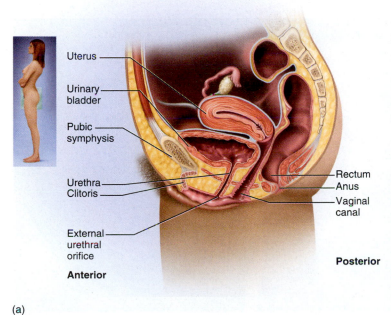

(a)

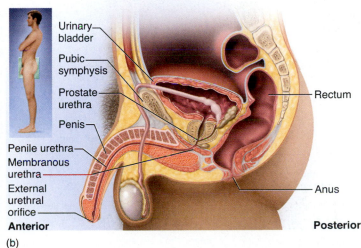

(b)

FIGURE 27.6

Midsagittal Section of the Female and Male Pelves (a) Female pelvis; (b) male pelvis.

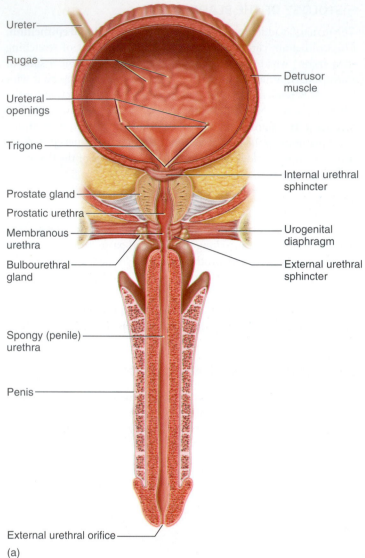

Ureter

Rugae

Ureteral openings

Trigone

Detrusor muscle

Prostate gland

Prostatic urethra

Membranous urethra

Bulbourethral gland

Internal urethral sphincter

Urogenital diaphragm

External urethral sphincter

Spongy (penile) urethra

Penis

External urethral orifice

(a)

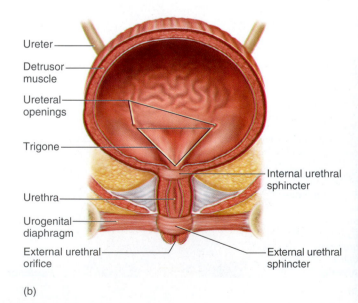

Ureter

Detrusor muscle

Ureteral openings

Trigone

Urethra

Urogenital diaphragm

External urethral orifice

Internal urethral sphincter

External urethral sphincter

(b)

FIGURE 27.7

Bladder and Urethra, Coronal Sections (a) Male; (b) female.

hilum, and the tubes are the **renal artery, renal vein,** and **ureter.** Make an incision in the sheep kidney a little off center in the coronal plane (fig. 27.9). Locate the **renal cortex, renal medulla, renal pyramids, renal papillae, minor** and **major calyces,** and **renal pelvis.** Lift the renal pelvis somewhat to pull it away from the **renal sinus.** When you are finished with the dissection, place the material in the proper waste container provided by your instructor.

Cat Anatomy

If you are using cats, turn to section 9, "Urinary Anatomy of the Cat," on page 443 of this lab manual.

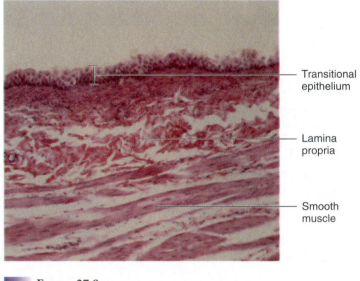

Transitional epithelium

Lamina propria

Smooth muscle

FIGURE 27.8

Histology of the Bladder (100×)

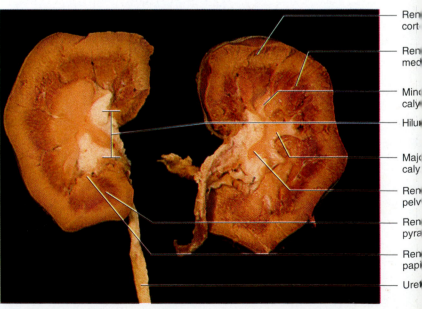

Renal cortex

Renal medulla

Minor calyx

Hilum

Major calyx

Renal pelvis

Renal pyramid

Renal papilla

Ureter

FIGURE 27.9

Dissection of a Sheep Kidney

Name _____ Date _____

1. What is the name of the outer region of the kidney deep to the renal capsule?

2. Describe the kidneys with regard to their position to the parietal peritoneum.

3. What takes urine directly to the urinary bladder?

4. Urinary bladder infections are more common in females than in males. How can you explain this in terms of microorganisms moving into the bladder from the urethra?

5. What is the functional value of the increased surface area caused by the microvilli of the proximal convoluted tubule?

6. On the posterior bladder there is a triangular region. What is it called?

7. What is the name of the cluster of capillaries in the kidney where filtration occurs?

8. Distal convoluted tubules flow directly into what structures?

9. What is a renal papilla?

10. Name the connective tissue covering a kidney.

11. Urine flows from the tip of a renal pyramid into what structure?

12. Blood in the glomerulus flows to what arteriole?

13. Which shows greater anatomic difference between the sexes: ureters, urinary bladder, or urethra?

14. Describe the pathway of the filtered material from the glomerulus to the urethra.

15. What histologic feature distinguishes a proximal convoluted tubule from a distal convoluted tubule?

16. Name the parts of the nephron.

17. What type of tissue lines the bladder?

18. Fill out the following illustration using the terms provided.

minor calyx	renal artery	renal vein
renal pelvis	major calyx	ureter
renal capsule	renal pyramid	renal cortex

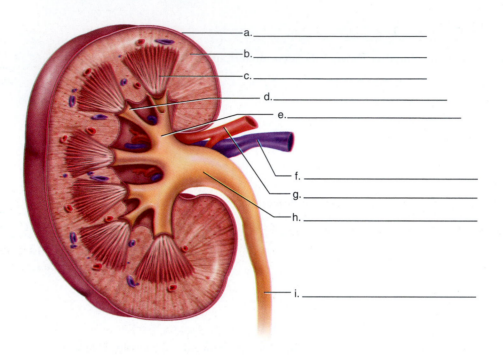

a. _____

b. _____

c. _____

d. _____

e. _____

f. _____

g. _____

h. _____

i. _____

The Male Reproductive System

Reproductive System

INTRODUCTION

The male reproductive system produces male **gametes (spermatozoa)**, transports the gametes to the female reproductive tract, and secretes the male reproductive hormone, **testosterone.** The gonad, or gamete-producing structure, of the male reproductive system is the *testis* (plural, *testes*). The testes are considered mixed glands in that they have both an exocrine and an endocrine function. The exocrine nature of the gland is the secretion of gametes through ducts or tubules. The endocrine function involves the ductless secretion of testosterone. In this exercise you examine the gross anatomy of the male reproductive system and the histology of the system.

LEARNING OBJECTIVES

At the end of this exercise you should be able to
1. identify the gamete-producing organ of the male reproductive system;
2. describe the anatomy of the major structures of the male reproductive system;
3. describe the formation of spermatozoa in the testis;
4. list the pathway that spermatozoa follow from production to expulsion;
5. describe the anatomy of the spermatic cord;
6. list the four components of semen;
7. name the three cylinders of erectile tissue in the penis.

MATERIALS

Charts, models, and illustrations of the male reproductive system
Microscopes
Prepared slides of a cross section of testis and sperm smear

PROCEDURE

Overview of the Gross Anatomy of the Male Reproductive System

Examine the models and charts of the male reproductive system available in the lab and locate the listed structures in figure 28.1. When you find the structures, place a check mark next to the name.

_____ Testis
_____ Epididymis
_____ Scrotum (scrotal sac)
_____ Ductus deferens
_____ Seminal vesicle
_____ Prostate gland
_____ Bulbourethral gland
_____ Penis

Testes

The **testes** are paired organs wrapped in a tough connective tissue sheath called the **tunica albuginea** (fig. 28.1) and covered with an extension of the peritoneum called the **tunica vaginalis.** They lie outside of the body cavity, where the temperature is somewhat cooler and are surrounded by the **scrotum (scrotal sac),** which envelops the testes. The testes are the site of **spermatozoa (sperm)** production, and this process must occur at about 35°C. The scrotum is lined with a layer of muscle called the **dartos muscle.** It is composed of smooth muscle fibers that contract when the testes are cold, thus bringing them closer to the body. When the environment around the testes is warm, the dartos muscles relax and the testes descend from the body, thus becoming cooler. **Cremaster muscles** attach to the testes. They contract, which brings the testes up, warming them, and relax to lower the testes. Examine charts and models of the testes and compare them with figures 28.1 and 28.2.

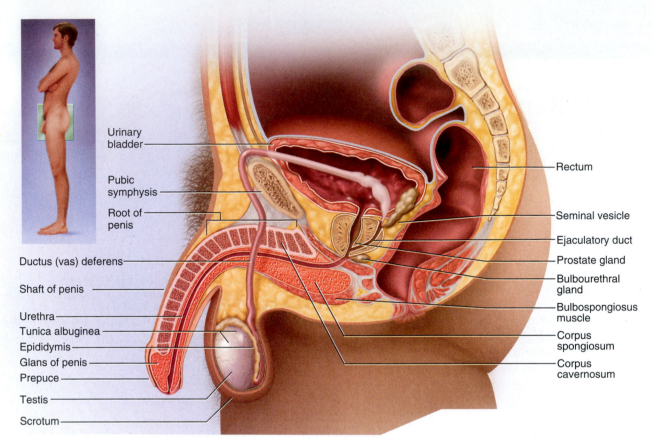

FIGURE 28.1
Male Reproductive System, Midsagittal View

HISTOLOGY OF THE TESTIS

The testis, as the gamete-producing organ of the male reproductive system, should be viewed in a prepared slide. Numerous tubules are seen in cross section. These are the **seminiferous tubules.** The gametes, or spermatozoa, are produced in seminiferous tubules in the testis (figs. 28.3 and 28.4).

Find the clusters of cells that frequently appear as triangles inbetween the tubules. These are called **interstitial cells.** They produce the male sex hormone, **testosterone.** Examine the seminiferous tubules under high magnification. You should be able to see the outer row of cells called the **spermatogonia.** These cells reproduce by mitosis to produce **primary spermatocytes. Sustentacular (Sertoli) cells** assist in the movement of the primary spermatocytes.

The primary spermatocytes undergo meiosis, or reduction division, to eventually produce the sex cells (spermatozoa). The primary spermatocytes divide to form **secondary spermatocytes,** which are found closer to the lumen. The secondary spermatocytes become **spermatids.** Spermatids lose their remaining cytoplasm and mature into **spermatozoa.** Examine a prepared slide of

testis and compare it with figures 28.3 and 28.4. Locate the spermatogonia, primary and secondary spermatocytes, spermatids, and spermatozoa.

SPERMATOZOA

The structure of an individual spermatozoa (sperm) consists of a **head** and a **tail.** The head contains the genetic information (DNA), as well as a cap known as the **acrosome.** The acrosome contains digestive enzymes, which digest the exterior covering of the female gamete. The midpiece of the tail of the sperm contains **mitochondria** that provide ATP to move the spermatozoa. The remainder of the tail of the sperm is a flagellum that propels the sperm forward. Examine a prepared slide of the sperm and compare it with figure 28.5.

EPIDIDYMIS

Spermatozoa from each testis travel from tubules in the testis to the **rete testis** and into the **epididymis,** where they are stored and mature. Each epididymis has a rounded **head,** an elongated **body,** and a tapering **tail** that leads to the ductus deferens. Sperm mat-

FIGURE 28.2

Scrotum, Testis, Spermatic Cord, and Penis, Anterior View

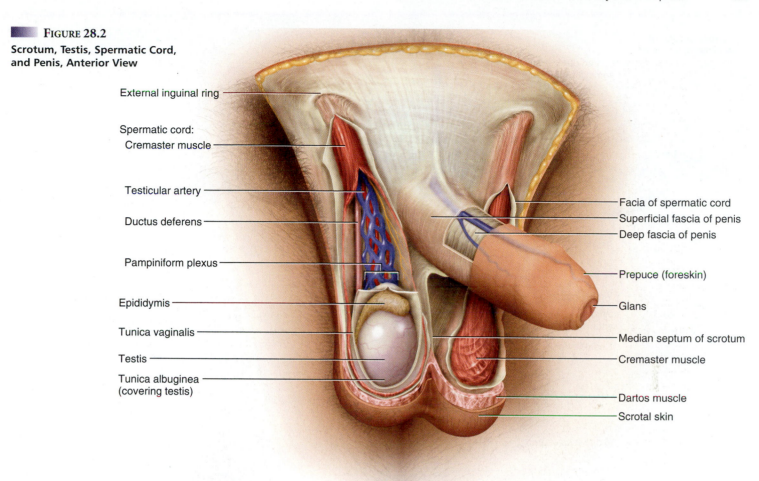

External inguinal ring

Spermatic cord:
Cremaster muscle

Testicular artery

Ductus deferens

Pampiniform plexus

Epididymis

Tunica vaginalis

Testis

Tunica albuginea
(covering testis)

Facia of spermatic cord
Superficial fascia of penis
Deep fascia of penis
Prepuce (foreskin)
Glans
Median septum of scrotum
Cremaster muscle
Dartos muscle
Scrotal skin

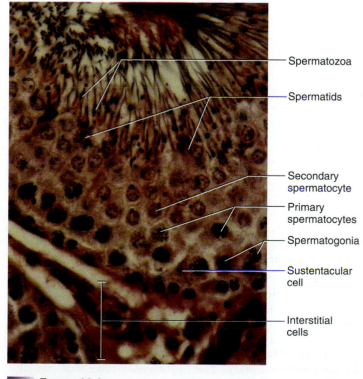

Spermatozoa

Spermatids

Secondary spermatocyte

Primary spermatocytes

Spermatogonia

Sustentacular cell

Interstitial cells

FIGURE 28.3

Histology of the Testis (400×)

uration occurs in the epididymis. If spermatozoa are removed from the testis proper, they are not capable of fertilizing the female oocyte (egg). Spermatozoa move slowly through coiled tubules of the epididymis, which lead to the ductus deferens. Examine a model or chart of the longitudinal section of a testis and epididymis and locate the structures by comparing them with figure 28.6.

SPERMATIC CORD AND ACCESSORY GLANDS

Spermatozoa travel from the epididymis into the **ductus deferens.** The ductus deferens is enclosed in the **spermatic cord,** which is a complex structure consisting of the ductus deferens, the **testicular artery** and **vein,** the **testicular nerves,** and the **cremaster muscle.** The cremaster muscle is a cluster of skeletal muscle fibers. The spermatic cord is longer on the left side than on the right; therefore, the left testis is lower than the right. Locate the structures of the spermatic cord in figure 28.6.

As the spermatic cords reach the **inguinal canals,** each ductus deferens travels around the posterior aspect of the urinary bladder (fig. 28.1). You can trace the course of the ductus deferens until they reach the inferior portion of the bladder. The ductus deferens enlarge somewhat here to form the **ampullae.** Each ductus deferens joins with a **seminal vesicle,** which is a gland that adds fluid to the spermatozoa. The union of the ductus deferens and the two seminal vesicles produces the **ejaculatory ducts.** Locate the

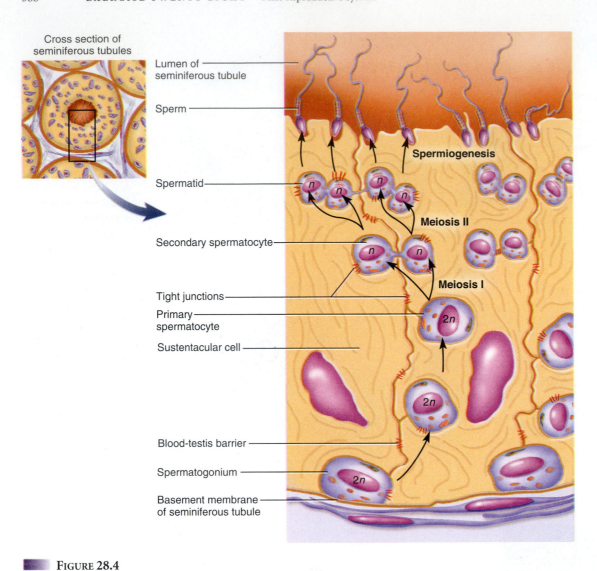

Cross section of
seminiferous tubules

Lumen of
seminiferous tubule

Sperm

Spermiogenesis

Spermatid

Meiosis II

Secondary spermatocyte

Meiosis I

Tight junctions

Primary
spermatocyte

Sustentacular cell

Blood-testis barrier

Spermatogonium

Basement membrane
of seminiferous tubule

FIGURE 28.4

Spermatogenesis in the Seminiferous Tubule Diploid cells (2*n*) reduce their chromosome number by half to haploid cells *(n)* in this process.

seminal vesicle and the ejaculatory duct in figure 28.7. The fluid from the seminal vesicles adds about 60% to the final volume of semen.

From this location the ejaculatory ducts lead to the inferior portion of the bladder and join with the urethra, which passes through the prostate gland. The **prostate gland** is located just inferior to the urinary bladder, and the urethra that passes through the gland is known as the **prostatic urethra.** The prostate gland adds a buffering fluid to the secretions of the testes and seminal vesicles. The prostate fluid makes up slightly less than 40% of the final semen volume. As the prostatic urethra exits the prostate gland it becomes the **membranous urethra** and passes through the body wall. Here the paired **bulbourethral (Cowper's) glands** are found, which add a lubricant to the seminal fluid. **Seminal fluid** consists of secretions from the seminal vesicles, prostate gland, and bulbourethral glands. **Semen** consists of seminal fluid plus spermatozoa from the testes. Spermatozoa make up much less than 1% of the total volume of semen. The urethra passes out of the body cavity

and becomes the **spongy,** or **penile, urethra** of the penis. Locate the portions of the urethra and the accessory glands in figure 28.7.

External Genitalia: Penis

The **penis** consists of an elongated **shaft** and a distally expanded **glans penis.** The glans is covered with the **prepuce,** or **foreskin,** which is removed in some males by a procedure called **circumcision.** At the inferior portion of the glans is a region richly supplied with nerve endings called the **frenulum.** The glans penis is a distal, expanded region that stimulates the vagina of the female. The erect penis is, on average, about 16 cm in length. The various structures of the penis can be seen in figures 28.1, 28.2, and 28.8.

The penis contains three cylinders of **erectile tissue.** The **corpus spongiosum** is the cylinder of erectile tissue that contains the **spongy (penile) urethra.** The two **corpora cavernosa** (singular, **corpus cavernosum**) are located anterior to the corpus spongiosum. Examine a model or chart of a cross section of penis and compare it

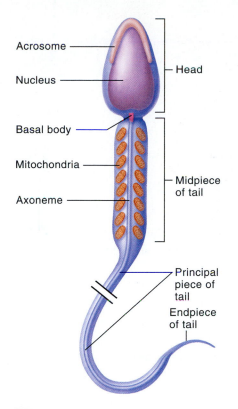

Acrosome —
Nucleus —
⎤ Head

Basal body —
Mitochondria —
Axoneme —
⎤ Midpiece of tail

Principal piece of tail
Endpiece of tail

FIGURE 28.5
Isolated Sperm

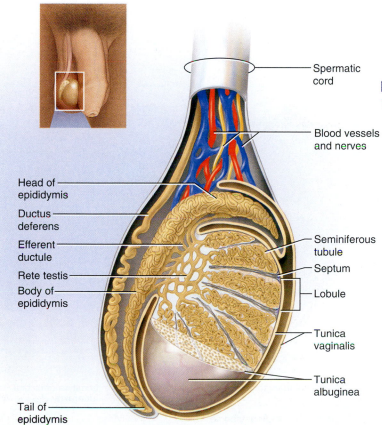

Spermatic cord

Blood vessels and nerves

Head of epididymis
Ductus deferens
Efferent ductule
Rete testis
Body of epididymis

Seminiferous tubule
Septum
Lobule
Tunica vaginalis
Tunica albuginea

Tail of epididymis

FIGURE 28.6
Testis and Epididymis, Lateral View

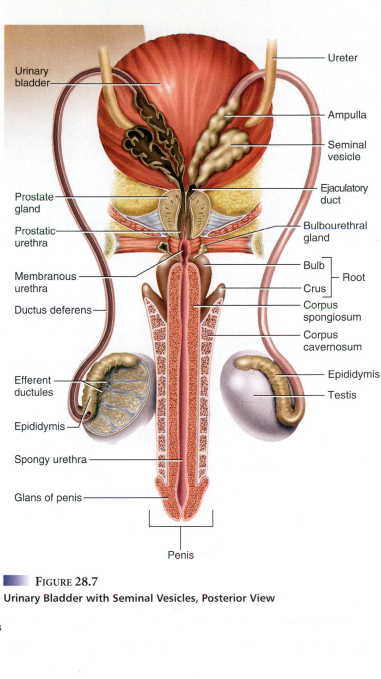

Urinary bladder —
Ureter
Ampulla
Seminal vesicle
Ejaculatory duct

Prostate gland —
Prostatic urethra —
Bulbourethral gland
Bulb ⎤
Crus ⎦ Root

Membranous urethra —
Corpus spongiosum
Ductus deferens —
Corpus cavernosum

Efferent ductules —
Epididymis
Testis

Epididymis —

Spongy urethra —

Glans of penis —

Penis

FIGURE 28.7
Urinary Bladder with Seminal Vesicles, Posterior View

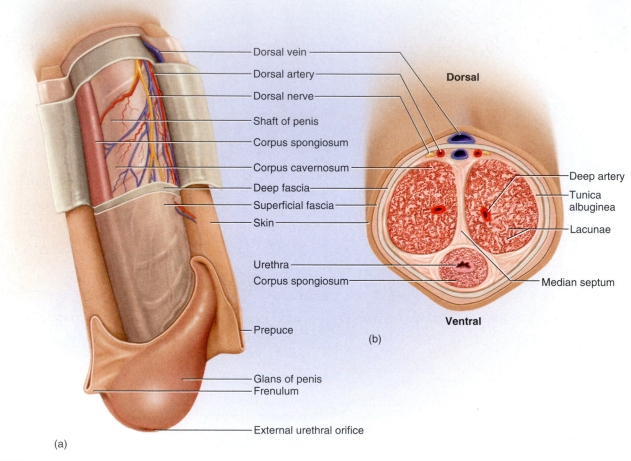

Dorsal vein

Dorsal artery

Dorsal nerve

Shaft of penis

Corpus spongiosum

Corpus cavernosum

Deep fascia

Superficial fascia

Skin

Urethra

Corpus spongiosum

Prepuce

Glans of penis

Frenulum

External urethral orifice

(a)

Dorsal

Deep artery

Tunica albuginea

Lacunae

Median septum

Ventral

(b)

FIGURE 28.8.

Penis (a) 3/4 view; (b) cross section.

with figure 28.8. The proximal parts of the cylinders of erectile tissue are anchored to the body. Locate the **crus,** which is an expansion of the corpora cavernosa. The **bulb** of the penis is an extension of the corpus spongiosum. The crus and the bulb form the **root** of the penis. The corpus spongiosum expands distally to form the glans penis. Note the **dorsal arteries** of the penis. These take blood to the penis. Locate the **dorsal vein** of the penis. When the arteries of the penis dilate, the erectile tissues engorge with blood and the penis becomes erect. The erection subsides as the arteries constrict, decreasing blood flow into the penis. Examine a model of the penis and find the features listed in figures 28.7 and 28.8.

The floor of the pelvis as seen from the outside is referred to as the **perineum.** It can be divided into a posterior **anal triangle** and an anterior **urogenital triangle.** The anal triangle surrounds the anus and the urogenital triangle encloses the penis and scrotum (fig. 28.9).

Cat Anatomy

If you are using cats, turn to section 10, "Reproductive Anatomy of the Cat," on page 444 of this Lab manual.

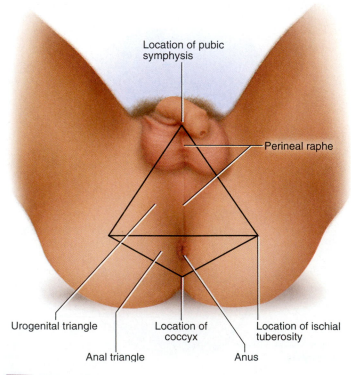

Location of pubic symphysis

Perineal raphe

Urogenital triangle

Location of coccyx

Location of ischial tuberosity

Anal triangle

Anus

FIGURE 28.9

Male Perineum

Name _____ Date _____

1. What is the gonad in the male reproductive system?

2. The testes are considered mixed glands because they have both an endocrine and an exocrine function. Describe the endocrine and exocrine products that come from the testes.

3. Proper sperm production must occur at what temperature?

4. Male sterility can result from excessively high temperatures around the testes. What mechanism occurs in the scrotum to counteract the effects of high temperature?

5. Trace the movement of sperm from the testes to where they exit the penis. List all of the vessels they come into contact with as they move through the system.

6. Where are spermatozoa produced in the testis?

7. What are the cells that initiate spermatozoa production?

8. Where is the cremaster muscle found?

9. List all the structures involved in producing semen.

10. Name the three organs that produce seminal fluid.

11. How do spermatozoa differ from seminal fluid?

12. A vasectomy is the cutting and tying of the two ductus deferens at the level of the spermatic cords. Review the percentage of spermatozoa that composes semen and determine what effect a vasectomy has on semen volume.

13. Which one of the seminal fluid glands is not a paired gland?

14. Where is the glans penis located?

15. What is the cylinder of erectile tissue inferior to the corpora cavernosa?

16. Label the following illustration using the terms provided.

seminal vesicle   prostate gland   bulbourethral gland

glans penis   prepuce   epididymis

bulb of penis   urinary bladder   corpus cavernosum

ductus deferens   testis   scrotum

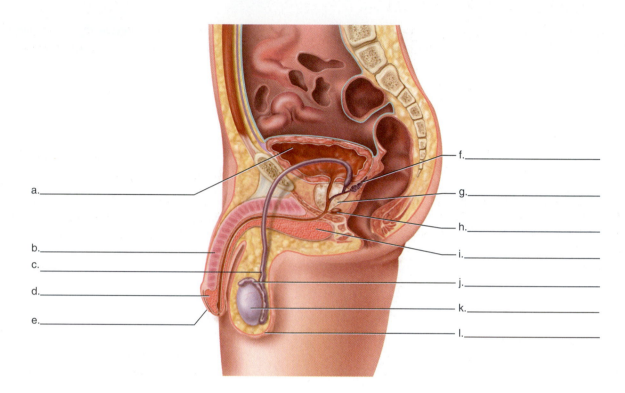

a._____

b._____

c._____

d._____

e._____

f._____

g._____

h._____

i._____

j._____

k._____

l._____

The Female Reproductive System and Development

Reproductive System

INTRODUCTION

The female reproductive system is functionally more complex than the male reproductive system. In the male, the reproductive system produces gametes and delivers them to the female reproductive system. The female reproductive system not only produces gametes and receives the gametes from the male but also provides space and nutrients for the developing **conceptus.** Finally, the female reproductive system delivers the child into the outer environment. In this exercise you learn about the structure of the female reproductive system and development.

LEARNING OBJECTIVES

At the end of this exercise you should be able to
1. identify the gamete-producing organ of the female reproductive system;
2. trace the pathway of a gamete from the ovary to the site of implantation;
3. list the structures of the vulva;
4. describe the function of each structure in the female reproductive system;
5. name the layers of the uterus from superficial to deep.

MATERIALS

Charts, models, and illustrations of the female reproductive system
Microscopes
Prepared slides of ovary and uterus

PROCEDURE

Overview of the Gross Anatomy of the Female Reproductive System

Examine a model or chart of the female reproductive system and locate the following major reproductive organs there and in figure 29.1. When you find the structure, check it off from the list.

_____ Ovary

_____ Uterine tube

_____ Uterus

_____ Vagina

_____ Clitoris

_____ Labia minora (singular, *labium minus*)

_____ Labia majora (singular, *labium majus*)

Ovaries

The ovaries are the gamete-producing organs of the female reproductive system. The ovaries are a mixed gland having both an endocrine and an exocrine function. The exocrine function is the production of oocytes, and the endocrine function is the production of the female sex hormones—estrogen (estradiol) and progesterone.

Each **ovary** is an ovoid organ approximately 3 to 4 cm long (fig. 29.1). The ovaries produce **oocytes,** which are shed from the outer surface of the ovary during **ovulation.** From here the oocytes (which may mature into ova, if fertilized) move into the **uterine (fallopian) tube.** The ovaries are not directly attached to the uterine tube, and the oocytes must move from the surface of the ovary into the uterine tube.

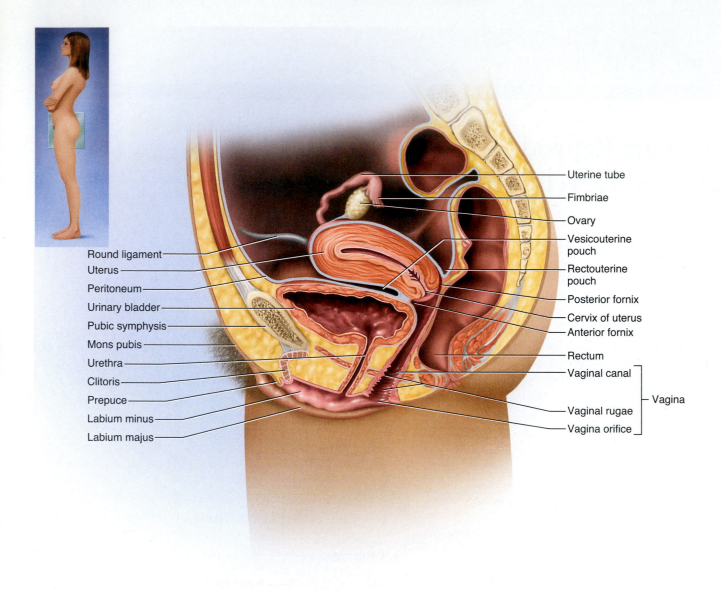

FIGURE 29.1
Female Reproductive System, Midsagittal View

Labels in figure:
- Round ligament
- Uterus
- Peritoneum
- Urinary bladder
- Pubic symphysis
- Mons pubis
- Urethra
- Clitoris
- Prepuce
- Labium minus
- Labium majus
- Uterine tube
- Fimbriae
- Ovary
- Vesicouterine pouch
- Rectouterine pouch
- Posterior fornix
- Cervix of uterus
- Anterior fornix
- Rectum
- Vaginal canal
- Vaginal rugae
- Vagina orifice
- Vagina

HISTOLOGY OF THE OVARY

Examine a prepared slide of the ovary (cat or human) under the microscope on low power. Locate the background substance of the ovary, which is known as the **stroma.** The stroma is divided into a superficial **cortex** and a deep **medulla.** Look for circular structures in the ovary. These are the **ovarian follicles,** which include an oocyte surrounded by follicular cells. The smallest of the follicles are the **primordial follicles.** Locate the primordial follicles in your slide and compare these with the follicles in figure 29.2.

You should also locate the enlarging **primary** and **secondary follicles.** Some of the follicles may contain **oocytes.** All follicles contain oocytes, but you may not see them if the sectioning plane did not pass through the oocyte. The largest follicles in the ovary are the **graafian follicles,** or **mature ovarian follicles,** and one may be present in your slide. During ovulation in humans usually one secondary oocyte is shed from the ovary. In cats many oocytes may be shed. Follicles produce estrogen, which contributes to the endocrine function of the ovary.

Examine your slide and compare it with figures 29.2 and 29.3. Draw what you see in your slide in the following space.

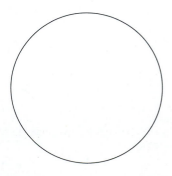

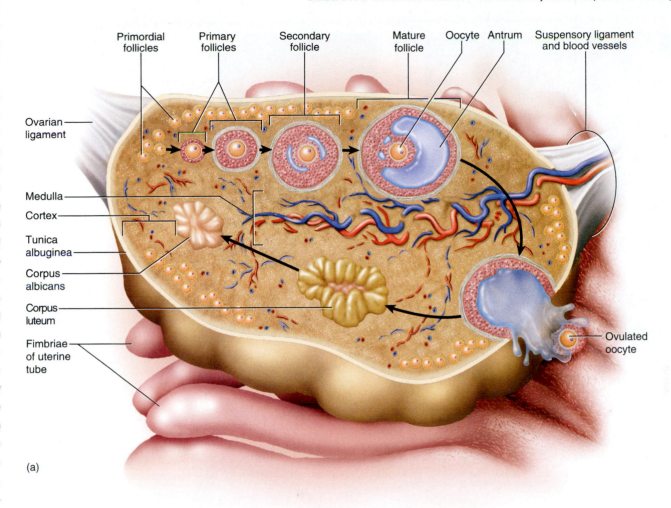

Primordial follicles

Primary follicles

Secondary follicle

Mature follicle

Oocyte Antrum

Suspensory ligament and blood vessels

Ovarian ligament

Medulla

Cortex

Tunica albuginea

Corpus albicans

Corpus luteum

Fimbriae of uterine tube

Ovulated oocyte

(a)

Free surface of ovary

Primordial follicle

Primary follicle

Antrum

Secondary follicle

Oocyte

Stroma

(b)

FIGURE 29.2

Histology of the Ovary (a) Diagram (arrows indicate a timeline of development from primordial follicles to the corpus albicans);
(b) photomicrograph (40×).

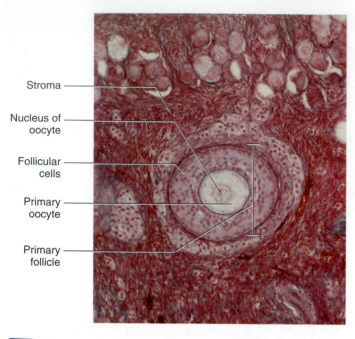

Stroma

Nucleus of
oocyte

Follicular
cells

Primary
oocyte

Primary
follicle

FIGURE 29.3
Primary Oocyte in Follicle (100×)

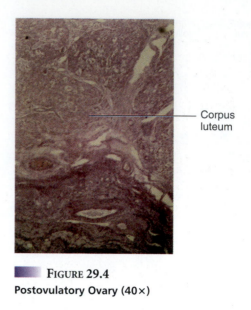

Corpus
luteum

FIGURE 29.4
Postovulatory Ovary (40×)

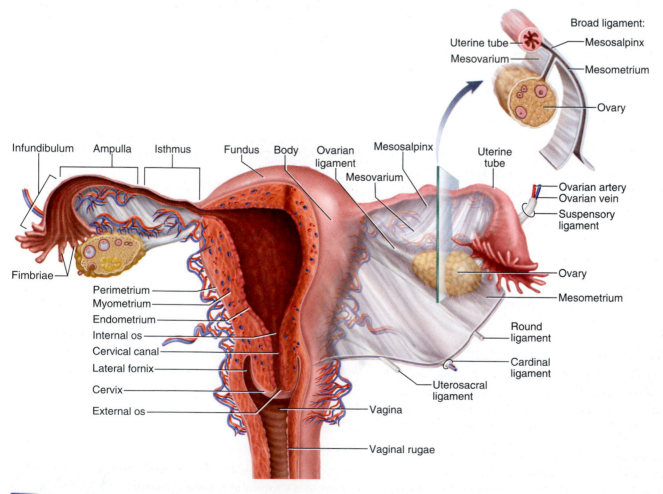

Broad ligament:

Uterine tube

Mesovarium

Mesosalpinx

Mesometrium

Ovary

Infundibulum   Ampulla   Isthmus   Fundus   Body   Ovarian
ligament

Mesosalpinx   Uterine
tube

Mesovarium

Ovarian artery
Ovarian vein
Suspensory
ligament

Fimbriae

Ovary

Perimetrium

Mesometrium

Myometrium

Endometrium

Internal os

Round
ligament

Cervical canal

Lateral fornix

Cardinal
ligament

Cervix

External os

Uterosacral
ligament

Vagina

Vaginal rugae

FIGURE 29.5
Female Reproductive System, Posterior View

After ovulation the remains of a mature ovarian follicle become a **corpus luteum,** which primarily secretes progesterone. If pregnancy does not occur, the corpus luteum decreases in size and becomes the **corpus albicans.** This progression is seen in figure 29.2*a*. Examine a prepared slide of the ovary with a corpus luteum or corpus albicans and compare it with figures 29.2 and 29.4.

Uterine Tubes

The uterine tube has a small fringe on the distal region known as the **fimbriae.** These are small, fingerlike projections attached to an expanded region known as the **infundibulum.** The uterine tube also has an enlarged region known as the **ampulla** and a narrower portion, the **isthmus,** toward the uterus. Oocytes move down the uterine tubes by ciliary movement in the uterine tube. Fertilization takes place in the uterine tubes. Examine a model or chart of the female reproductive system and compare it with figure 29.5.

Uterus

The **uterus** is a pear-shaped organ with a domed **fundus,** a **body,** and a circular, inferior end called the **cervix.** The uterine tubes enter the uterus at approximately the junction of the fundus with the uterine body. A constricted portion of the inferior uterus is called the **isthmus.** The isthmus is the upper third of the cervix.

The uterine wall is composed of three layers. The outer (*superficial*) surface of the uterus is called the **perimetrium.** This is located near the body cavity. The majority of the uterine wall consists of the **myometrium,** a thick layer of smooth muscle, and the innermost (deepest) layer of the uterus is the **endometrium.** Examine the models or charts in the lab and locate the structures in figure 29.5.

HISTOLOGY OF THE UTERUS

Examine a prepared slide of the uterus and locate the outer **perimetrium,** the smooth muscle of the **myometrium,** and the inner **endometrium.** Compare your slide with figures 29.5 and 29.6.

Now examine the two layers of the endometrium of the uterus under higher magnification. The stratum functionalis is the one that is shed during menstruation. It is composed of **spiral arterioles** and **uterine glands,** which appear as wavy lines toward the edge of the tissue. The stratum basalis is deeper and contains **straight arterioles.** Deep to the endometrium is the myometrium, which can be distinguished by the presence of smooth muscle.

OVARIAN AND MENSTRUAL CYCLES

The endometrium undergoes dynamic changes during the **menstrual cycle. Luteinizing hormone (LH)** and **follicle-stimulating hormone (FSH),** from the anterior pituitary, and subsequently **estrogen** and **progesterone,** from the ovary, have significant effects on the endometrium. FSH stimulates the ovarian follicles to secrete estrogens and some progesterone. Elevated levels of these hormones promote the thickening of the endometrium, as indicated in figure 29.7.

After ovulation, progesterone levels increase and the endometrial cells remain intact. Toward the end of a woman's menstrual cycle, estrogen and progesterone levels drop and this causes

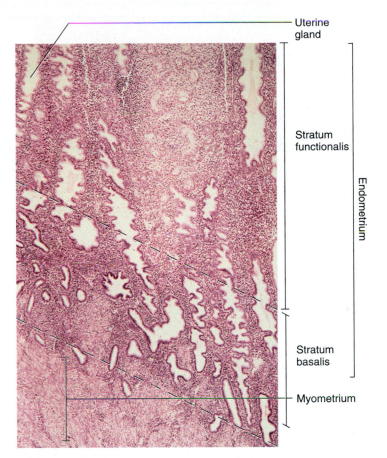

FIGURE 29.6
Histology of the Uterus (40×)

the functional layer of the endometrium to slough off. The loss of the endometrial layer is the beginning of a woman's period, or **menstruation.**

These events are briefly outlined in figure 29.7. Examine the figure and note how the endothelial layer begins to decrease when estrogen and progesterone levels fall (about day 25 in figure 29.7).

Ligaments

The uterus and ovaries are suspended in the pelvic cavity by a number of connective tissue sheaths called ligaments. The **broad ligament** anchors the uterus to the anterior pelvic wall. The **round ligament** attaches the uterus to the anterior pelvic wall at the region of the inguinal canal. The **ovarian ligament** directly attaches the ovary to the uterus, and the **suspensory ligament** attaches the ovaries to the lumbar region. Locate these structures in models or charts in the lab and in figure 29.5.

Vagina

The **vagina** consists of the **vaginal canal** and the **vaginal orifice.** The uterus joins with the vaginal canal at the cervix. The vagina is a tough, muscular tube with an anterior and a posterior recessed region around the cervix known as the **fornix.** The vaginal canal is

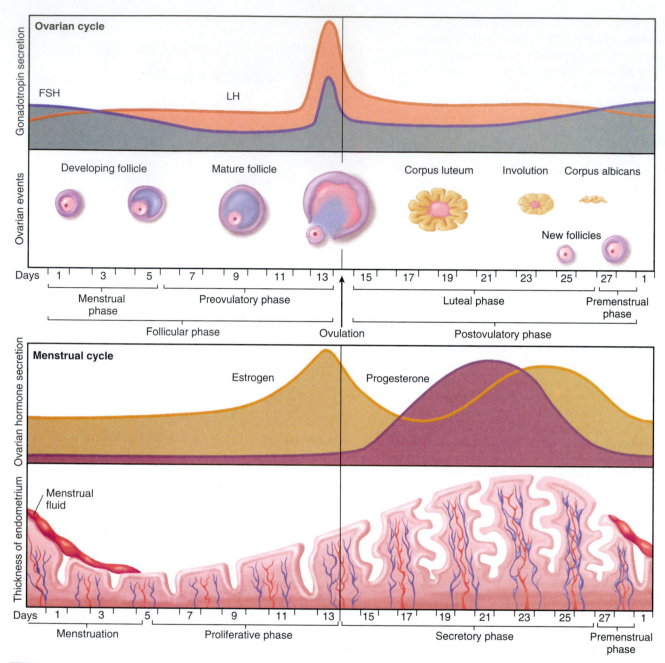

FIGURE 29.7

Ovarian and Menstrual Cycles

poorly supplied with nerves, and the wall of the vagina has cross ridges called **rugae.** The vagina expands greatly during the delivery of a child. Locate the vaginal canal, fornix, rugae, and vaginal orifice in figure 29.1.

External Genitalia

As in the male, the floor of the pelvis can be divided into the **anal triangle** and the **urogenital triangle.** The anal triangle contains the **anus,** and the urogenital triangle contains the **external female re-**

productive structures, or **genitalia.** Examine charts or models in lab and compare the structure of the genitalia with figure 29.8. The external female reproductive structures are collectively referred to as the **vulva.** The **mons pubis** is the anterior-most structure of the vulva and is an adipose pad that overlies the symphysis pubis. Posterior to the mons pubis is the **clitoris,** which is a cylinder of erectile tissue embedded in the body wall that terminates as the **glans clitoris.** The body of the clitoris is a curved structure, illustrated in figure 29.1, and the glans clitoris is the superficial portion. The glans clitoris is anterior to the **external urethral orifice.** The clitoris has an embryonic origin similar to that of the penis in males, and

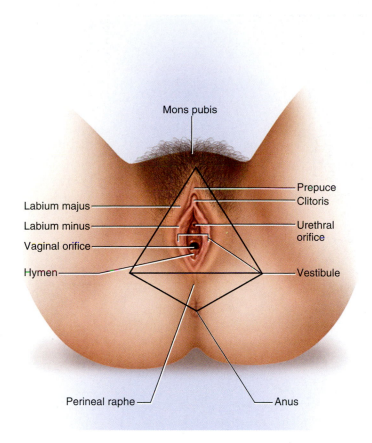

FIGURE 29.8
Female Perineum and External Genitalia

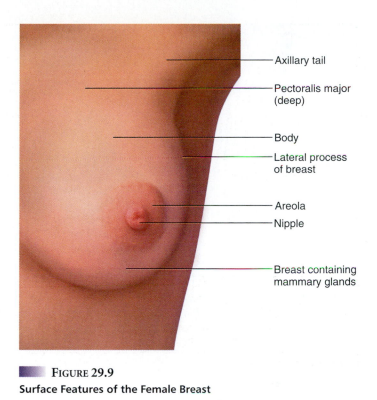

FIGURE 29.9
Surface Features of the Female Breast

like the penis it has erectile tissue and is richly supplied with nerve endings. The anterior edge of the clitoris is enclosed by the **prepuce,** which is an extension of the labia minora.

Posterior to the clitoris is the external urethral orifice and posterior to this is the **vaginal orifice.** The vaginal orifice is partially enclosed by a mucous membrane structure known as the **hymen.** The hymen is variable anatomically and has historically (and sometimes incorrectly) been used as an indicator of virginity. Lateral to the vaginal orifice are the **labia minora** (singular, **labium minus**). These are commonly known as the inner vaginal lips. The space between the labia minora is known as the **vestibule,** and located laterally and posteriorly to the vestibule are the **greater vestibular glands (Bartholin's glands).** These glands provide lubrication to the vagina during intercourse. Lateral to the labia minora are the paired **labia majora** (singular, **labium majus**). Locate these structures on models or charts in the lab and in figure 29.8.

Anatomy of the Breast

The human breast is an integumentary structure, yet the female breast is discussed as a reproductive structure due to its importance as a source of nourishment for offspring. The major structures of the external breast are the pigmented **areola,** the protruding

nipple, the **body** of the breast, and the **axillary tail (tail of Spence).** The axillary tail is of clinical importance in that breast tumors frequently occur there. Examine the surface features of the breast in figure 29.9 and locate the structures listed.

Compare models or charts in the lab and locate the internal structures of the breast as seen in figure 29.10. Note that the breast is anchored to the pectoralis major muscle and dermis of the skin with **suspensory ligaments.** Much of the breast is composed of **adipose tissue,** and embedded in the adipose tissue are the **mammary glands.** The mammary glands are responsible for the production of milk in lactating females. The glands are clustered in **lobes,** and there are about 15 to 20 lobes in each breast. In lactation the mammary glands increase in size and lead to **lactiferous ducts,** which subsequently lead to **lactiferous sinuses (ampullae).** These sinuses exit via the nipple. Humans have several sinuses leading to each nipple. The mammary glands in females begin to undergo changes prior to puberty and become functional glands after delivery of a child. Note the features of the breast in figure 29.10.

Stages of Development

EARLY DEVELOPMENT

The union of a sperm and an egg in a process known as **fertilization** initiates a remarkable phenomenon of growth and differentiation from the single-celled **zygote** to the adult human. The sperm and egg pronuclei fuse and the genetic information from the mother and father form the genes of the **conceptus.** Fertilization usually occurs in the uterine tube, and the zygote divides into 2 cells, then 4, 8, 16, and

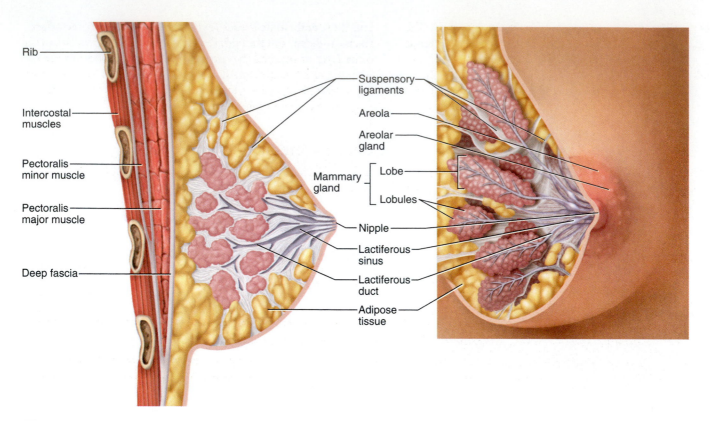

FIGURE 29.10

Interior of the Female Breast

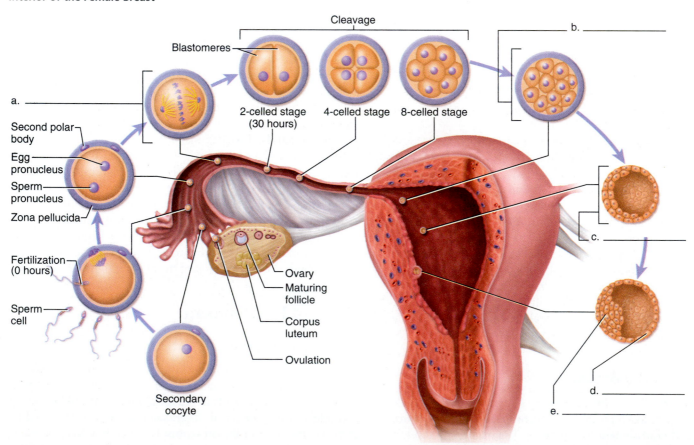

FIGURE 29.11

Development of the Conceptus Use the following terms to fill in this illustration: blastocyst, zygote, embryoblast, morula, trophoblast.

so on until a solid cluster of cells, called a **morula,** is formed. The morula is so named because it looks like a mulberry (*Morus* = genus name for mulberry). The morula continues to divide until it becomes a hollow ball of cells known as the **blastocyst.** The covering of cells on the outside of the blastocyst is called the **trophoblast,** and the cluster of cells on the inside is known as the **embryoblast,** or **inner mass.** Examine charts or models in lab or illustrations in a text and the preceding description to fill in figure 29.11.

EMBRYONIC TISSUES

The blastocyst is the stage that is involved in implantation into the endometrium of the uterus. The conceptus continues to progress and three embryonic tissues form. These tissue are the **ectoderm,** the **mesoderm,** and the **endoderm.** The ectoderm gives rise to the outer layer of skin and the nervous tissue, the mesoderm gives rise to bones and muscles, and the endoderm gives rise to many internal organs, such as digestive and respiratory organs. The derivations from the embryonic tissues are outlined in table 29.1.

Cat Anatomy

If you are using cats, turn to section 10, "Reproductive Anatomy of the Cat," on page 444 of this lab manual.

TABLE 29.1	
Derivations of Embryonic Tissues	
Embryonic Tissues	**Adult Derivatives**
Ectoderm	Epidermis and most of its derivatives, nervous system, outer surface of the eye, outer and inner ear, and epithelia of the mouth, nose, and anus
Mesoderm	Bone; bone marrow; skeletal, cardiac, and most of the smooth muscle; dermis of the skin; and blood
Endoderm	Most of the digestive and respiratory epithelium, most of the digestive glands (except salivary glands), and urinary bladder epithelium

Name _____ Date _____

1. What is the gonad in the female reproductive system?

2. Where is the fornix in the female reproductive system?

3. What is the background substance of the ovary called?

4. What is the inner layer of the uterus called?

5. The ovaries attach to the uterus by what structure?

6. What happens to estrogen and progesterone levels just prior to menstruation?

7. What three hormones are at elevated levels just prior to ovulation?

8. Which is more anterior, the urethral opening or the clitoris?

9. What is the name for the expulsion of the oocyte from the ovary?

10. What is the layer of the endometrium closest to the myometrium called?

11. Ectopic pregnancies are those that occur outside of the endometrial layer of the uterus. Explain how pregnancies can occur in the uterine tube (thus, a tubal pregnancy) or in the abdominopelvic cavity.

12. Trace the pathway of milk from the mammary glands to expulsion.

13. What is the name of the part of the breast nearest to the shoulder?

14. What are the milk-producing glands of the breast called?

15. A zygote is formed from the fusion of what two cells?

16. Embryonic tissue consists of three layers. What are these called?

17. What is a morula?

18. Label the following illustration using the terms provided.

cervix   labium minus   fundus

fornix   vaginal orifice   vaginal canal

clitoris   rugae   urinary bladder

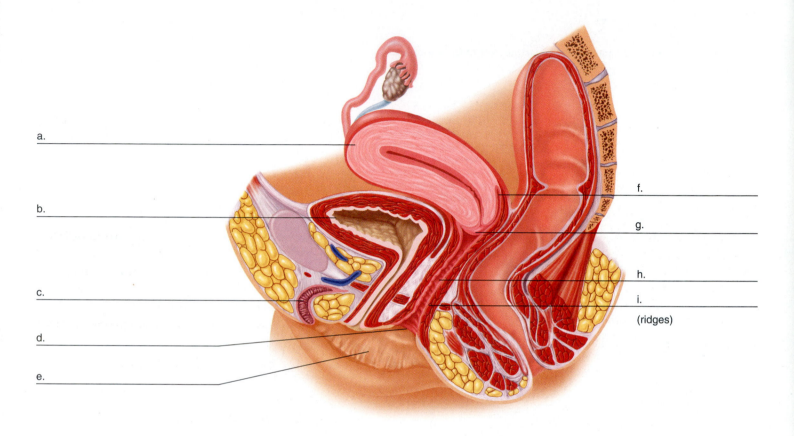

a.

b.

c.

d.

e.

f.

g.

h.

i.

(ridges)

Cat Anatomy

The cat is a reasonable study specimen for understanding human anatomy. The fact that cats and humans have many of the same bones, muscles, and organs reflects the evolutionary mammalian pattern. There are some noticeable differences between cats and humans, though. As cats are quadrupeds, they have a skeletal system and muscles adapted to walking on all fours.

Because this is the study of human anatomy, human terms will be used to describe anatomic features of the cat as a general rule in these lab exercises. For example, the *anterior vena cava* in cats is homologous to the *superior vena cava* in humans. There are some structures in the cat, such as the *epitrochlearis* muscle, that are not found in the human. These structures will retain the cat name.

The exercises in the main portion of the lab manual direct you to the appropriate cat anatomy section. You can find the appropriate section in the following contents description.

MATERIALS FOR CAT DISSECTION

Each time you dissect your cat you should have the appropriate area and tools for dissection. Remember that many people are sensitive to the dissection of animals, so please be aware of this concern as you do your work. A list of materials for dissection follows:

Cats
Large plastic bag to store cat
String to keep bag closed
Tag to identify specimen
Paper towels or clean cloth (an old T-shirt works well) to keep cat moist during storage
Dissection trays—large enough to hold cat and keep materials in place
Scalpel and two or three extra blades
Gloves (household latex gloves work well for repeated use)
Blunt (Mall) probe
Pins to hold parts down or for identification
Forceps
Sharp scissors
Animal waste disposal container

ADDITIONAL MATERIALS TO HAVE ON HAND DURING DISSECTION LABS

First aid kit in lab or prep area
Sharps container for used scalpel blades

DISSECTION CONCERNS

Wear an apron or an old overshirt, gloves, and safety glasses as you dissect. Dissection instruments are sharp, and you must be careful when using scalpels. Do not cut *down* into the specimen but, rather, lift structures gently and try to make incisions so the scalpel blade cuts *laterally*. Cut *away* from yourself and your lab partners. *If you do cut yourself, notify the instructor immediately!* Wash the cut with antimicrobial soap, cover the wound with a sterile bandage, and seek medical advice to reduce the chance of infection.

The laboratory should be well ventilated, but if your eyes burn and you develop a headache, get some fresh air for a moment. If you dissect without having your face directly over the specimen, it may help. Occasionally, students have allergic reactions to formaldehyde. This usually consists of a feeling of restriction of breath either during or after lab. If you have this experience, notify your instructor.

At the end of the exercise wash your dissection equipment with soap and water, taking extra precaution with the scalpel blade. Place all used blades or sharp material in the **sharps container** in lab. Remove all the excess animal material on the dissection trays and put it in the appropriate animal waste container. *Do not dump any animal material in the lab sinks!* Wash the dissection trays and place them in the appropriate area to dry.

CAT CARE

Keep your cat in a plastic bag. It is recommended that you retain the skin of the cat as a protective wrapping when you are finished with the day's dissection or that you take a small towel (or an old T-shirt) and wrap the specimen in the cloth, soaking it in a cat wetting solution before you place it back in the bag. Usually, one lab period is required to skin the cat and remove the superficial fascia over the muscles, preparing the specimen for further study.

CLEANUP

When you are done with your dissection, carefully place your cat back in the plastic bag. Place all waste material in the appropriate animal waste disposal container, not down the sinks in lab.

SECTION 1: SKELETAL ANATOMY OF THE CAT

The cat skeleton differs from the human skeleton in several ways. Humans are bipedal, and the vertebral column reflects this. In humans the lower vertebrae are larger than the upper vertebrae. The vertebrae in cats are more evenly matched in size because cats are quadrupeds and the weight on the vertebral column is distributed more evenly. Humans have a coccyx, while cats have tail vertebrae that are fully functional.

Other skeletal differences can be seen in that humans walk on the entire inferior surface of the foot, while cats walk on their toes. Examine a cat skeleton in the lab and note the general features of the skeleton (fig. S1.1)

SECTION 2: DISSECTION, OVERVIEW, AND FORELIMB MUSCLES OF THE CAT

The objective in using a cat for the exercise on musculature is to provide a study specimen for dissection and application to the hu-

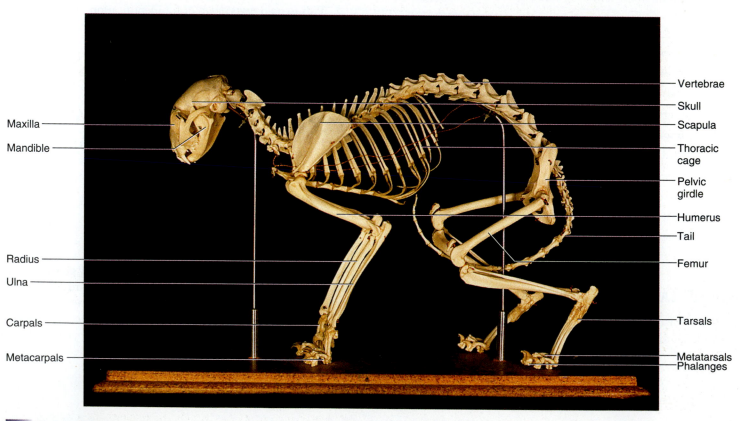

Maxilla

Mandible

Radius

Ulna

Carpals

Metacarpals

Vertebrae

Skull

Scapula

Thoracic cage

Pelvic girdle

Humerus

Tail

Femur

Tarsals

Metatarsals
Phalanges

■ **FIGURE S1.1**
Cat Skeleton, Lateral View

man system. Differences occur between cat musculature and human musculature, but you should focus on similar structures in order to gain an appreciation of human musculature. Read all the material in the exercise prior to beginning the dissection. Dissection is a skill in which you try to separate the overlying structures from those underneath while keeping intact as much material as you can.

External Features

Take a dissection tray and a cat specimen to your table. Remove the cat from the plastic bag and place it on the tray. You may have excess fluid in the plastic bag, and this should be disposed of properly as directed by your instructor. Place the cat on its back and determine whether you have a male or a female. Both sexes have multiple **teats** (nipples), yet the males have a **scrotum** near the base of the tail with a **prepuce** (penile foreskin) ventral to the scrotum. Females have a **urogenital opening** anterior to the anus without a scrotum and prepuce. Once you have identified the sex of your specimen, compare yours with others in the class, so that you can identify the sexes externally. If you have difficulty determining the sex, ask your instructor for help.

Examine the cat and notice the **vibrissae** (vib-RISS-ee), or whiskers, in the facial region. Other variations from humans are the presence of **claws** and **friction pads** on the extremities and a **tail**.

Review the planes of the body as illustrated in laboratory exercise 1 before you begin the dissection. The terms used in that exercise will be of great importance in the dissection procedures.

Removal of the Skin

Begin your dissection by lifting the skin in the pectoral region with a forceps and making a small cut with a scalpel or sharp scissors in the midline. Work a blunt probe gently into the cut, so that you free the skin from the underlying fascia and muscle somewhat. Be careful—the muscles are close to the skin. Insert your scalpel, blade side up, and make small incisions in the skin, cutting away from the underlying muscle, as illustrated in figure S2.1.

Make a cut that runs up the midline of the sternal region until you reach the neck. Likewise, cut posteriorly until you reach an area anterior to the genital region. Leave the genitals intact and carefully make an incision that runs perpendicular to your first cut. Likewise, make a perpendicular incision along the upper thoracic region (fig. S2.2). Continue your caudal incision along the medial aspect of the thigh. Watch out for superficial blood vessels in this area (especially the great saphenous vein) and stop when you reach the knee. Cut the skin around the knee and begin removing it as a layer from the dorsal side of the cat. Cut the skin from the base of the tail and remove the skin from the back.

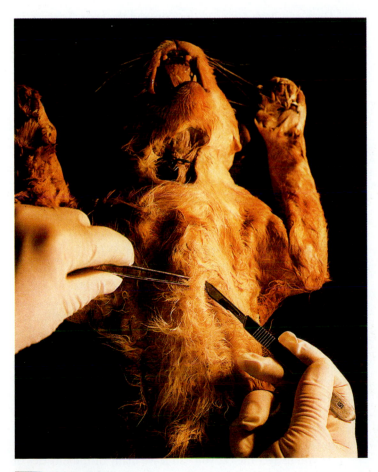

FIGURE S2.1
Lateral Cutting with a Scalpel

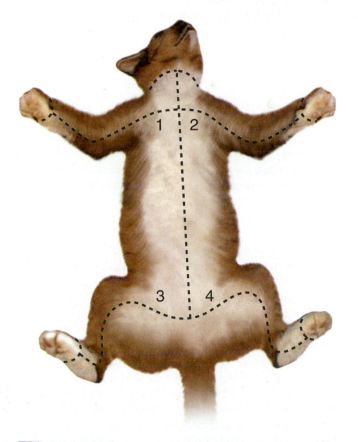

FIGURE S2.2
Removal of the Skin of the Cat After cutting down the midline, make incisions into the limbs (follow the numbers) and gently remove the skin from underlying structures.

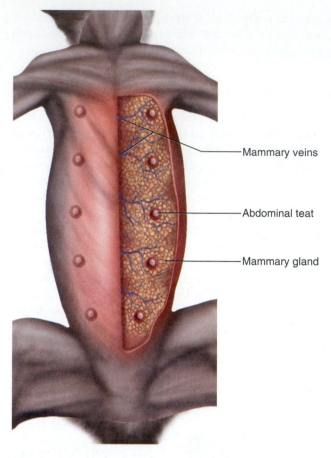

Mammary veins

Abdominal teat

Mammary gland

Mammary Glands of the Cat

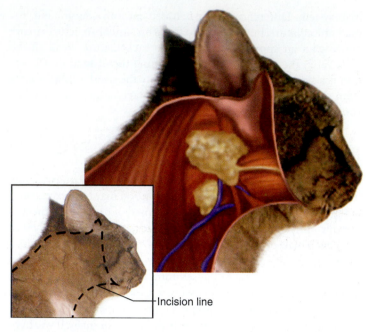

Incision line

Removal of Skin from the Head of a Cat

Once you reach the shoulders, turn the cat back to the ventral side and remove the skin on the medial side of the arm until you reach the elbow. Stop at this level and remove the skin from the lateral aspect of the arm, working back toward the shoulder. Be careful not to damage the superficial veins on the lateral side of the forearm. Continue your removal of the skin from the ventral body region. If your cat is a female, locate the mammary glands, which are elongated, beige, lobular tissue on each side of the midline on the ventral side (fig. S2.3).

Make an incision on the ventral side of the neck along the midline. Be careful—there are numerous blood vessels along the neck. Do *not* cut through these blood vessels. Continue up into the face and look for beige lumps of glandular tissue in the region of the mandible. These are the **salivary glands.** Cut the skin carefully from the face and remove it from the head by cutting around the ears (fig. S2.4).

You should be able to remove all of the skin from the cat at this time. Examine the undersurface of the skin and note the superficial muscles that cause the skin to move. These are the **cutaneous** (que-TAY-nee-us) **maximus** and the **platysma** (plah-TISS-mah).

Return to the cat and remove as much fat as you can. **Subcutaneous fat** is variable from cat to cat, and your specimen may have little or may have significant amounts. The muscle also is covered by fascia, a connective tissue wrapping. Remove the fascia from the muscle, so that the fiber direction is apparent.

Once you have removed the skin and the superficial fascia, you should identify the major muscles of the cat. Look for the large latissimus dorsi muscle of the back and the external abdominal oblique muscle. You should also find the deltoids and triceps brachii muscles of the shoulder region, as well as the gluteus and biceps femoris muscles of the hip and thigh region. Compare your cat with figure S2.5.

Individual Muscles of the Cat

Begin your **dissection** of the muscles by understanding that the term "dissect" means to separate. When you isolate one muscle from another, locate the tendons of that muscle. The **tendon** is the attachment point of the muscle to a bone. Broad, flat tendons are known as **aponeuroses** (AH-poh-nyeu-ROH-seez).

Cat Dissection

If you did not already remove the skin from the forelimb of the cat, you should do so now. This can be accomplished by making a longitudinal incision along the length of the forelimb. Be very careful not to cut the tendons, blood vessels, or nerves. Pull the skin off as if you were removing a pair of knee socks. As you get to the tips of the digits, cut the skin from the digits. Pay particular attention and keep the tendons intact. It is a good procedure to dissect only one side of the cat at a time. If you make an error on one side, you will have the other side to dissect.

As you dissect the cat you can leave many of the muscles of the forelimb intact. Deeper muscles can generally be seen by moving the more superficial muscles off to the side.

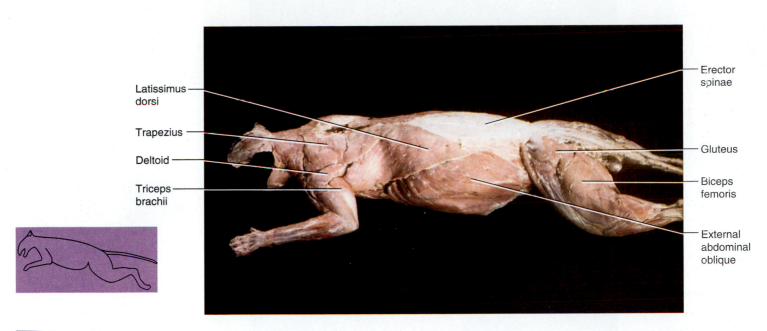

Latissimus dorsi

Trapezius

Deltoid

Triceps brachii

Erector spinae

Gluteus

Biceps femoris

External abdominal oblique

FIGURE S2.5
Major Muscles of the Cat

Once you locate the muscle, tug gently on it to locate its **origin** and **insertion.** If you pull too hard, you may rip the muscle. The outer wrapping of the muscle is known as the **fascia,** and it should be removed to find the fiber direction of the muscle. The main part of the muscle is known as the **belly.** You may need to cut a muscle to locate deeper muscles. This is done by **transecting** the muscle, which is to cut the muscle into two sections perpendicular to the fiber direction. Lift the muscle with a probe to transect it. Do not cut underlying muscles. After transecting the muscle you may want to **reflect** it, or pull it toward its attachment site. When separating two muscles, you may find a cottonlike material in between them. This is loose connective tissue that forms part of the fascia.

THORACIC MUSCLES

Pectoral Muscles of the Cat   There are four major muscles in a superficial view of the pectoral region of the cat. These are the pectoantebrachialis, pectoralis major, pectoralis minor, and xiphihumeralis. The **pectoantebrachialis** has no corresponding muscle in the human. It originates on the sternum and inserts on the forearm. Transect and reflect the pectoantebrachialis to see the pectoralis major. The **pectoralis major** originates on the sternum and inserts on the upper humerus. The **pectoralis minor** is a large muscle in cats; it inserts on the upper humerus. The **xiphihumeralis** (ZIE-fee-HEU-mur-AL-is) is another cat muscle that has no corresponding human muscle and it originates on the sternum and inserts on the proximal humerus, along with the pectoralis major and minor. Locate these muscles on your cat and in figure S2.6.

In humans there is a single trapezius on each side of the body. In cats each trapezius consists of three muscles. Examine the cat from the dorsal side and locate the **clavotrapezius,** the **acromiotrapezius,** and the **spinotrapezius.** All these muscles originate on the vertebral column, with the clavotrapezius also

originating on the occipital bone. The clavotrapezius inserts on the clavicle, the acromiotrapezius on the acromion process of the scapula, and the spinotrapezius on the spine of the scapula. Compare these muscles with figure S2.7.

The deltoid muscle in cats also consists of three muscles, unlike the single deltoid in humans. The deltoid muscles are named for their bony attachments. The **clavodeltoid** (clavobrachialis) originates on the clavicle, the **acromiodeltoid** on the acromion process, and the **spinodeltoid** on the spine of the scapula. Insertions of these muscles are on the arm or forelimb. Locate these muscles on the cat and compare them with figure S2.7.

Two other muscles of the region are the **latissimus dorsi** and the **levator scapulae ventralis.** These two muscles are similar to those in the human. Locate these muscles on the cat and compare them with figure S2.7.

Scapular Muscles That Act on the Humerus   The deep muscles of the scapula can be seen by reflecting the overlying muscles. The **supraspinatus, infraspinatus, subscapularis, teres major,** and **teres minor** are roughly equivalent to the same muscles in the human. Locate the supraspinatus, infraspinatus, and teres major on the cat and find them in figure S2.8.

Forelimb Muscles   The muscles that have an action on the forelimb of the cat typically either flex the forelimb or extend the forelimb as their primary actions. The **epitrochlearis** is a muscle that does not have a corresponding muscle in humans (fig. S2.9). The epitrochlearis is on the medial side of the humerus and extends the forelimb. The **biceps brachii** is also a medial muscle, and it flexes the forelimb. The **triceps brachii, anconeus, brachioradialis,** and **brachialis** are lateral or posterior muscles. The triceps brachii and the anconeus extend the forearm, while the brachialis flexes the forearm. Find the lateral muscles, using figures S2.9 and S2.10 as a guide.

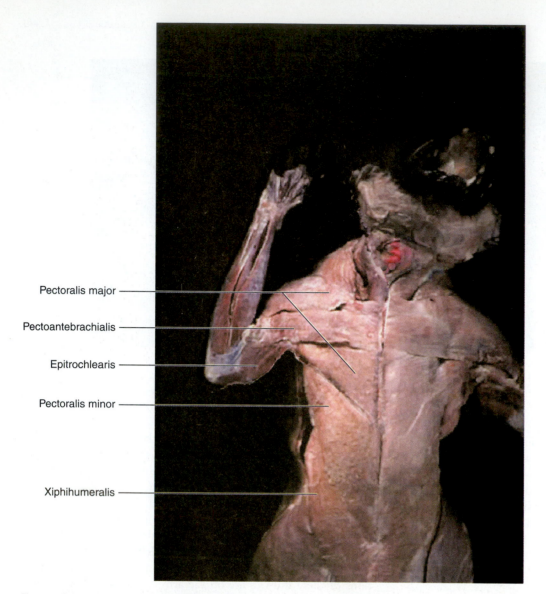
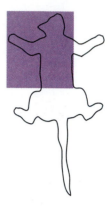

Pectoralis major

Pectoantebrachialis

Epitrochlearis

Pectoralis minor

Xiphihumeralis

FIGURE S2.6.
Muscles of the Pectoral Region of the Cat

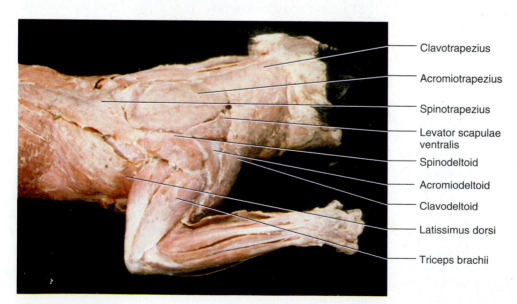

Clavotrapezius

Acromiotrapezius

Spinotrapezius

Levator scapulae ventralis

Spinodeltoid

Acromiodeltoid

Clavodeltoid

Latissimus dorsi

Triceps brachii

FIGURE S2.7
Superficial Muscles of the Shoulder of the Cat

414

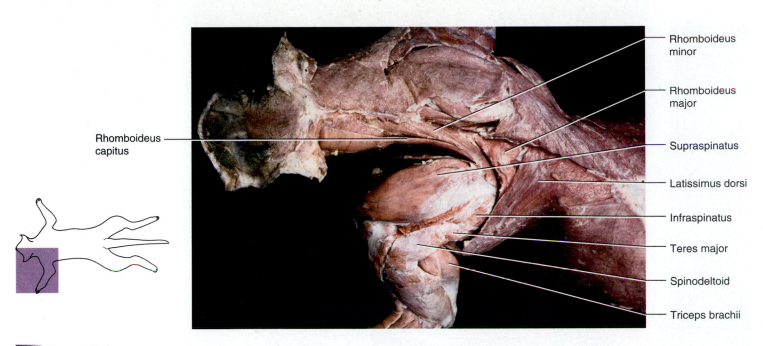

Rhomboideus
minor

Rhomboideus
major

Supraspinatus

Latissimus dorsi

Infraspinatus

Teres major

Spinodeltoid

Triceps brachii

Rhomboideus
capitus

FIGURE S2.8
Deep Muscles of the Scapula of the Cat

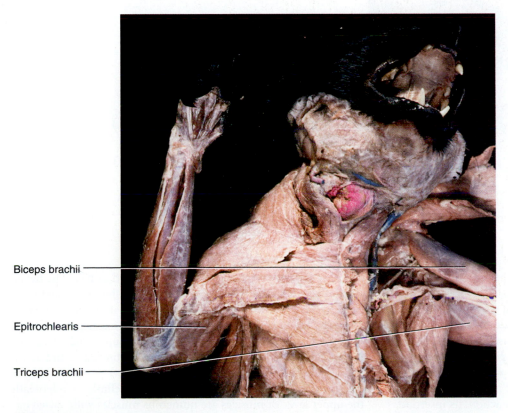

Biceps brachii

Epitrochlearis

Triceps brachii

FIGURE S2.9
Muscles of the Proximal Forelimb of the Cat, Ventral View

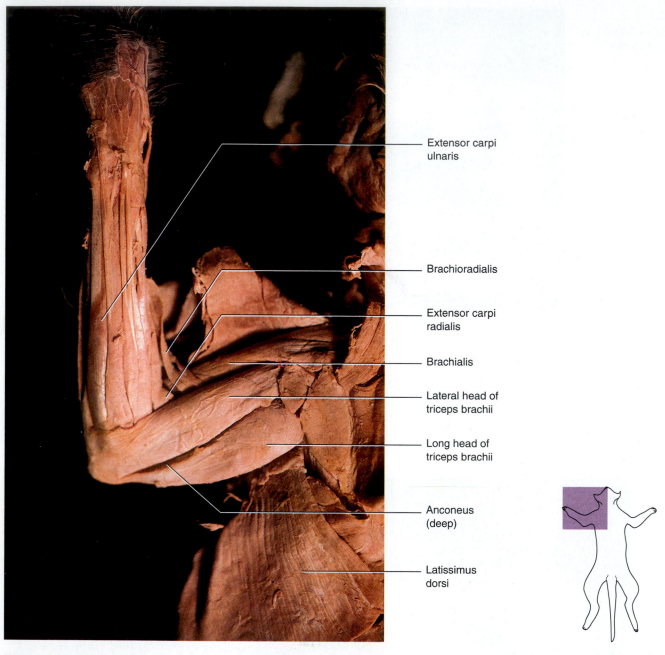

Extensor carpi
ulnaris

Brachioradialis

Extensor carpi
radialis

Brachialis

Lateral head of
triceps brachii

Long head of
triceps brachii

Anconeus
(deep)

Latissimus
dorsi

FIGURE S2.10
Muscles of the Proximal Forelimb of the Cat, Dorsal View

Superficial Muscles on the Medial Aspect of the Forelimb   Most of the muscles of the forelimb of the cat run parallel to the radius and ulna. An exception to this is the **pronator teres,** which is a small slip of muscle that runs obliquely down the forelimb. It pronates the wrist. The **palmaris longus** is a broad, flat muscle, superficially located on the forelimb with insertions into the digits. In humans the palmaris longus terminates in the fascia of the palm.

The **flexor carpi radialis** is a thin muscle inserting on the second and third metacarpals. It is named for its action, its insertion, and its location. The **flexor carpi ulnaris** originates on the humerus

and ulna and inserts on medial metacarpals and carpals. The **flexor digitorum superficialis** (or flexor digitorum sublimis in cats) is located in the forelimb as a middle-level muscle, underneath the palmaris longus. It originates on fascia of other forearm muscles, as opposed to originating on bone. Examine the superficial muscles of the medial forelimb and compare them with figures S2.11 and S2.12.

Deep Muscles on the Medial Aspect of the Forelimb   Underneath the upper layer of muscles are numerous muscles with varied actions. The **supinator** is a deep muscle that runs diagonally from the

Figure S2.11

Superficial Muscles of the Right Medial Forelimb of the Cat

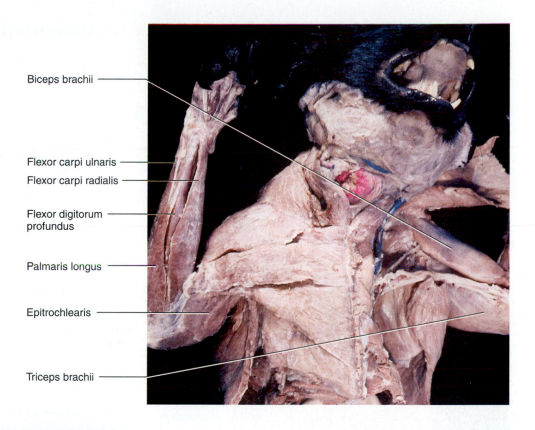

Biceps brachii

Flexor carpi ulnaris

Flexor carpi radialis

Flexor digitorum profundus

Palmaris longus

Epitrochlearis

Triceps brachii

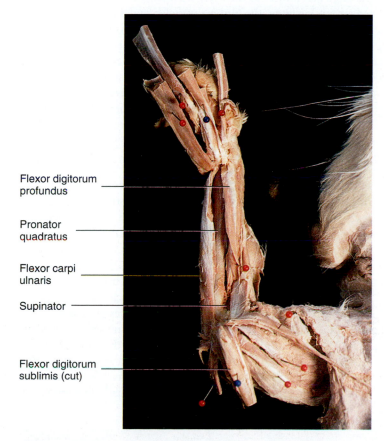

Flexor digitorum profundus

Pronator quadratus

Flexor carpi ulnaris

Supinator

Flexor digitorum sublimis (cut)

Figure S2.12

Deep Muscles of the Right Medial Forelimb of the Cat

lateral epicondyle of the humerus to the proximal radius. It is the deepest of the proximal forelimb muscles.

The **flexor digitorum profundus** is an extensive muscle that inserts on the first through fifth digits. It replaces the flexor pollicis longus for the thumb flexion, since this muscle is absent in cats. The **pronator quadratus** is a square muscle located between the radius and ulna deep to the flexor digitorum profundus. Find the deep muscles in the cat and compare them with figure S2.12.

Muscles on the Lateral Aspect of the Forelimb   The lateral muscles of the forelimb are the extensor group; the **extensor carpi radialis longus** muscle is deep to the brachioradialis and inserts on the second metacarpal. The **extensor carpi radialis brevis** is underneath the extensor carpi radialis longus and inserts on the third metacarpal.

Locate the **extensor carpi ulnaris,** which is next to the extensor digitorum lateralis, inserting on the fifth metacarpal. The **extensor digitorum communis** is on the lateral aspect of the forelimb. It is a broad muscle inserting by tendons on the second through fifth digits. Locate these muscles on the cat and in figure S2.13.

The **extensor digitorum lateralis** is specific to the cat and inserts with the tendons of the extensor digitorum communis to the digits. The **extensor pollicis brevis** is a well-developed muscle in the cat, while the abductor pollicis longus is absent in cats.

After you have examined these muscles on the cat in lab, make sure that you put your cat away in a plastic bag and clean up your lab station. Put all waste animal material in the appropriate container (not the trash can or down the sink) and clean off your desk. After you have spent the time dissecting your cat, you will probably realize why it is important not to set food down on the lab tables.

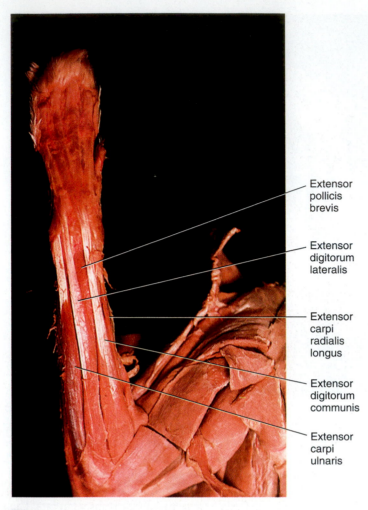

Extensor
pollicis
brevis

Extensor
digitorum
lateralis

Extensor
carpi
radialis
longus

Extensor
digitorum
communis

Extensor
carpi
ulnaris

FIGURE S2.13
Lateral Muscles of the Left Forelimb of the Cat

SECTION 3: HINDLIMB MUSCLES OF THE CAT

If you have not done so already, remove the skin from the lower portion of the hind limb of the cat. Be careful cutting the skin from the distal portions of the leg, so that you do not cut through the tendons of the foot. Be careful in the dissection of the cat muscles that you *leave the blood vessels and nerves intact* as you dissect your specimen. You will be looking at these structures in later sections. Pay particular attention to the **great saphenous vein,** which runs under the skin and should be preserved. Remove any excess fat and fascia from the muscles as you dissect the material. If you need to cut a muscle, bisect it perpendicular to the fiber direction, so that half of it is attached to the origin side and half of it is attached to the insertion side.

Cat Musculature

The cat has many muscles of the hip and thigh that serve as good models for studying human muscles. Because cats are quadrupedal, some of the thigh muscles are different in size or placement than those of humans.

MUSCLES OF THE THIGH

Medial Muscles of the Thigh of the Cat   Two major muscles on the medial aspect of the thigh in the cat are the **sartorius** and the **gracilis muscles.** Locate these muscles in figure S3.1 and note that they are much broader in the cat than in the human.

Cut through the sartorius and gracilis and reflect the ends of these muscles. You should be able to see the deeper muscles of the thigh, including the **tensor fasciae latae,** the **vastus medialis,** the **adductor femoris** (a large muscle specific to the cat), and the **semimembranosus.** The distal portion of the **semitendinosus** may also be seen from this view. Examine figure S3.1 for the deep muscles of the thigh and locate the **rectus femoris, vastus medialis,** and **vastus lateralis.** Bisect the rectus femoris in order to see the **vastus intermedius** muscle. The **adductor longus** is a thin muscle anterior to the **adductor femoris,** and it can be seen as a small muscle on the medial aspect of the thigh.

Lateral Muscles of the Thigh of the Cat   On the lateral aspect of the thigh are numerous muscles, including the **biceps femoris,** the **tensor fasciae latae,** the **gluteus** muscles, and a muscle specific to the cat, the **caudofemoralis.** The biceps femoris is the largest and most lateral muscle of the thigh. The **gluteus medius** is larger than the **gluteus maximus** in cats due to the lengthening of the pelvic girdle. The **semimembranosus** is a large muscle in cats (much larger than the semitendinosus), and it can be seen from the lateral side of the cat. The adductor magnus and adductor brevis are not found in the cat. Examine these muscles in figures S3.2 and S3.3.

MUSCLES OF THE LEG

Posterior Muscles   Examine the large **gastrocnemius** on the posterior aspect of the leg. The **soleus** is deep to the gastrocnemius and inserts with the gastrocnemius on the calcaneus. In cats the soleus has only one point of origin, the fibula, while in humans the soleus originates on the tibia and fibula. Examine figure S3.4 with your dissection. Cut through the calcaneal tendon and lift the gastrocnemius and soleus, so that you can study the underlying muscles.

The **popliteus** is a small, triangular muscle that crosses the knee joint. Do not damage the nerves and blood vessels that pass over the popliteus because you will study them in the subsequent sections.

The **flexor digitorum longus** is a muscle that runs along the medial side of the hind limb and flexes the digits of the cat. It joins with the **flexor hallucis longus,** which inserts on all the digits of the hind limb. The **tibialis posterior** is a narrow muscle that inserts on

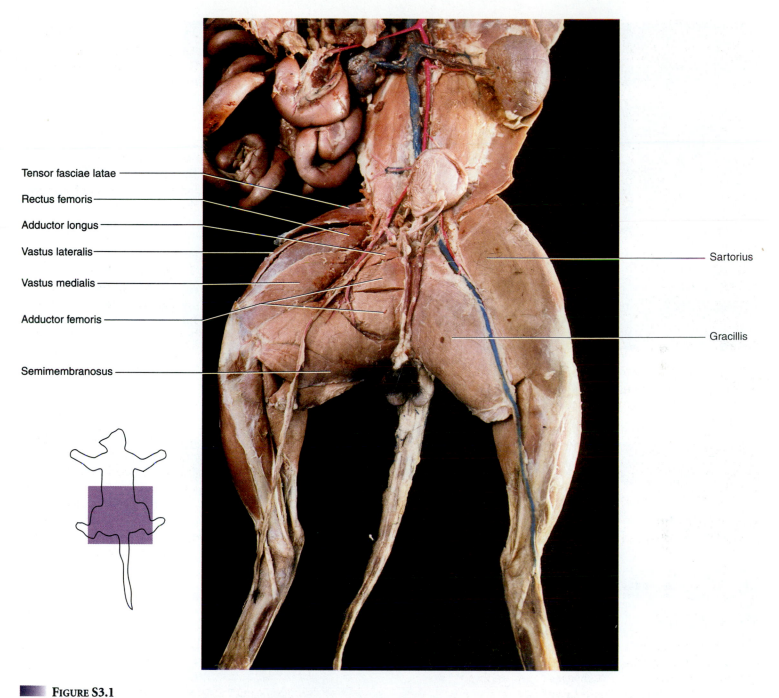

Tensor fasciae latae

Rectus femoris

Adductor longus

Vastus lateralis

Vastus medialis

Adductor femoris

Semimembranosus

Sartorius

Gracillis

FIGURE S3.1

Thigh Muscles of the Cat Superficial muscles (left side of cat); deep muscles (right side of cat).

the tarsal bones of the foot. Locate these muscles in the cat and compare them with figure S3.5.

Anterior Muscles   The **tibialis anterior** is a large muscle of the leg and inserts on the dorsum of the foot. The **extensor digitorum longus** originates on the femur in cats and inserts on the distal phalanges in all the digits in the cat. The extensor hallucis longus is not found in cats. These muscles can be seen in figure S3.6. Examine these muscles and isolate them in your dissection.

The **fibularis longus** extends along the length of the fibula with the **fibularis brevis.** The fibularis longus inserts at the base of the metatarsals, while the fibularis brevis inserts on the fifth metatarsal. The **fibularis tertius** inserts into the tendon of the extensor digitorum muscle. These muscles can be seen in figure S3.7.

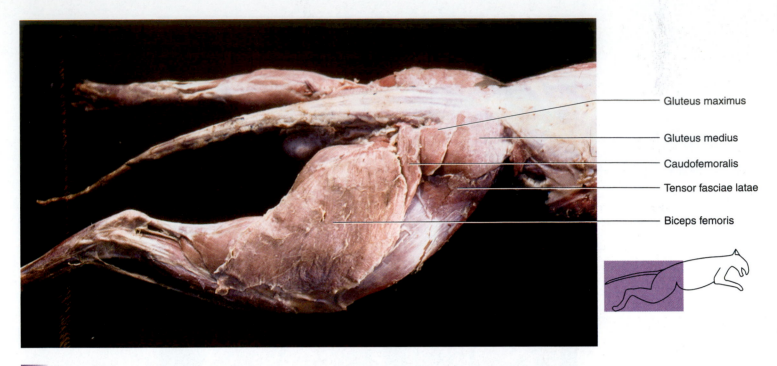

Gluteus maximus

Gluteus medius

Caudofemoralis

Tensor fasciae latae

Biceps femoris

FIGURE S3.2

Lateral Thigh Muscles of the Cat Superficial muscles of the right side.

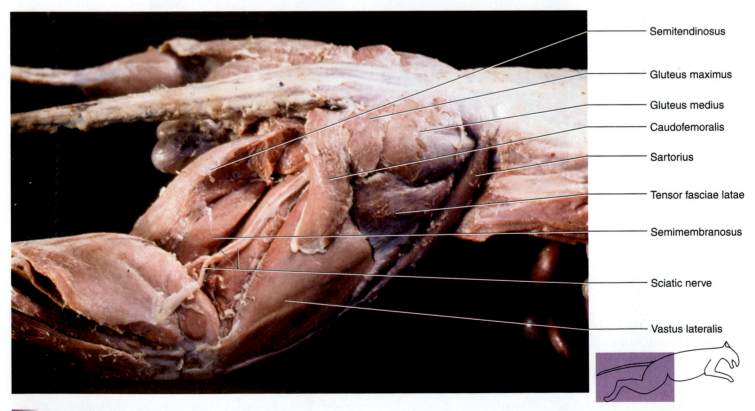

Semitendinosus

Gluteus maximus

Gluteus medius

Caudofemoralis

Sartorius

Tensor fasciae latae

Semimembranosus

Sciatic nerve

Vastus lateralis

FIGURE S3.3

Lateral Thigh Muscles of the Cat Deep muscles of the right side.

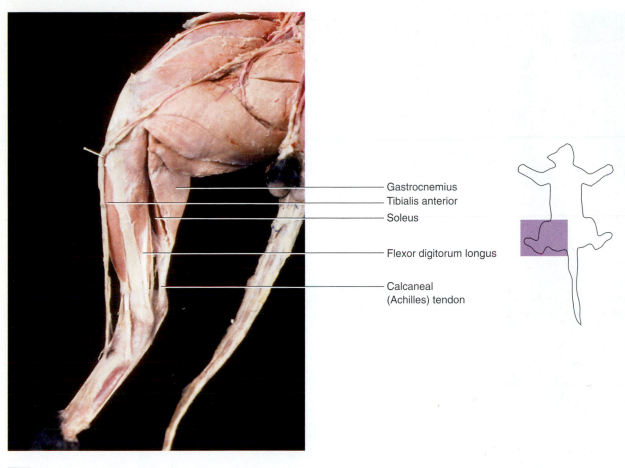

Gastrocnemius

Tibialis anterior

Soleus

Flexor digitorum longus

Calcaneal
(Achilles) tendon

FIGURE S3.4

Superficial Muscles of the Right Leg of the Cat, Medial View

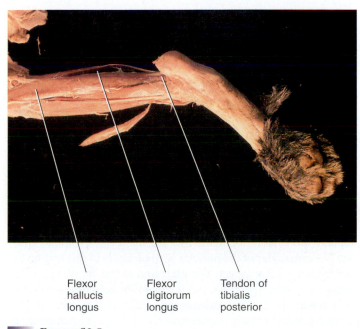

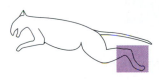

Flexor
hallucis
longus

Flexor
digitorum
longus

Tendon of
tibialis
posterior

FIGURE S3.5

Deep Muscles of the Left Leg of the Cat, Lateral View

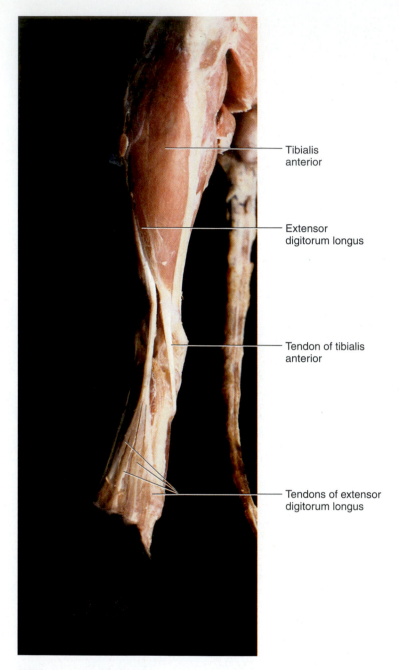

Tibialis anterior

Extensor digitorum longus

Tendon of tibialis anterior

Tendons of extensor digitorum longus

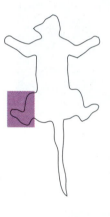

FIGURE S3.6

Muscles of the Right Leg of the Cat, Anterior View

SECTION 4: HEAD AND NECK MUSCLES OF THE CAT

In dissecting the muscles of the head and neck, take care not to cut through the salivary glands, or the veins and arteries. You will study these structures in later sections. Dissect only one side of the head, leaving the structures on the other intact for study of the digestive system. A dorsal neck muscle of the cat is the **levator scapulae,** which can be found deep to a shoulder muscle known as the trapezius. You will need to cut a part of the trapezius, the clavotrapezius,

in order to see the levator scapulae. In cats there is an additional muscle called the **levator scapulae ventralis,** which inserts on the scapular spine. Examine figure S4.1 for the levator scapulae muscles. The remainder of the muscles you will study in this section are seen from a ventral aspect. The **platysma** in the cat was removed during the skinning process, and it will not be seen unless you kept the skin with the cat. The **scalenes** can be dissected by reflecting the pectoralis minor muscle. Notice how the scalenes are composed of separate slips of muscle that run from the ribs to the neck. In humans the muscle is more lateral than in cats. Compare your specimen with figure S4.1.

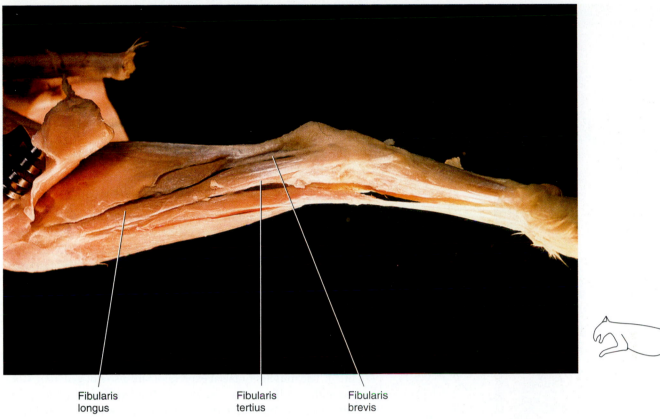

Fibularis longus

Fibularis tertius

Fibularis brevis

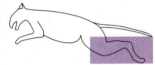

FIGURE S3.7

Muscles of the Left Leg of the Cat, Lateral View

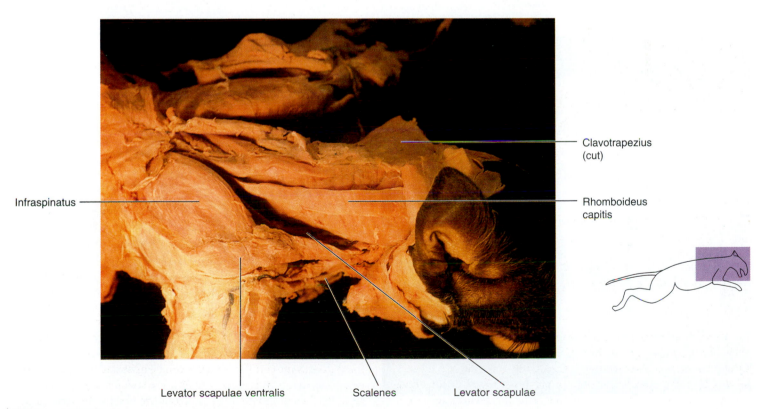

Infraspinatus

Clavotrapezius (cut)

Rhomboideus capitis

Levator scapulae ventralis

Scalenes

Levator scapulae

FIGURE S4.1

Muscles of the Neck of the Cat, Dorsolateral View

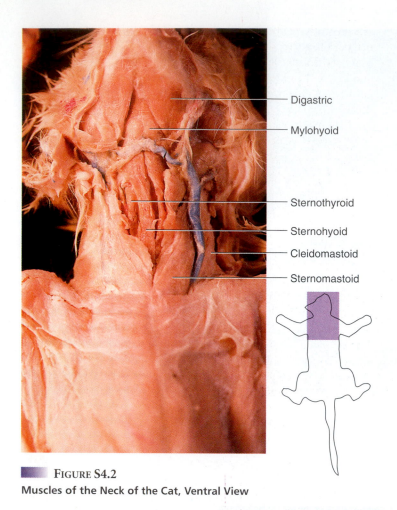

In the cat the **sternocleidomastoid** consists of two muscles, the **sternomastoid** and the **cleidomastoid.** The sternomastoid extends from the sternum to the mastoid process of the skull, and the cleidomastoid runs from the clavicle to the mastoid process. Underneath the sternomastoid is the most medial muscle of the neck group, the **sternohyoid.** The **sternothyroid** is deeper and more lateral than the sternohyoid. These muscles can be seen in figure S4.2.

The **digastric** muscle runs parallel to the lower edge of the mandible and underneath the submandibular gland. Deep to the digastric is the **mylohyoid,** a broad muscle that runs perpendicular to the direction of the digastric. Compare your dissection with figure S4.2.

The **occipitalis** is a posterior head muscle in the cat that attaches to the **galea aponeurotica.** The **frontalis** attaches anteriorly to the galea. Lateral to these muscles are the temporalis and masseter muscles. The **temporalis** is located more dorsally than the masseter and can be dissected by removing the skin and fascia anterior to the ear. The **masseter** is a large, well-developed muscle in the cat that originates on the zygomatic arch and inserts on the lateral surface of the mandible. Examine the temporalis and masseter muscles in figure S4.3.

The **pterygoids** are usually not dissected in the cat because you have to cut through the ramus of the mandible to examine them. The muscles of facial expression are also not generally studied in the cat. These muscles are small and are frequently removed with the skin. You should study these muscles on human models or a cadaver (if available).

Digastric

Mylohyoid

Sternothyroid

Sternohyoid

Cleidomastoid

Sternomastoid

FIGURE S4.2
Muscles of the Neck of the Cat, Ventral View

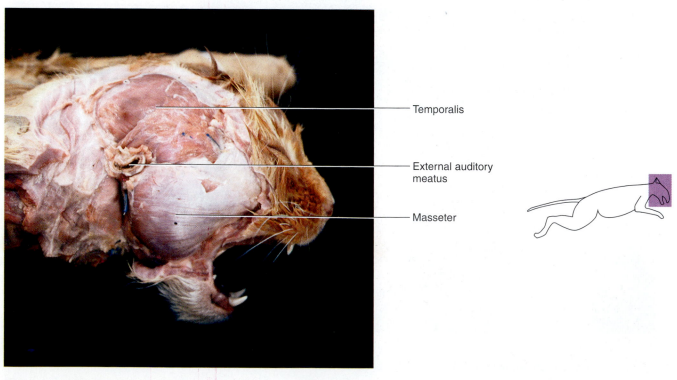

Temporalis

External auditory meatus

Masseter

FIGURE S4.3
Muscles of the Head of the Cat, Lateral View

SECTION 5: TORSO MUSCLES OF THE CAT

Abdominal Muscles

Place the cat on its back and examine the abdominal muscles. The abdominal muscles in the cat are similar to those in the human in that the **external abdominal oblique** is a broad superficial muscle on the anterior abdomen. Carefully cut through the external abdominal oblique to reveal the **internal abdominal oblique,** as illustrated in figure S5.1. Deep to this is the **transversus abdominis,** and it can be seen by carefully dissecting the internal abdominal oblique. If you cut too deeply, you will enter the abdominal cavity, so be careful in this part of the dissection. The **rectus abdominis** is a muscle that runs from the pubic region to the sternum.

Move to the thoracic region and examine the muscle of the lateral thorax dorsal to the xiphihumeralis. This is the **serratus anterior** (actually, serratus ventralis in the cat), and you should see the scalloped edges of the muscle. Carefully separate this muscle from the others and follow where it inserts on the scapula. To see the intercostal muscles you will have to bisect the superficial chest muscles. If you have not done so already, cut through the middle of the belly of the pectoral muscles, exposing the ribs of the cat. Carefully remove the outer layer of fascia from the muscle between the ribs and locate the **external intercostal** muscle. You should be able to cut part of this muscle away and expose the **internal intercostal** muscle. Note how the fibers run perpendicular to one another. Do not look for the diaphragm at this time. You can see it in section 7, "Respiratory Anatomy of the Cat," when you examine the lungs. Examine these thoracic muscles in figure S5.2.

Deep Muscles of the Back

Place the cat so that you can examine the dorsal surface. You will need to dissect the trapezius carefully to see the **rhomboideus** muscles. These muscles originate on the vertebral column and insert on the scapula. If you examine the muscles of the neck and head, you should see the **splenius** and the **semispinalis** muscles. These are located in figure S5.3. You will need to bisect the latissimus dorsi and the posterior portion of the external abdominal oblique to see the **erector spinae** muscles. The relative position of the cat erector spinae can be seen in figure S5.4. Compare this figure with your dissection. Locate the **iliocostalis, longissimus,** and **spinalis** of the erector spinae along with the **multifidus** in the cat.

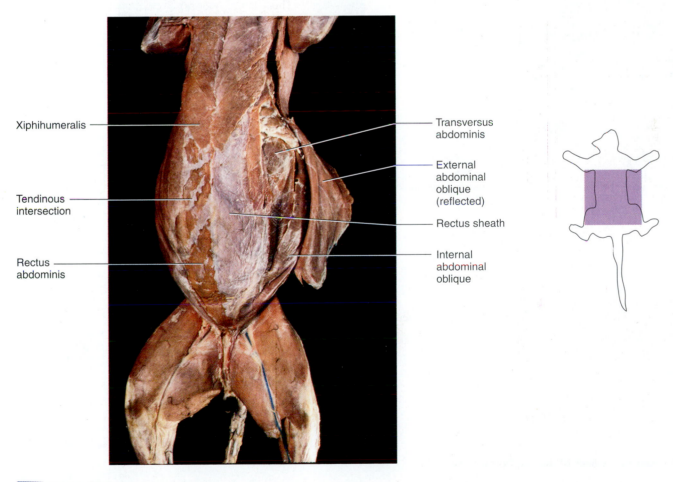

Xiphihumeralis

Tendinous intersection

Rectus abdominis

Transversus abdominis

External abdominal oblique (reflected)

Rectus sheath

Internal abdominal oblique

FIGURE S5.1

Muscles of the Abdomen of the Cat, Ventral View

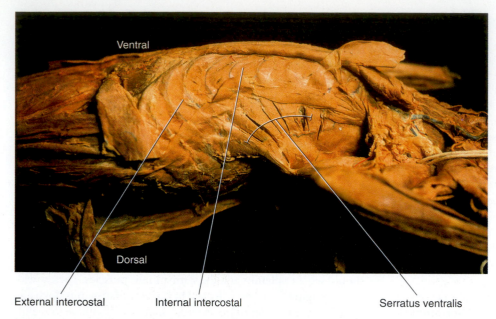

Ventral

Dorsal

External intercostal   Internal intercostal   Serratus ventralis

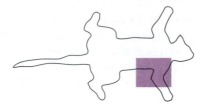

Figure S5.2

Muscles of the Thorax of the Cat, Lateral View

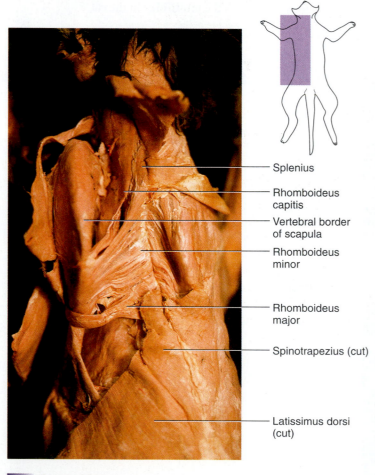

Splenius

Rhomboideus capitis

Vertebral border of scapula

Rhomboideus minor

Rhomboideus major

Spinotrapezius (cut)

Latissimus dorsi (cut)

Figure S5.3

Anterior, Deep Muscles of the Back of the Cat, Dorsal View

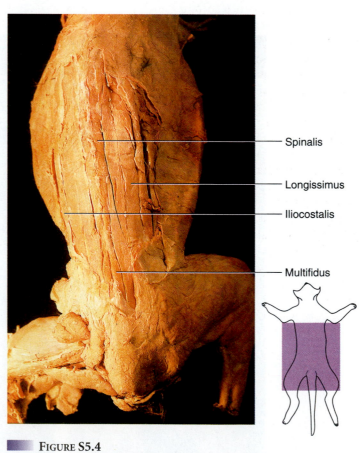

Spinalis

Longissimus

Iliocostalis

Multifidus

Figure S5.4

Posterior, Deep Muscles of the Back of the Cat, Dorsal View

SECTION 6: BLOOD VESSEL ANATOMY OF THE CAT—BLOOD VESSELS AND LYMPH SYSTEM

Open the thoracic and abdominal regions of the cat if you have not done so already. Be careful as you make the first incision into the body cavity of the cat. Cut with a sharp scissors or, if using a scalpel, use the blade facing to one side or, even better, use a lifting motion and cut upwards through the tissue. If you use a downwards sawing motion you may damage the internal organs of the cat.

Make an incision into the body wall of the cat in the belly, slightly lateral to midline. Cut into the muscle and then grasp the tissue with a forceps and lift the body wall away from the viscera. Continue cutting anteriorly through the belly. Stop at the level of the diaphragm, and then continue cutting a little off center as you cut through the costal cartilages. Use a scalpel or scissors to cut through the thoracic region.

You can now make a transverse incision in the lower region of the belly, so that you have an inverted **T**-shaped incision (fig. S6.1). Lift the tissue near the ribs and expose the diaphragm. Gently and carefully snip or cut the diaphragm away from the ventral region by cutting as close to the ribs as possible.

Make the third incision by cutting transversely across the upper part of the chest. Be careful not to cut the blood vessels of the neck that were exposed during the original skinning process. Open the body cavity by pulling the two flaps laterally, exposing the thoracic and abdominal cavities. You may want to cut through the ribs at their most lateral point to facilitate the opening of the thoracic cavity.

Carefully expose the thoracic region of the cat and locate the two lungs and the heart. Note that the heart is enclosed in the pericardial sac. Look for the fatty material, called the **greater omentum,** that covers the **intestines** in the abdominal region. As you dissect the arteries of the cat, pay careful attention to the veins and nerves that travel along with the arteries. *Do not dissect other organs.* You will study these systems (e.g., respiratory, digestive, and urogenital systems) later.

As you open the abdominal cavity, you will see the space that contains the internal organs, known as the **coelom,** or **body cavity.** The lining on the inside of the body wall is the **parietal peritoneum,** and the lining that wraps around the outside of the intestines is the **visceral peritoneum.**

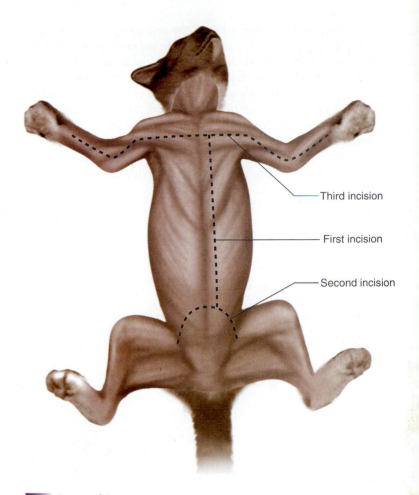

Third incision

First incision

Second incision

FIGURE S6.1
Incisions to Open the Thoracic and Abdominal Cavities

Arteries of the Cat

The blood vessels in the cat have been injected with colored latex. If the cat is doubly injected, the **arteries are red** and the **veins are blue.** If the cat has been triply injected, the hepatic portal vein is typically injected with yellow latex. Be careful locating the arteries in the cat. You can tease away some of the connective tissue with a dissection needle to see the vessel more clearly. Note the difference in the aortic arch arteries in the cat, compared with that of the human. Look for the **coronary arteries,** which take blood from the ascending aorta to the heart tissue.

You can begin your study of the arteries of the cat by examining the ascending aorta. Note that the **aorta** is composed of the **ascending aorta,** which bends and leads to the **aortic arch** and then posteriorly to form the **descending aorta.** The descending aorta is composed of the **thoracic aorta** anterior to the diaphragm and the **abdominal aorta** inferior to the diaphragm. The pattern of arteries that leave the aortic arch is somewhat different in cats than in humans. In the cat, typically only two large vessels leave the aortic arch—the large **brachiocephalic artery** and the **left subclavian artery.** The brachiocephalic artery further divides into the **left com-**

mon carotid artery, the **right common carotid artery,** and the **right subclavian artery.** Examine figure S6.2 for the pattern in the cat. How is the pattern in the cat different from that in the human?

Several arteries arise from the **subclavian artery.** The **vertebral artery** takes blood to the head, and the **subscapular artery** takes blood to the pectoral girdle muscles, along with the **ventral thoracic artery** and **long thoracic artery.** The ventral thoracic and long thoracic arteries supply the latissimus dorsi and pectoralis muscles with blood. The **left** and **right subclavian arteries** become the **left** and **right axillary arteries,** which in turn

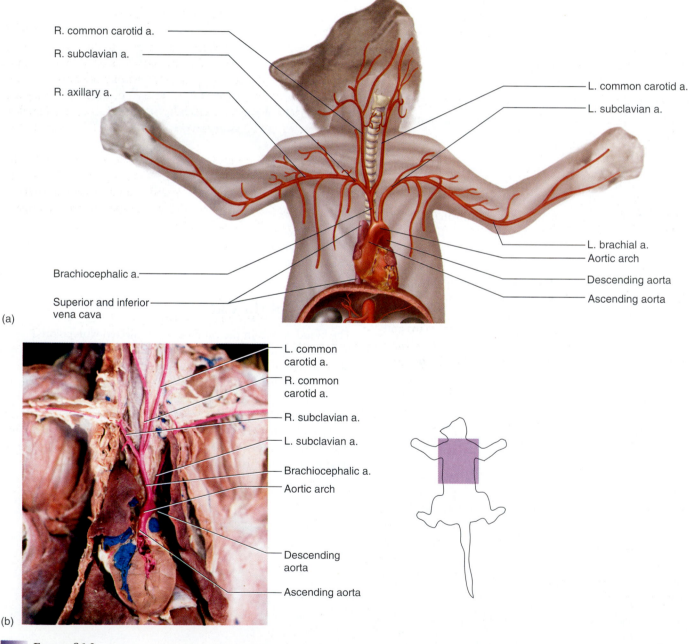

(a)

(b)

FIGURE S6.2

Aortic Arch Arteries of the Cat (a) Diagram; (b) photograph.

take blood to the brachial arteries. The **brachial artery** takes blood to the **radial** and **ulnar arteries.** Additional arteries that branch from the subclavian artery are the **internal mammary artery,** which takes blood to the ventral body wall, and the **thyrocervical artery,** which takes blood to the region of the shoulder and neck. The **costocervical artery** reaches the deep muscles of the back and neck (fig. S6.3).

The **common carotid artery** takes blood along the ventral surface of the neck to feed the small **internal carotid artery,** which supplies the brain. The common carotid artery also empties into the **external carotid artery,** which takes blood to the external portions of the head. The **occipital artery** receives blood from the common carotid artery and supplies the muscles of the neck.

The thoracic aorta not only takes blood to the abdominal aorta but also supplies the rib cage with blood from the **intercostal arteries. Esophageal arteries** also branch from the thoracic aorta, supplying the esophagus with blood. The thoracic aorta continues as the **abdominal aorta,** as it passes through the diaphragm.

Make sure you do not cut the blood vessels away from the organs that will be studied later. Note the major arteries of the caudal region of the cat in figure S6.4. Refer to this figure and the photographs for the remainder of this section.

Abdominal Arteries

The determination of the abdominal arteries is best done by looking for where these arteries leave the aorta and where they enter into the organs they supply. Locate the major organs of the abdominal region, such as the **stomach, spleen, liver,** and **small** and **large in-**

testines. Examine the specimen from the left side, gently lifting the stomach, spleen, and intestines toward the ventral right side as you look for the abdominal arteries. The **descending aorta** is located on the left side of the body, while the **inferior vena cava** is located on the right side. Locate the short **celiac trunk** (celiac artery), which quickly divides into three major vessels—the **splenic artery,** the **left gastric artery,** and the **hepatic artery.** The splenic artery takes blood to the spleen, the hepatic artery reaches the liver, and the left gastric artery supplies the stomach, as seen in figure S6.4.

The **superior mesenteric artery** is the next major vessel to take blood from the abdominal aorta. The superior mesenteric artery travels to the small intestine and the proximal portion of the large intestine. It branches into the **jejunal arteries,** the **ileal arteries,** and the **colic arteries.** Compare your dissection with figures S6.4 and S6.5.

The paired **adrenolumbar arteries** branch on each side of the abdominal aorta and take blood to the adrenal glands, parts of the body wall, and the diaphragm. Just caudal to the adrenolumbar arteries are the paired **renal arteries,** which take blood to the kidneys. The **gonadal arteries** are the next set of paired arteries, and these take blood to the testes in male cats, in which case they are called the **testicular arteries.** If the genital arteries occur in females, they are called the **ovarian arteries** because they take blood to the ovaries (fig. S6.4).

A single **inferior mesenteric artery** takes blood from the abdominal aorta to the terminal portion of the large intestine. The inferior mesenteric artery can be found caudal to genital arteries. Locate the lower abdominal arteries in your specimen and compare them with figures S6.4, S6.5, and S6.6.

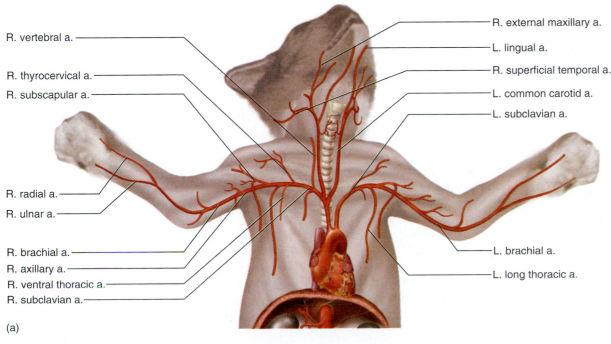

R. vertebral a.

R. thyrocervical a.

R. subscapular a.

R. radial a.

R. ulnar a.

R. brachial a.

R. axillary a.

R. ventral thoracic a.

R. subclavian a.

R. external maxillary a.

L. lingual a.

R. superficial temporal a.

L. common carotid a.

L. subclavian a.

L. brachial a.

L. long thoracic a.

(a)

FIGURE S6.3

Arteries of the Shoulder and Upper Extremity of the Cat (a) Diagram.

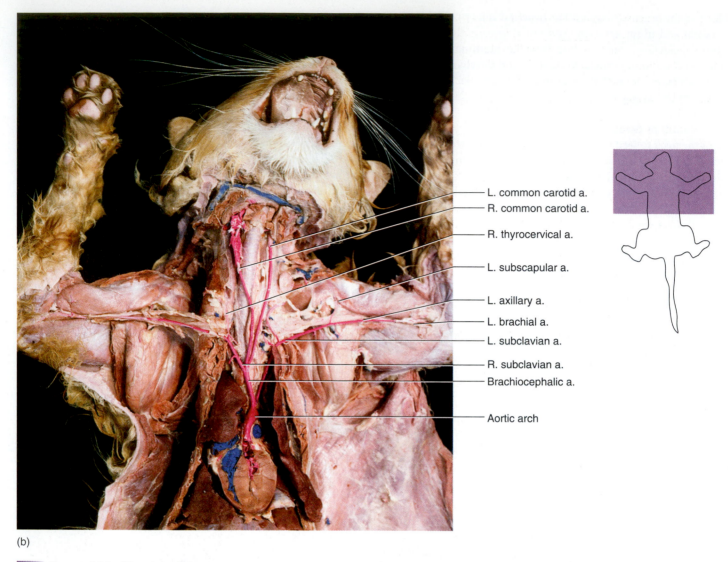

(b)

■ FIGURE S6.3 (Continued)
Arteries of the Shoulder and Upper Extremity of the Cat (b) Photograph.

The pelvic arteries are somewhat different in cats than in humans. The abdominal aorta splits caudally into the **external iliac arteries,** and a short section of the **aorta** continues on and then divides to form the two **internal iliac arteries** and the **caudal artery.** There is no common iliac artery in cats as there is in humans. In cats the caudal artery takes blood to the tail. The internal iliac artery takes blood to the urinary bladder, rectum, external genital organs, uterus, and some thigh muscles. As the external iliac artery exits from the pelvic region and enters the thigh, it becomes the **femoral artery,** which supplies blood to the thigh, leg, and foot. The femoral artery becomes the **popliteal artery** behind the knee, which further divides to form the **tibial arteries.** Locate the pelvic and lower extremity arteries in your cat and compare them with figures S6.4*a*, S6.6, and S6.7.

Veins and Lymph System

The blood vessels in the cat have been injected with colored latex. If the cat has been doubly injected, the arteries are red and the veins are blue. If the cat has been triply injected, the hepatic portal vein is also injected, typically with yellow latex.

VEINS OF THE UPPER LIMB

Begin your dissection by examining the veins of the upper limb. The basilic vein does not occur in cats, but you should be able to find the **ulnar vein** and see that it joins the **radial vein** to form the **brachial vein.** The brachial vein is next to the brachial artery. Locate the **median cubital vein** and the **cephalic vein.** The **axillary vein** and **subscapular vein** (from the shoulder) join to form the

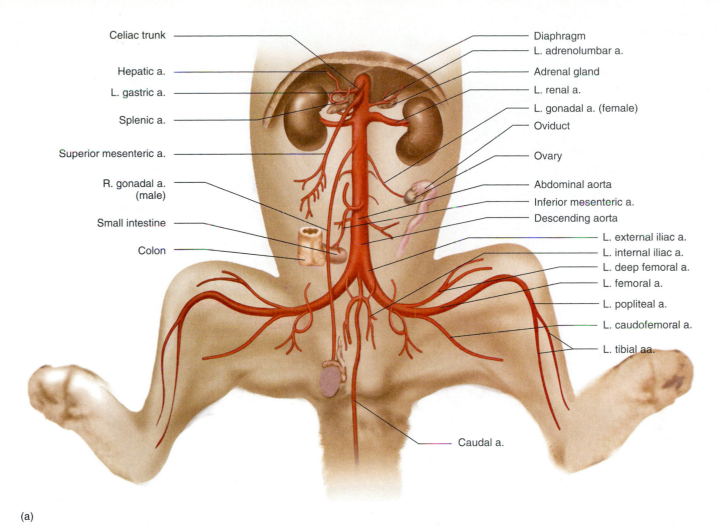

(a)

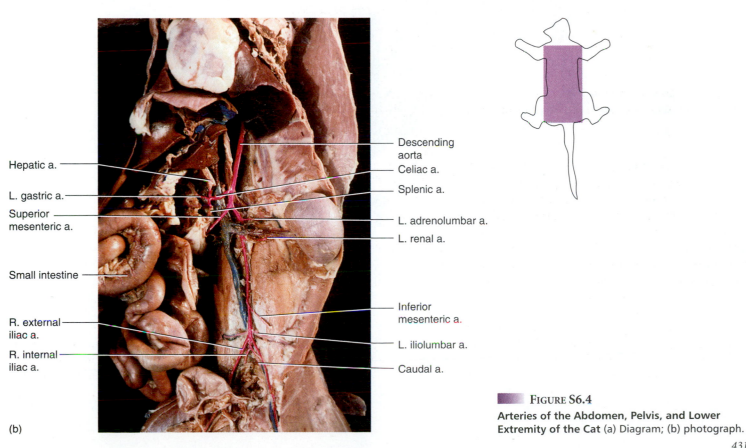

(b)

FIGURE S6.4

Arteries of the Abdomen, Pelvis, and Lower Extremity of the Cat (a) Diagram; (b) photograph.

431

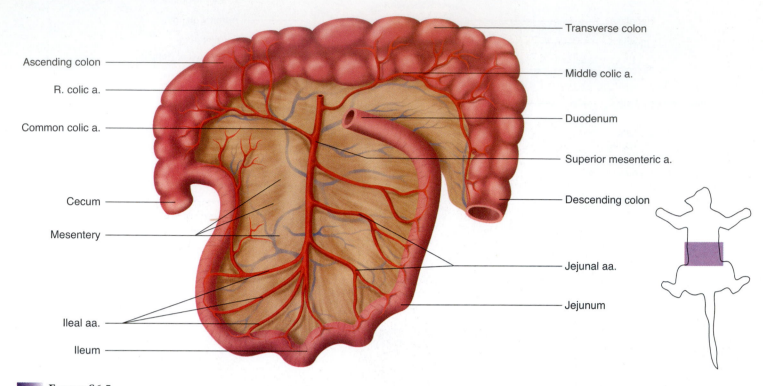

Ascending colon

R. colic a.

Common colic a.

Cecum

Mesentery

Ileal aa.

Ileum

Transverse colon

Middle colic a.

Duodenum

Superior mesenteric a.

Descending colon

Jejunal aa.

Jejunum

FIGURE S6.5

Branches of the Superior Mesenteric Artery of the Cat

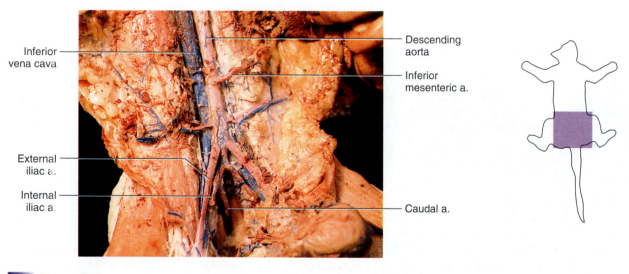

Inferior
vena cava

External
iliac a.

Internal
iliac a.

Descending
aorta

Inferior
mesenteric a.

Caudal a.

FIGURE S6.6

Pelvic Arteries of the Cat

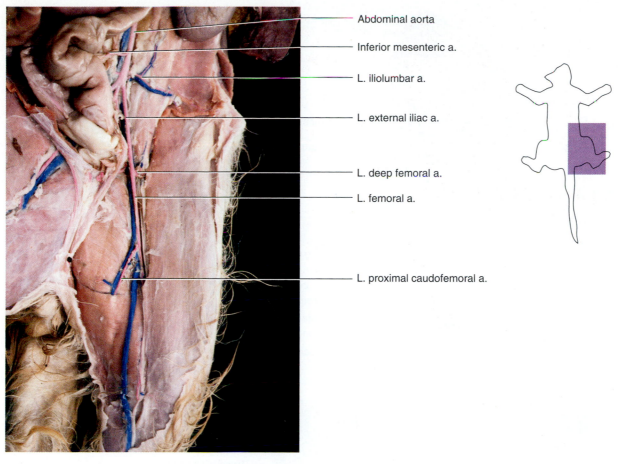

Abdominal aorta

Inferior mesenteric a.

L. iliolumbar a.

L. external iliac a.

L. deep femoral a.

L. femoral a.

L. proximal caudofemoral a.

FIGURE S6.7
Arteries of the Left Lower Extremity of the Cat

subclavian vein, which takes blood to the **brachiocephalic vein.** The cephalic vein continues on into the **transverse scapular vein,** which takes blood to the **external jugular vein.** Locate these veins in figure S6.8.

THORACIC VEINS

Anterior to the diaphragm are the veins of the thorax. The internal mammary vein (sternal vein) may be difficult to locate, since it is frequently cut while opening up the thoracic cavity. The **internal mammary vein** receives blood from the ventral body wall. Another vein of the thoracic region is the **azygos vein.** This vessel is found on the right side of the body and takes blood from the **esophageal, intercostal,** and **bronchial veins.** Compare the veins in your cat with figure S6.8.

VEINS OF THE HEAD AND NECK

The veins in this region of the cat have a few variations from those in the human. In the cat the **external jugular vein** is larger than the **internal jugular vein.** The reverse is true in humans. What anatomic difference between the cat and human might explain the difference in the volume of blood carried by these two vessels?

The **transverse jugular vein,** which is present in cats but absent in humans, connects the two external jugular veins. The jugular veins, along with the **costocervical veins** and the subclavian

veins, unite to form the brachiocephalic veins (fig. S6.8). The **vertebral veins** take blood from the head of the cat and empty into the subclavian veins.

VEINS OF THE LOWER LIMB

Locate the superficial **great saphenous vein** in the cat on the medial side of the lower limb. The great saphenous vein joins the **femoral vein** just above the knee. Note the long **tibial veins** of the leg that join and form the **popliteal veins.** Find the femoral vein, which is found with the femoral artery. The vessels of the distal part of the lower limb take blood to the femoral vein, which turns into the **external iliac vein** as it passes into the body cavity at the level of the inguinal ligament. Find these veins in the cat and compare them with figure S6.9.

VEINS OF THE PELVIS

The **external iliac vein** and the **internal iliac vein** join to form the **common iliac vein.** The two common iliac veins lead to the **inferior vena cava** along with the **caudal vein,** which takes blood from the tail. Be careful as you dissect these structures in the cat. There are numerous ducts that cross the external iliac veins, such as the ureter and, in males, the ductus deferens. Look for these structures in figures S6.9 and S6.10 and do not cut them. You will study them later.

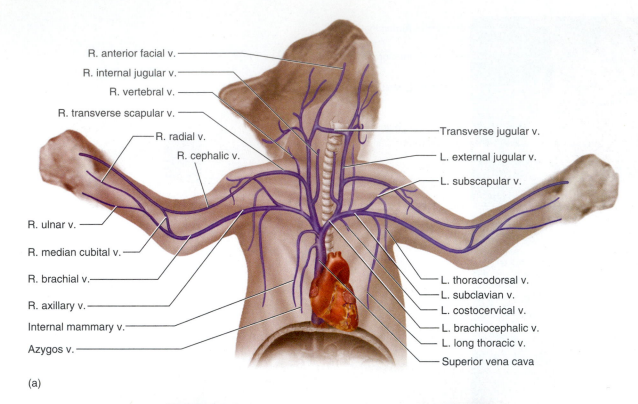

R. anterior facial v.

R. internal jugular v.

R. vertebral v.

R. transverse scapular v.

R. radial v.

R. cephalic v.

R. ulnar v.

R. median cubital v.

R. brachial v.

R. axillary v.

Internal mammary v.

Azygos v.

Transverse jugular v.

L. external jugular v.

L. subscapular v.

L. thoracodorsal v.

L. subclavian v.

L. costocervical v.

L. brachiocephalic v.

L. long thoracic v.

Superior vena cava

(a)

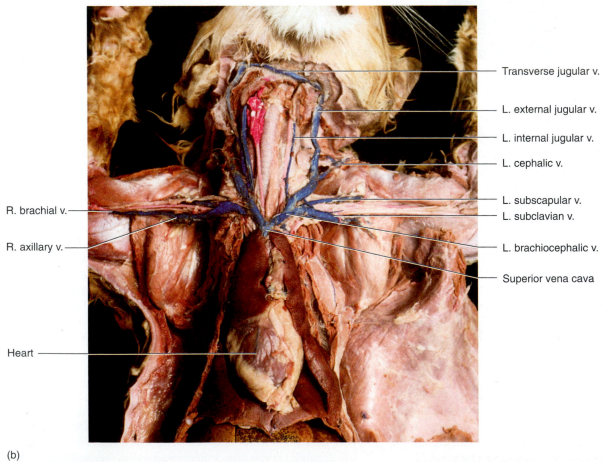

Transverse jugular v.

L. external jugular v.

L. internal jugular v.

L. cephalic v.

L. subscapular v.

L. subclavian v.

L. brachiocephalic v.

Superior vena cava

R. brachial v.

R. axillary v.

Heart

(b)

FIGURE S6.8

Anterior Veins of the Cat (a) Diagram; (b) photograph.

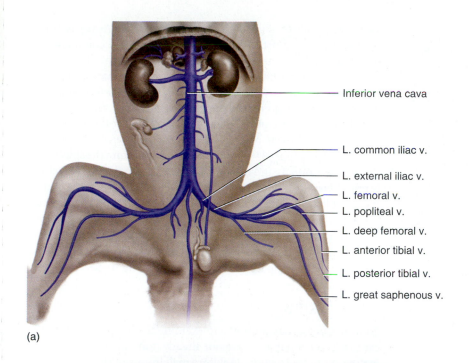

Inferior vena cava

L. common iliac v.

L. external iliac v.

L. femoral v.

L. popliteal v.

L. deep femoral v.

L. anterior tibial v.

L. posterior tibial v.

L. great saphenous v.

(a)

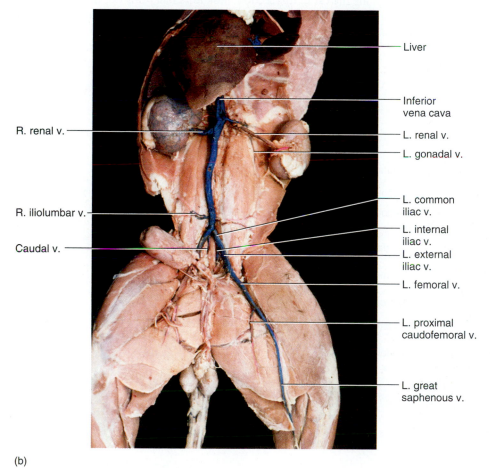

Liver

Inferior vena cava

R. renal v.

L. renal v.

L. gonadal v.

R. iliolumbar v.

L. common iliac v.

L. internal iliac v.

Caudal v.

L. external iliac v.

L. femoral v.

L. proximal caudofemoral v.

L. great saphenous v.

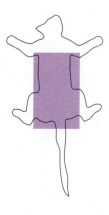

(b)

FIGURE S6.9

Veins of the Lower Extremity of the Cat (a) Diagram; (b) photograph.

ABDOMINAL VEINS

Make sure you identify the major organs in the digestive tract before you proceed with the study of the veins of this region. Pay particular attention to the stomach, liver, spleen, kidneys, adrenal glands, small intestine, and large intestine. The inferior vena cava is a large vessel that travels up the right side of the body in the cat. You may have to gently move some of the digestive organs to the side to find the inferior vena cava. Examine where the common iliac veins join to form the inferior vena cava. The caudal vein joins the inferior vena cava at this point. This vein is absent in humans. The gonadal veins receive blood from the testes or ovaries. The right gonadal vein takes blood to the inferior vena cava and the left gonadal vein takes blood to the left renal vein, which subsequently leads to the inferior vena cava. The paired renal veins receive blood from each kidney and empty into the inferior vena cava. The iliolumbar veins take blood from the body wall and return it via the inferior vena cava. Other vessels that drain the body wall are the adrenolumbar veins, which receive blood from the adrenal glands in addition to the body wall. Compare the veins in your cat with figure S6.10.

HEPATIC PORTAL SYSTEM

The veins that flow into the liver before returning to the heart belong to the hepatic portal system. These vessels take blood from a number of abdominal organs and transfer it to the liver, where a significant amount of metabolic processing occurs. As you examine the lower portion of the digestive tract (transverse colon and descending colon), locate the inferior mesenteric vein. The superior mesenteric vein receives blood from the small intestine and part of the large intestine, as well as from the inferior mesenteric vein. The gastrosplenic vein joins the superior mesenteric vein and takes blood to the hepatic portal vein, along with other digestive veins. The gastrosplenic vein receives blood from the spleen and the stomach. The hepatic portal vein thus receives blood from the spleen and the digestive organs and transports that blood to the liver. After the liver receives the blood, it is transferred to the inferior vena cava by the short hepatic vein. This vein is usually difficult to dissect, since it is at the dorsal aspect of the liver and joins the inferior vena cava at about the junction of the diaphragm. Examine the hepatic portal system of the cat while trying to maintain the integrity of the digestive organs. Compare your dissection with figure S6.11.

LYMPH SYSTEM

Your instructor may want you to examine the lymph system in the cat in this section if you have not already done so. If you do examine the cat in this section, look for small lumps of tissue in the axilla. These are the **lymph nodes** of the region. Note the large **spleen,** which is an approximately 6-inch elongated organ on the left side of the abdominal cavity, and the **thymus,** which is anterior to the heart. Compare what you find in the cat with figure S6.12.

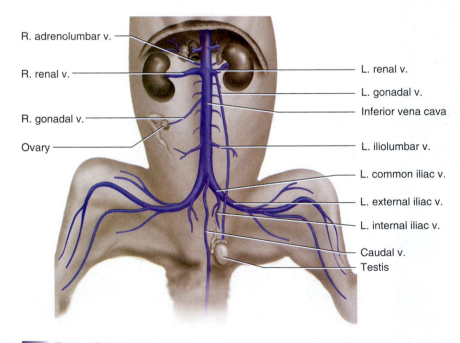

R. adrenolumbar v.
R. renal v.
R. gonadal v.
Ovary

L. renal v.
L. gonadal v.
Inferior vena cava
L. iliolumbar v.
L. common iliac v.
L. external iliac v.
L. internal iliac v.
Caudal v.
Testis

FIGURE S6.10
Veins of the Pelvis and Abdomen of the Cat

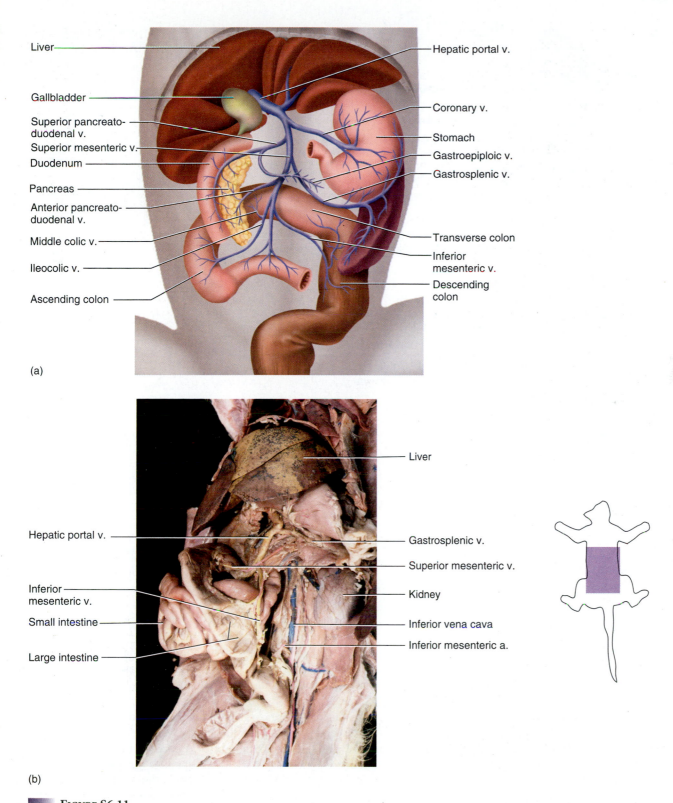

Liver

Gallbladder

Superior pancreato-
duodenal v.

Superior mesenteric v.

Duodenum

Pancreas

Anterior pancreato-
duodenal v.

Middle colic v.

Ileocolic v.

Ascending colon

Hepatic portal v.

Coronary v.

Stomach

Gastroepiploic v.

Gastrosplenic v.

Transverse colon

Inferior
mesenteric v.

Descending
colon

(a)

Hepatic portal v.

Inferior
mesenteric v.

Small intestine

Large intestine

Liver

Gastrosplenic v.

Superior mesenteric v.

Kidney

Inferior vena cava

Inferior mesenteric a.

(b)

FIGURE S6.11

Hepatic Portal System of the Cat (a) Diagram; (b) photograph.

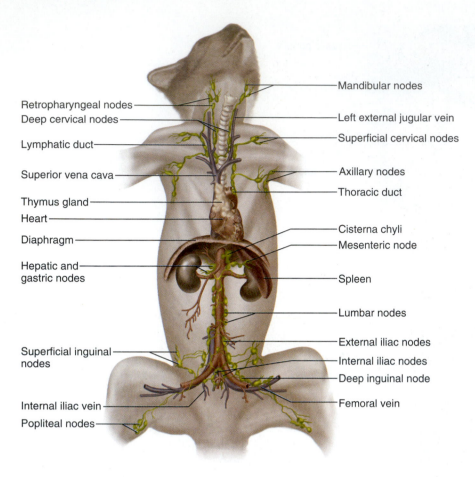

Retropharyngeal nodes

Deep cervical nodes

Lymphatic duct

Superior vena cava

Thymus gland

Heart

Diaphragm

Hepatic and
gastric nodes

Superficial inguinal
nodes

Internal iliac vein

Popliteal nodes

Mandibular nodes

Left external jugular vein

Superficial cervical nodes

Axillary nodes

Thoracic duct

Cisterna chyli

Mesenteric node

Spleen

Lumbar nodes

External iliac nodes

Internal iliac nodes

Deep inguinal node

Femoral vein

FIGURE S6.12
Lymph System of the Cat

SECTION 7: RESPIRATORY ANATOMY OF THE CAT

If you have not opened the chest cavity in your study of the cat, you should do so now. Removal of the skin is discussed in section 2 and the procedure for opening the thoracic cavity is described in section 6 of "Cat Anatomy."

Locate the **larynx** of the cat above the **trachea.** Notice the broad, wedge-shaped structure made of hyaline cartilage. This is the **thyroid cartilage.** Make a midsagittal incision through the thyroid cartilage and continue carefully cutting until you have cut completely through the larynx. To examine the larynx more completely, you may wish to continue your midsagittal cut partway down through the trachea. Open the larynx and find the **epiglottis** of the cat. Notice how the elastic cartilage of the epiglottis is lighter in color than that of the thyroid cartilage. Find the **cricoid cartilage,** the **arytenoid cartilages,** and the **true** and **false vocal folds** (fig. S7.1).

Examine the trachea as it passes from the larynx and into the thoracic cavity. Ask your instructor for permission before you cut

the trachea in cross section. If you do, you should see the **tracheal cartilages** and the **posterior tracheal membrane** (trachealis muscle). The trachea is ventral to the esophagus.

Notice how the trachea splits into the two **bronchi,** which then enter the lungs. The lobes of the lungs are different in the cat than in the human. The right lung in the cat has four lobes, while the left lung has three lobes. How does this compare with the pattern in humans? Examine the structure of the lungs in your specimen and compare it with figure S7.2.

The lungs are covered with a thin serous membrane called the **visceral pleura,** while the outer covering (on the deep surface of the ribs and intercostal muscles) is called the **parietal pleura.** The space between these two membranes is the **pleural cavity.** Cut into one of the lungs of the cat and examine the lung tissue. Note how the lungs look like a very fine mesh sponge. The alveoli of the lungs are microscopic.

At the inferior portion of the thoracic cavity is the diaphragm. As it contracts, the pressure in the thoracic cavity decreases and air fills the lungs. Examine the diaphragm in your specimen and compare it with figure S7.2.

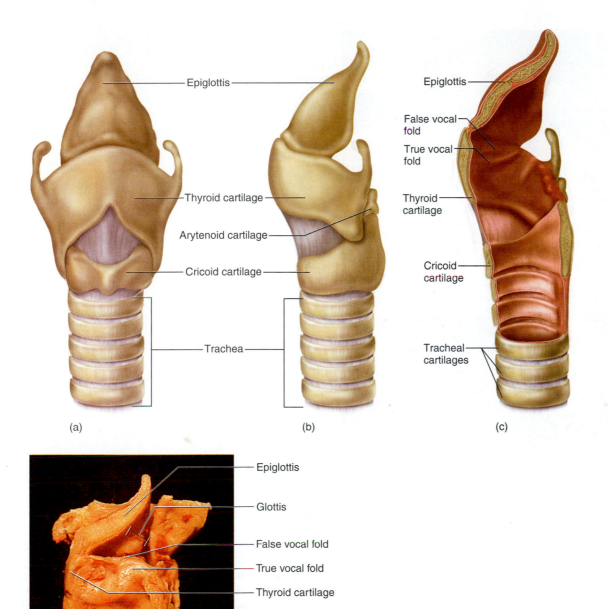

Epiglottis

Thyroid cartilage

Arytenoid cartilage

Cricoid cartilage

Trachea

(a)

(b)

Epiglottis

False vocal fold

True vocal fold

Thyroid cartilage

Cricoid cartilage

Tracheal cartilages

(c)

Epiglottis

Glottis

False vocal fold

True vocal fold

Thyroid cartilage

Cricoid cartilage

(d)

FIGURE S7.1

Larynx of the Cat Diagram (a) anterior view; (b) left lateral view; (c) midsagittal view. Photograph (d) midsagittal view.

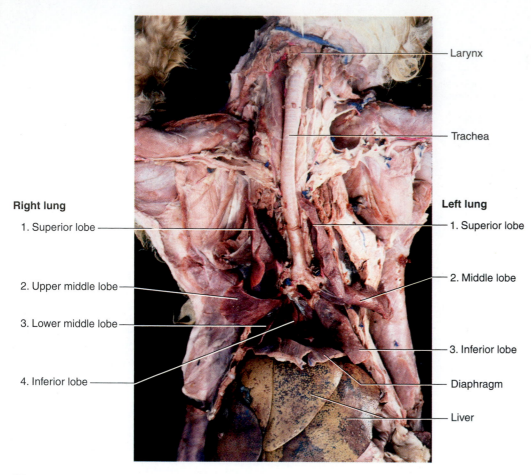

Right lung

1. Superior lobe

2. Upper middle lobe

3. Lower middle lobe

4. Inferior lobe

Larynx

Trachea

Left lung

1. Superior lobe

2. Middle lobe

3. Inferior lobe

Diaphragm

Liver

FIGURE S7.2
Respiratory Structures of the Cat

SECTION 8: DIGESTIVE ANATOMY OF THE CAT

Begin your study of the digestive system of the cat by locating the large **salivary glands** around the face. You must first remove the skin from in front of the ear. If you have dissected the face for the musculature, you may have already removed the **parotid gland,** which is a spongy, cream-colored pad anterior to the ear. You can also locate the **submandibular gland** as a pad of tissue slightly anterior to the angle of the mandible and lateral to the digastric muscle. The **sublingual gland** is just anterior to the submandibular gland and is an elongated gland that parallels the mandible. Use figure S8.1 to help you locate these glands.

Examine the **tongue** of the cat and note the numerous papillae on the dorsal surface. These are **filiform papillae** and have both a digestive and a grooming function. You can lift up the tongue and examine the **lingual frenulum** on the ventral surface.

Dissection of the head of the cat should be done only if your instructor gives you permission to do so. To dissect the head, you should first use a scalpel and make a cut in the midsagittal plane from the forehead of the cat to the region of the occipital bone. Cut through any overlying muscle. Then use a small saw to gently cut through the cra-

nial region of the skull, being careful to stay in the midsagittal plane. Once you make an initial cut through the dorsal side of the head, cut the mental symphysis of the cat, thus separating the mandible.

You can then use a long knife (or scalpel) to cut through the softer regions of the skull, brain, and tongue. Use a scalpel to cut through the floor of the mouth to the hyoid bone. It is best to use a scissors or small bone cutter to cut through the hyoid bone. This should allow you to open the head and examine the structures seen in a midsagittal section, as shown in figure S8.2. Locate the hard palate, tongue, mouth, and oropharynx. Note how the larynx in the cat is closer to the tongue than in humans.

Look for the muscular tube of the **esophagus** by carefully lifting the trachea ventrally away from the neck. You should be able to insert your blunt dissection probe into the mouth and gently wiggle it into the esophagus. The tongue and esophagus are represented in figure S8.2.

Abdominal Organs

To get a better view of the abdominal organs, it is best to cut the **diaphragm** away from the body wall. Carefully cut the lateral edges of the diaphragm from the ribs. You should see a fatty drape

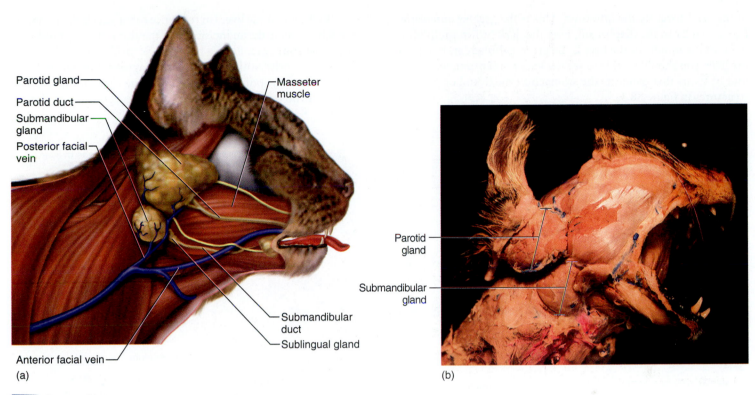

Parotid gland
Parotid duct
Submandibular gland
Posterior facial vein
Masseter muscle
Submandibular duct
Sublingual gland
Anterior facial vein
(a)

Parotid gland
Submandibular gland
(b)

FIGURE S8.1

Salivary Glands of the Cat (a) Diagram; (b) photograph.

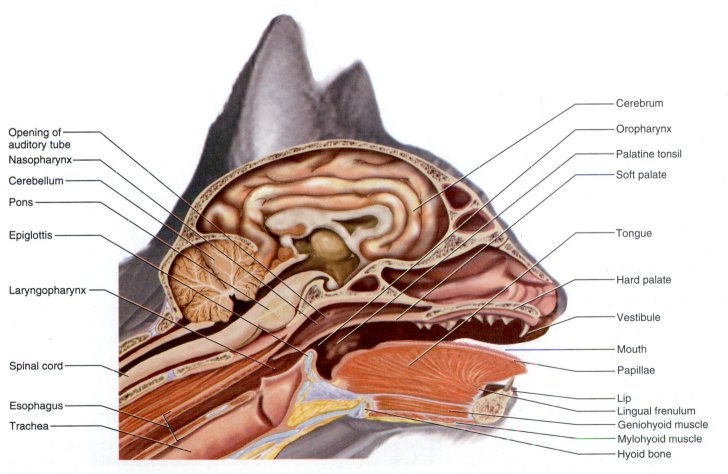

Opening of auditory tube
Nasopharynx
Cerebellum
Pons
Epiglottis
Laryngopharynx
Spinal cord
Esophagus
Trachea

Cerebrum
Oropharynx
Palatine tonsil
Soft palate
Tongue
Hard palate
Vestibule
Mouth
Papillae
Lip
Lingual frenulum
Geniohyoid muscle
Mylohyoid muscle
Hyoid bone

FIGURE S8.2

Midsagittal Section of the Head and Neck of a Cat

of material covering the intestines. This is the **greater omentum.** Just posterior to the diaphragm, note the dark brown multilobed **liver.** In the middle of the liver is the green **gallbladder.** If you lift the liver, you should be able to see the **lesser omentum,** which is a fold of tissue that connects the stomach to the liver. Locate these structures in figure S8.3.

To the left of the liver (in reference to the cat) is the J-shaped **stomach.** If you make an incision into the stomach you should see folds known as **rugae.** Place a blunt probe inside the stomach and move it anteriorly. Notice how the **esophageal sphincter** makes it difficult to put the pointer into the esophagus anterior to the diaphragm. If you do move the probe into the esophagus you should

■ FIGURE S8.3

Abdominal Organs of the Cat (a) Diagram; (b) photograph.

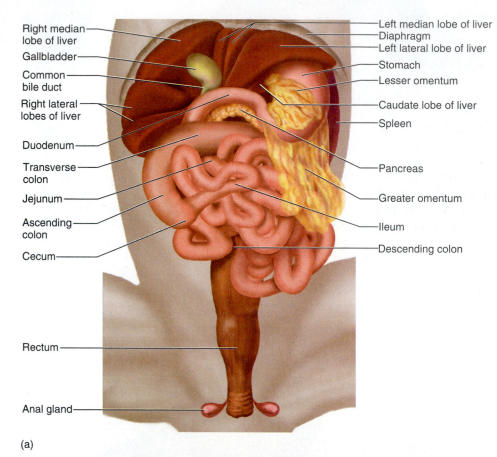

Right median lobe of liver
Gallbladder
Common bile duct
Right lateral lobes of liver
Duodenum
Transverse colon
Jejunum
Ascending colon
Cecum

Left median lobe of liver
Diaphragm
Left lateral lobe of liver
Stomach
Lesser omentum
Caudate lobe of liver
Spleen
Pancreas
Greater omentum
Ileum
Descending colon

Rectum

Anal gland

(a)

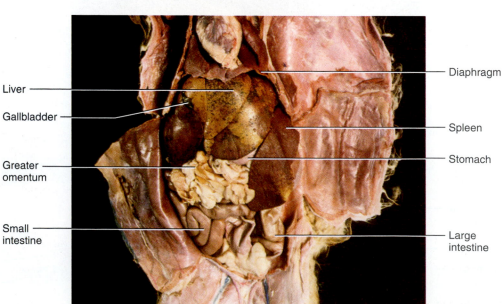

Liver
Gallbladder
Greater omentum
Small intestine

Diaphragm
Spleen
Stomach
Large intestine

(b)

be able to see it as you look at the esophagus from the thoracic cavity.

The region of the stomach near the esophagus is called the **cardiac region.** A small part of it forms an anterior dome or pouch. This is the **fundus** of the stomach. The stomach is curved, with the left side of the stomach forming the **greater curvature** and the right side forming the **lesser curvature.** The main region of the stomach is the **body** and the narrowed section of the stomach, near the small intestine, is the **pyloric region.** Locate the parts of the stomach on your cat.

Cut lengthwise into the pyloric region of the stomach and then through the duodenum. You should be able to locate a tight sphincter muscle between the stomach and the duodenum. This is the **pyloric sphincter.** As you move into the duodenum, you may have to scrape some of the chyme away from the wall of the small intestine to be able to see the fuzzy texture of the intestinal wall. This texture is due to the presence of **villi.** As in the human, the small intestine is composed of three regions, a proximal **duodenum,** the middle **jejunum,** and a terminal **ileum.** Compare the stomach and small intestine with figure S8.3.

If you elevate the greater omentum, you should be able to see the **pancreas,** along the caudal side of the stomach. The pancreas appears granular and brown. The **tail** of the pancreas is near the **spleen,** which is a brown, elongated organ on the left side of the body.

The small intestine is an elongated, coiled tube about the diameter of a wooden pencil. Note the **mesentery,** which holds the small intestine to the posterior body wall near the vertebrae. The small intestine is extensive in the cat and rapidly expands into the large intestine. The appendix is absent in the cat.

The **large intestine** in the cat is a fairly short tube with a diameter slightly larger than your thumb. The first part of the large intestine is a pouch called the **cecum.** The remainder of the large intestine can be further divided into the **ascending, transverse,** and **descending colon** and the **rectum.** Compare these with figure S8.3. Examine the **parietal peritoneum** along the inner surface of the body wall and the **visceral peritoneum** that envelops the intestines.

SECTION 9: URINARY ANATOMY OF THE CAT

The kidneys are on the dorsal body wall of the cat and are located dorsal to the parietal peritoneum. Examine the kidneys and the structures that lead to and from the hilum of the kidney. You should locate the renal veins that take blood from the kidney. The veins are larger in diameter than the renal arteries, and they are attached to the inferior vena cava of the cat that runs along the ventral, right side of the vertebral column. Find the renal arteries that take blood from the aorta to the kidneys. The aorta lies to the left of the inferior vena cava. You should be able to locate the ureters that run posteriorly from the kidney to the urinary bladder. Do not dissect the urethra at this time. You can locate the urethra during the dissection of the reproductive structures in section 10. Compare your dissection with figure S9.1.

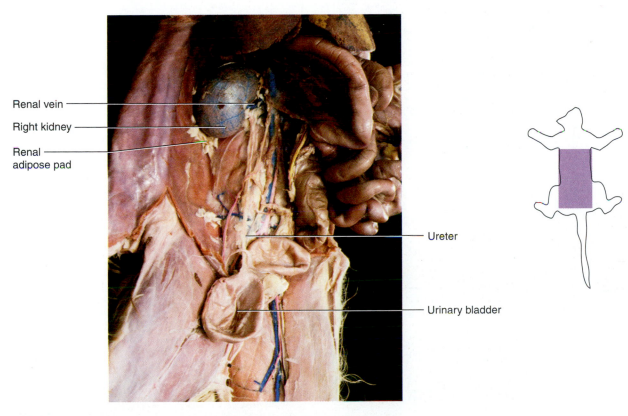

Renal vein

Right kidney

Renal adipose pad

Ureter

Urinary bladder

FIGURE S9.1

Urinary System in the Cat

SECTION 10: REPRODUCTIVE ANATOMY OF THE CAT

Check to see whether your cat is male or female. The penis may be retracted in your specimen, so look for the opening of the penile urethra and the scrotum. Team up with a lab partner or group that has a cat of a different sex than your specimen, so that you can learn both male and female reproductive systems. Once you are sure you have a male cat, locate the **scrotum** and paired **testes.** Make an incision on the lateral side of the scrotum and locate the testis inside the scrotal sac. If your cat was neutered, you will not be able to locate the testis. If the testes are present, cut through the connective tissue of the scrotum (the **tunica vaginalis**) and observe both the testis and the **epididymis.** The testis is covered by a tough connective tissue membrane called the **tunica albuginea.** Use figure S10.1 as a guide. Spermatozoa move from the testis and into the epididymis, where the spermatozoa mature.

Once you have located the epididymis, proceed in an anterior direction and trace the thin **ductus deferens** from the epididymis into the **spermatic cord.** The spermatic cord traverses the body wall on the exterior and enters the body of the cat at the **inguinal canal** (an opening in the inguinal ligament). You may want to gently insert a blunt probe into the inguinal canal, so that you can locate the ductus deferens as it passes into the **coelom.** Locate the ductus deferens as it enters the body cavity and notice how it arches around the ureter on the dorsal side of the urinary bladder. You may have cut one of the ductus deferens in an earlier exercise, so if you cannot find it on one side, look for it on the other side.

You may want to look for the accessory organs of the male reproductive system, but this takes some significant dissection. *Check with your instructor before cutting through the pelvis of your cat.* If your instructor directs you to do so, begin by cutting through the musculature of the cat at the level of the **symphysis pubis.** Make a midsagittal incision through the groin muscles, and carefully cut through the cartilage of the symphysis pubis. You should now be able to open the pelvic cavity and locate the single **prostate gland** and the paired **bulbourethral glands** (fig. S10.2). Much of the anatomy of the cat is similar to that of the human except that there are no seminal vesicles in the cat. Trace the ductus deferens from the posterior surface of the bladder through the prostate gland and

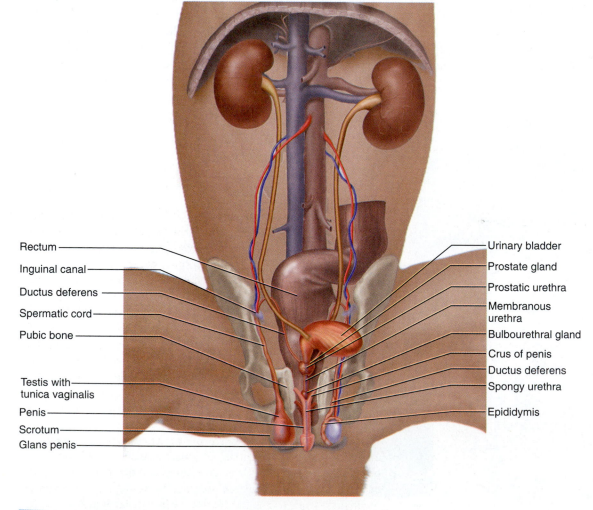

Rectum
Inguinal canal
Ductus deferens
Spermatic cord
Pubic bone
Testis with tunica vaginalis
Penis
Scrotum
Glans penis

Urinary bladder
Prostate gland
Prostatic urethra
Membranous urethra
Bulbourethral gland
Crus of penis
Ductus deferens
Spongy urethra
Epididymis

FIGURE S10.1

Male Reproductive Organs of the Cat (a) Diagram.

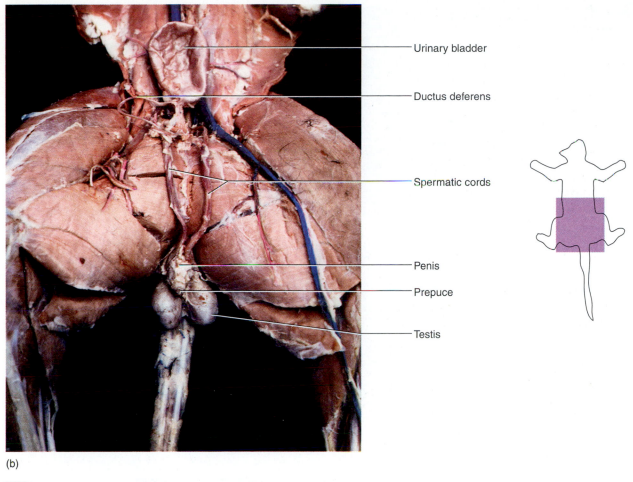

(b)

FIGURE S10.1 *(Continued)*
Male Reproductive Organs of the Cat (b) Photograph.

Labels for figure S10.1(b):
- Urinary bladder
- Ductus deferens
- Spermatic cords
- Penis
- Prepuce
- Testis

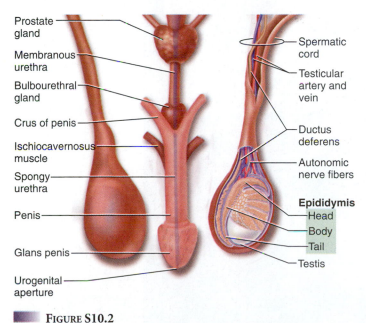

Labels for figure S10.2:
- Prostate gland
- Membranous urethra
- Bulbourethral gland
- Crus of penis
- Ischiocavernosus muscle
- Spongy urethra
- Penis
- Glans penis
- Urogenital aperture
- Spermatic cord
- Testicular artery and vein
- Ductus deferens
- Autonomic nerve fibers
- Epididymis
  - Head
  - Body
  - Tail
- Testis

FIGURE S10.2
Testis, Epididymis, Spermatic Cord, and Penis of the Cat

the penis. You can either make a longitudinal section through the penis to trace the urethra in the erectile tissue or make a cross section of the penis to see the three cylinders of erectile tissue—the **corpus spongiosum** and the **two corpora cavernosa.**

You probably removed the multiple mammary glands of the female cat during the removal of the skin and study of the muscles. Examine the external genitalia for the **urogenital orifice.** *Check with your instructor before cutting through the pelvis of your cat.* If your instructor directs you to do so, begin by cutting through the musculature of the cat at the level of the **symphysis pubis.** Continue to cut in the midsagittal plane through the cartilage of the symphysis pubis and then cranially through the abdominal muscles as described in the dissection of the male cat. This should expose the reproductive organs (fig. S10.3).

Unlike the human female, the cat has a **horned (bipartite) uterus.** The uterus of a cat has an appearance of a Y—with the upper two branches being the **uterine horns** and the stem of the Y the **body** of the uterus. Humans normally have single births from a pregnancy, while the expanded uterus in cats facilitates multiple births. If your cat is pregnant, the uterus will be greatly enlarged, and you may find many fetuses inside.

Female Reproductive Organs of the Cat (a) Diagram; (b) photograph.

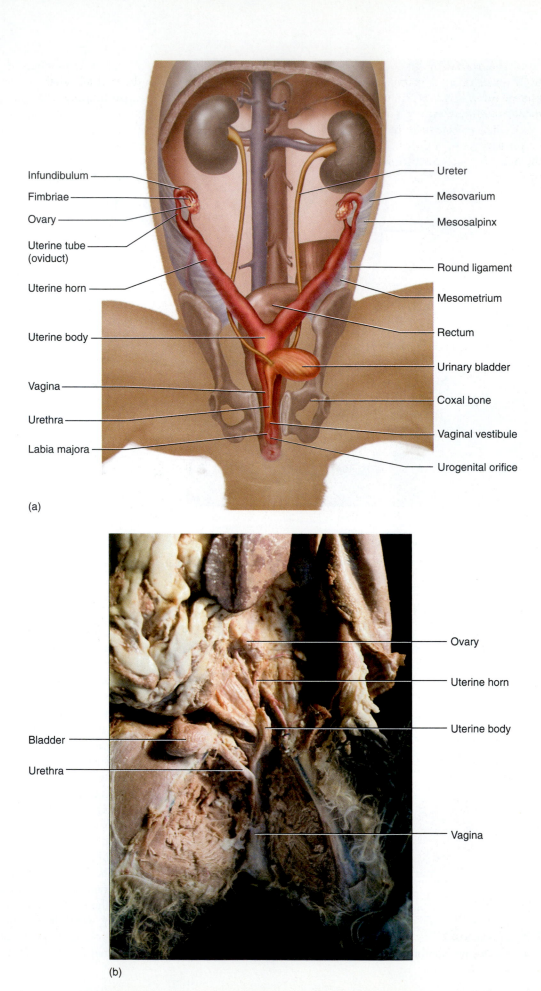

Infundibulum

Fimbriae

Ovary

Uterine tube (oviduct)

Uterine horn

Uterine body

Vagina

Urethra

Labia majora

Ureter

Mesovarium

Mesosalpinx

Round ligament

Mesometrium

Rectum

Urinary bladder

Coxal bone

Vaginal vestibule

Urogenital orifice

(a)

Ovary

Uterine horn

Uterine body

Bladder

Urethra

Vagina

(b)

The **ovaries** in cats are caudal to the kidneys and are relatively small organs. Examine the paired ovaries and the short **uterine tubes (oviducts)** in the cat. If you have difficulty locating them, trace the uterus towards the uterine horns and locate the ovaries. The uterine tubes in the human female are proportionally longer than those in the cat. The opening of the uterine tube near the ovary is called the **ostium,** and it receives the oocytes during ovulation.

At the termination of the uterus is the **cervix,** which leads to the **vaginal canal.** The vagina in cats is different than in humans in that the **urethral opening** is internally enclosed in the vaginal canal. This region where the vagina and the urethral opening are located is called the **vaginal vestibule.** Thus, the opening to the external environment is a common urinary and reproductive outlet called the **urogenital orifice.** Locate these structures in figure S10.3.

Preparation of Materials

The following solutions are designed for a lab of 24 students. The preparations are listed alphabetically and the solutions are the approximate volume for the needs of a class of 24. The number of the laboratory exercise or cat anatomy section follows the solution description for crossreferencing.

BLEACH SOLUTION (10%)

Mix 100 mL household bleach (sodium hypochlorite) with 900 mL tap water. (Laboratory exercise 20)

CAT WETTING SOLUTION

Numerous formulations are available for keeping preserved specimens moist. Some commercial preparations are available that reduce the exposure of students to formalin or phenol. You may not need any wetting solution at all if the cats are kept in a plastic bag that is securely closed. You can make a wetting solution by putting 75 mL formalin, 100 mL glycerol, and 825 mL distilled water in a 1-Liter squeeze bottle. Another mixture consists of equal parts Lysol and water. (Cat anatomy sections 2–10)

PHOSPHATE BUFFER SOLUTION

Add 3.3 g potassium phosphate (monobasic) and 1.3 g sodium phosphate (dibasic) to 500 mL water. Place in squeeze bottles. (Laboratory exercise 20)

WRIGHT'S STAIN

Wright's stain is available as a commercially prepared solution from a number of biological supply houses. (Laboratory exercise 20)